Game Theory for Economists

Game Theory for Economists

Jürgen Eichberger
Department of Economics
University of Melbourne
Parkville, Victoria
Australia

Academic Press, Inc.
A Division of Harcourt Brace & Company

San Diego New York Boston London Sydney Tokyo Toronto

This book is printed on acid-free paper. ♾

Academic Press, Inc.
1250 Sixth Avenue, San Diego, California 92101-4311

United Kingdom Edition published by
Academic Press Limited
24–28 Oval Road, London NW1 7DX

Library of Congress Cataloging-in-Publication Data

Eichberger, Jürgen. Date
Game theory for economists /Jürgen Eichberger.
p. cm.
Includes bibliographical references and index.
ISBN 0-12-233620-8
1. Game theory. 2. Equilibrium (Economics) I. Title.
HB144.E34 1993
330'.01'5193–dc20 92-40743
CIP

PRINTED IN THE UNITED STATES OF AMERICA
93 94 95 96 97 98 QW 9 8 7 6 5 4 3 2 1

für meine Eltern

Contents

Preface

Game theory has become an increasingly important approach for theoretical analysis in the social sciences. In economics in particular, theory is often conducted in terms of game-theoretic concepts. This is reflected in the increasing number of books and articles that apply these concepts to model interactions of economic agents. Such applications can be found in fields ranging from industrial organization and public finance to macroeconomics and trade theory. There are at least two reasons for the growing importance of game theory in economics: game theory provides a unifying framework for economic analysis in many fields, and it structures the process of modeling economic behavior.

A fundamental postulate of economic theory maintains that economic agents maximize some objective function. In fact, *rational* behavior is usually defined in this way. A second, though less universal, postulate requires consistency in the beliefs that agents hold in regard to other agents' behavior. This second assumption becomes particularly important in models in which an agent's utility depends directly on other agents' behavior. Such consistent beliefs are often referred to as *rational* expectations. Most equilibrium concepts used in economics are based on both postulates and, therefore, characterize behavior and beliefs that are *rational*.

Game theory shares these two basic postulates with economics, but it applies them in the broader context of the analysis of social situations. It assumes that players choose strategies to maximize their payoffs and it imposes consistency requirements for beliefs about opponents' behavior. The similarity of approaches makes game theory appear a natural extension of the economic approach to other social sciences.

One can describe game theory as a method for analyzing the choice of strategies of agents in social situations. The concept of strategy as a complete plan of action is more general than the notion of activities such as commodity purchases or sales that has dominated, and often constrained, economic analysis in the past. Game theory enables economists to approach problems that, thirty years ago, seemed to be beyond formal modeling. Strategies provide game theory with a concept for modeling behavior that takes informational as well as dynamic characteristics of economic situations into account. In this sense, game theory is more than an extension of existing economic thinking because it offers guidelines for the modeling of economic problems. Macroeconomists, for example, would probably not have noted the importance of commitment possibilities for the effectiveness of government policy without the analysis of credible threats, which was made possible by the notion of a strategy. Or, in contract theory, the role of actions as signals of private information would most likely have been overlooked without the concept of a strategy that is contingent on the information that an agent holds. This strategic way of thinking makes the game-theoretic approach particularly valuable for economists. To a considerable extent, the rapid spread of game-theoretic ideas among economists during the last two decades can be attributed to their usefulness as a modeling tool.

Objectives of the Book

The advantage of a formal structure for analyzing many economic and social situations has as its price the need to learn a formal grammar. The rigor imposed on arguments by a formal language often makes it difficult for the student to follow arguments and, in particular, to argue in the same way. Most economists without a background in mathematics or logic find it hard to appreciate concepts proposed in game theory and to understand the conditions for and the meaning of game-theoretic results. This difficulty is compounded by the rapidly growing number of concepts and results developed in game theory—a typical phenomenon for fields under active research. This book tries to overcome these difficulties by emphasizing what needs to be specified to describe an economic situation as a game and how to read and apply existing results.

Given these objectives, I devote considerable space to introducing the formal description of games and strategies. Particular emphasis is placed on the relationship between different game forms and strategy concepts. To make results easier to appreciate and apply, this book focuses on central propositions and it makes no effort to present them in their most general form. Examples and counter-examples are used extensively to illustrate the meaning of assumptions and theorems. This procedure implies some loss of elegance in expositions and length of proofs but it will be helpful for readers who are not sufficiently familiar with mathematical reasoning to fill in "obvious" steps. In selecting material an attempt was made to include those concepts and results that have found widespread application in economics and to omit those that are of interest mainly for a systematic study of game theory.

Other Recent Books on Game Theory

On the whole, one can distinguish two groups of introductory books[1] on game theory. On the one side, there are informal treatments that, based on case studies, try to familiarize the reader with strategic thinking. The books by Dixit and Nalebuff (1991) and by McMillan (1992) represent excellent examples in this category. Both books provide a fascinating introduction to game-theoretic reasoning in the context of "real world" examples. They are certain to stimulate interest in the particular mode of approaching problems advocated by game theory, but they do not provide the techniques that are necessary to formalize problems for rigorous analysis.

At the other extreme, one finds elegant and more or less comprehensive treatments of game theory like Binmore (1992), Friedman (1990), Fudenberg and Tirole (1991), and Myerson (1991), and, with a more restricted choice of topics, Moulin (1986) and Van Damme (1991). The mathematical level of these books is quite high and it requires that the reader be familiar with general mathematical results. All of the authors have been at the forefront of game-theoretic research during the past decade, and these books include the latest developments of their research. Though invaluable for the researcher, they are difficult reading for the beginner and they require considerable prior knowledge of game theory to appreciate the importance of the results and the elegance of the proofs.

In this book I try to find the middle ground between the two levels. I have in mind a reader who is motivated to study game theory using a book from the first group and who feels reasonably comfortable with microeconomics at the level of Varian (1984). Though some familiarity with sets and functions in real-number spaces is required, no knowledge of general mathematical theorems is assumed. There are few sections of the book where such results are needed, and in those cases an intuitive diagrammatic presentation of the required result is provided. In addition, no parts of proofs or examples have been omitted as easy or obvious, except where the same steps have been made before. This will help students gain some proficiency with the kind of checks that proofs of formal statements require.

The game-theoretic concepts that are presented in this book and the level at which they are presented was guided by their importance for applications in economics. This criterion led to the inclusion of concepts such as subgame perfect equilibrium, Bayesian equilibrium, stable equilibrium, and cooperative concepts such as core and Shapley value, and to the

[1] There is a third group of books that cover some game-theoretic material in the context of other topics, for example, Kreps (1990), Rasmusen (1989), and Tirole (1988).

exclusion of correlated equilibrium, proper equilibrium, and many of the cooperative concepts. This is not meant to deny the importance of these concepts for a systematic study of game theory or for applications outside economics; rather, it reflects the fact that these concepts have not found economic applications so far.

How to Use This Book

The material presented here has been used for a one-semester course in game theory in economics. The experience from teaching game theory to economic students has influenced the format of the book. One of the least inspiring tasks of such a course is the exposition of the large amount of notation that is necessary for the description of games in extensive form. Familiarity with these concepts is, however, indispensable for the understanding of many issues that arise in the context of the refinements of Nash equilibrium.

Chapter 1 is a very careful and detailed exposition of the elements that need to be specified in describing extensive form games. The chapter uses many examples to illustrate the relationship between the graphical representation in tree diagrams and the formal description. The different concepts of strategy in extensive and strategic form games are discussed and their interrelationship is explored. Chapter 1 allows teachers to be more concise in the exposition of this material during lectures. Students can acquaint themselves with its content and can refer back to it in later sections.

In Chapter 2 the simple structure of two-person zero-sum games can be used to motivate the notion of a solution of a game. The value of a game is introduced and its equivalence with the equilibrium concept in this class of games is demonstrated. This material can be treated as optional. Though it helps in understanding the alternative ways of thinking about the solution of games, the coherence of the exposition is not broken by proceeding directly to Chapter 3.

Chapters 3, 4, 5, and 6 contain the core material on noncooperative games. In Chapter 3, dominant strategy equilibrium and iterated dominance ideas are considered. Chapter 4 focuses on the Nash equilibrium concept. A discussion of existence and uniqueness of Nash equilibrium is provided; this can be treated as optional. This chapter concludes with a section on the (trembling-hand) perfect equilibrium as an equilibrium refinement for normal form games.

The treatment of incomplete information by specifying types is introduced in Chapter 5. Special effort is devoted to the exposition of Bayesian updating, which students usually find difficult. It is shown that the expansion of a game to finite sets of types for the players makes it possible to

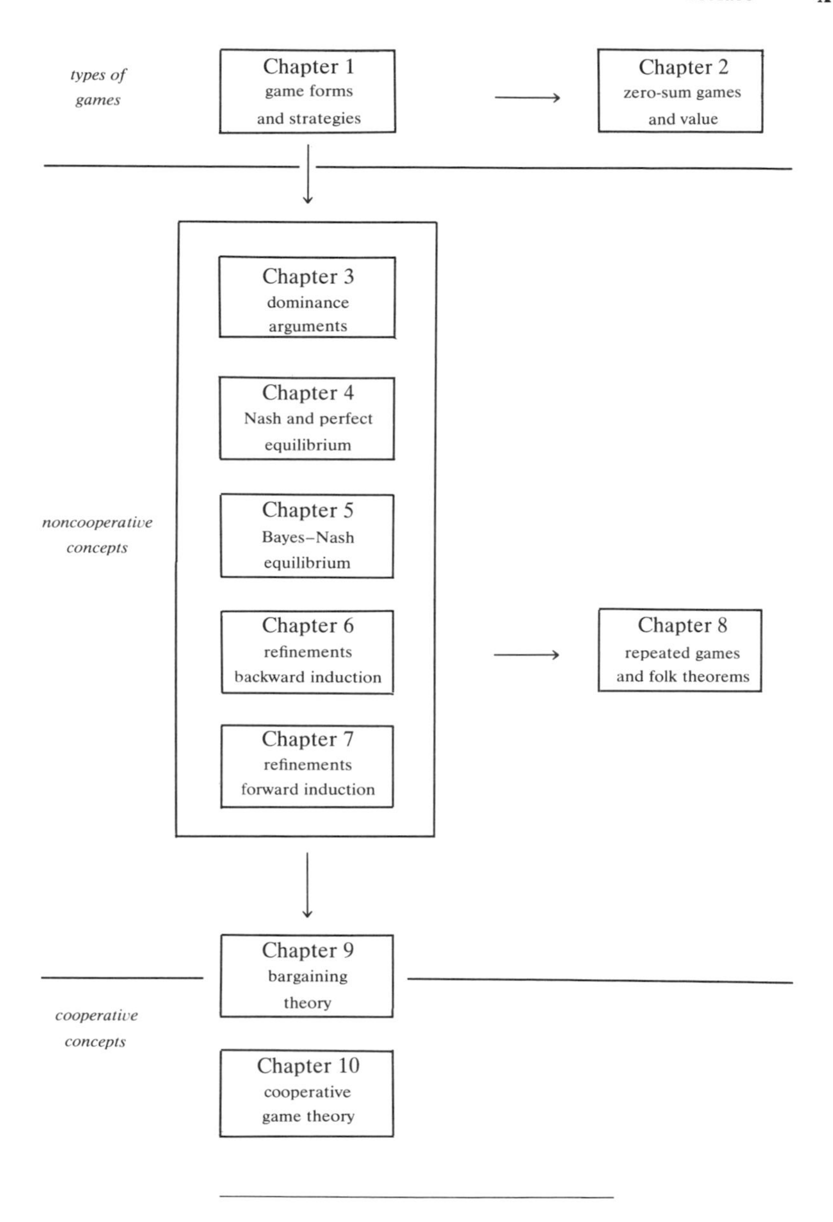

analyze the game either as an expanded normal form game or as an extensive form game where nature is the first mover.

Chapters 6 and 7 include considerations about equilibrium refinements in the context of extensive form games. The backward induction notion is used to motivate the exclusion of equilibria that contain incredible threats and promises (Chapter 6), and forward induction provides the intuition for the intuitive criterion and stability (Chapter 7).

The main results on repeated games are presented in Chapter 8. This chapter can be viewed as an extension of the previous chapters, but it may be treated independently once the notion of subgame perfection has been introduced. By restricting players' choices to pure strategies in each stage game, a great simplification of arguments is achieved at the small cost of folk theorems that support all individually rational outcomes of the stage game but do not support the full convex hull.

The bargaining problem is of particular relevance for economic analysis and it provides an excellent opportunity to introduce the axiomatic method that finds its main application in cooperative game theory. Bargaining theory is investigated in Chapter 9 in the noncooperative and the cooperative mode. This makes it possible to discuss the Nash program and bridges the gap between noncooperative and cooperative game theory. Chapter 10 presents an introduction to cooperative game theory with transferable utility. The core and the Shapley value are studied carefully as the two concepts that are most widely used in economics.

The diagram on page xv summarizes the book's organization.

Notation

The mathematical notation in this book is similar to that used in most microeconomic textbooks. The only exception is the use of "max" and "argmax," which is commonly found in game-theoretic books because of its economical representation of an optimization problem. Consider an optimization problem:

$$\text{Choose } x \text{ to maximize } f(x,a) \text{ subject to } x \in X(a)$$

The solution of this problem will be a maximum value $M(a)$, which depends in general on the parameter a, and one or more maximizer $m(a)$, which may depend on a as well. If there is more than one maximizer, say $m(a)$ and $m'(a)$, then $M(a) = f(m(a), a) = f(m'(a), a)$ must hold. Many economic and game-theoretic problems involve optimization problems of this kind. The following notation is therefore useful:

$$M(a) = \max\{f(x,a) \mid x \in X(a)\}$$

denotes the maximum value of the optimization problem and

$$m(a) \in \operatorname{argmax}\{f(x,a) \mid x \in X(a)\}$$

denotes a maximizer in the set of (possibly) many maximizers

$$\operatorname{argmax}\{f(x,a) \mid x \in X(a)\}.$$

Sometimes it may be useful to indicate the choice variables underneath the "max" or "argmax," as in

$$M(a) = \max_x \{f(x,a) \,|\, x \in X(a)\},$$

but mostly this will be unnecessary because writing $M(a)$ and $m(a)$ for the maximum and the maximizer indicates that a is a parameter that leaves only x as choice variables. Special cases of this notation are

$$M = \max\{z_1, z_2, \ldots, z_n\} = \max\{z_i | i \in N\}$$

for $N = \{1, 2, \ldots, n\}$, and

$$M(a) = \max\{f_1(a), \ldots, f_n(a)\} = \max\{f_i(a) | i \in N\},$$

where the maximizer is implicitly given by the index numbers in N. For an economic application one can write

$$f(p, y) = \operatorname{argmax}\{u(x) \,|\, p \cdot x \leq y, x \geq 0\}$$

for the demand function of a consumer with strictly concave utility function $u(x)$. This notation makes the derivation of the demand function from a utility maximization problem transparent. Note that one can use an equality with "argmax" if one is certain that there is a unique maximizer.

Acknowledgments

In 1986, when I was asked to teach a course on game theory at the Australian National University in Canberra, my prior knowledge of the subject was very limited. I had learned some cooperative game theory from Volker Böhm, and my interest in the subject had been stimulated by a series of lectures on bargaining theory that Ariel Rubinstein had given at the University of Western Ontario the year before. In organizing this course, I found that there was little literature on game theory available at that time that one could easily use as a text for students. James Friedman's (1986) *Game Theory with Applications to Economics* and Eric Van Damme's (1983) *Refinements of the Nash Equilibrium Concept*, together with journal articles, became the first reference material for myself and for my students. I have learned a good deal of game theory from these two books and that makes it appropriate to begin with a statement of my intellectual debt to these authors. A second major source of influence that helped to shape this book, though it was not clear at the time, was the interaction with students during the following years. The selection of topics and of examples in this book has been influenced by that interaction with them.

The idea to write *Game Theory for Economists* dates back to 1990 when I joined the University of Melbourne. The University of Melbourne supported this book with two teaching relief grants from the Faculty Research fund, which reduced my teaching load during two semesters. This book was written in a first draft as eight discussion papers that were circulated in 1991 and 1992. Many people have read the book or parts of it during the two years of its production phase. They provided helpful comments from various perspectives. In particular, I would like to thank Vincent Crawford and Hans Haller for reading the whole manuscript and for providing me with many suggestions that helped to improve the exposition. Frank Milne and David Kelsey used part of the book for their own game theory course. Their constructive proposals helped me revise Chapters 1–5 and Chapter 8. Volker Böhm observed some inaccuracies in Chapter 10, and Simon Grant saved me from some errors in Chapter 6. Jeff Borland's careful reading of the first four chapters helped to improve my English style and grammar. I am very grateful for the many comments and suggestions on different chapters by Richard Arnott, Ngo Van Long, Ian McDonald, John McMillan, Bob Marks, and Li-Anne Woo. My final word of thanks, however, goes to my wife for her support during many months of work beyond the usual hours.

1

Formal Representations of Games

Game theory studies the behavior of *rational players* in interaction with other *rational players*. Players act in an environment where other players' decisions influence their payoffs. Players are considered to be *rational* if they maximize their objective functions given their beliefs about the environment.

To specify the environment and how the players perceive it, a considerable amount of terminology is necessary to obtain unambiguously clear descriptions of

- the conditions under which an interaction takes place
- the state of information of the players
- the motivation of the participating individuals

Though this chapter may appear somewhat tedious, it is the foundation on which all the following propositions rest. Having a clear understanding of the terminology and notation used in game theory will make seemingly difficult theorems obvious later. The time it takes to study this first chapter is, therefore, a good investment for understanding later chapters.

There are essentially three ways to represent a social interaction as a game.

(i) The most complete description is called *extensive form*. It details the various stages of the interaction, the conditions under which a player has to move, the information an agent holds at different stages, and the motivation of agents.

(ii) More abstract is the *strategic form* (or *normal form*) representation of a game. Here one notes all possible strategies of each agent together with the payoff that results from strategy choices of the agents. In the strategic form, many details of the extensive form have been omitted. This allows one to concentrate on strategic aspects of a game, but neglects the dynamic structure of the game.

(iii) Whereas the strategic form can be viewed as a reduced version of the extensive form, the *characteristic function form* (or *coalitional form*) of a game represents more than just a further abstraction from details of the game. It is a description of social interactions where binding agreements can be made and enforced. Binding agreements allow groups of players, or *coalitions*, to commit themselves to actions that may be against the interest of individual players once the agreement is carried through. This representation of a game is particularly useful for analyzing distributional questions. The distribution of payoffs from the joint outcome among the members of a coalition will determine what kind of agreement can be reached and what kind of coalitions will form.

1.1 Games in Extensive Form

The extensive form is the most explicit description of a game. It notes the sequence of moves, all possible states of information, and the choices at different stages for all players of the game. Not surprisingly, therefore, it requires extensive notation.

1.1.1 The Game Tree

Most board games allow players to observe the moves of opponents. Players move in an order specified by the rules. This sequence of possible moves can be represented as a *game tree* describing the structure of the game.

EXAMPLE 1.1 (*Chess*). Consider a board game like chess. It has as basic elements

- a set of two players, say player 1 and player 2
- a set of rules (i.e., about which player is to move next in a given situation and what moves are feasible for this player)
- a description of those situations where the game ends

The only piece of information missing in order to be able to talk about "good" or "bad" strategies is each player's evaluation of the final outcome, which in chess is a win, a loss, or a draw. ■

These elements will now be described more precisely. The set of players is denoted by I. In most applications, it is a finite set described by listing its elements, $I = \{1, 2, \ldots, I\}$.[1] An arbitrary player from the set I is denoted by i, that is, $i \in I$.

Since the moves open to a player may depend on the situation, one has to specify what "situation" means. Each possible situation in a game is referred to as a *node*. Hence, a game is characterized by a set of nodes N. In chess, the initial position of the figures would be represented formally by a node. After the first player has moved, the play position reached would be represented by another node.

There is a natural sequence of positions (nodes) in games like chess. A *move* (or *action*) of a player leads from one node to the next node. The set of possible actions in a game is denoted by A, no matter which player takes them. There may be more nodes in a game than actions because, as will be discussed in detail, in some games positions have to be distinguished according to the information that players have. There are, however, always at least as many nodes as there are actions in a game.

Example 1.2. If $a, b, c, \ldots$ denote different nodes (positions) and $\alpha, \beta, \chi, \ldots$ denote different actions, then one can specify various orders of nodes indicating how one position arises from another.

In this example, the set of nodes is $N = \{a, b, c, d, e, f, g, h\}$ and the set of actions is $A = \{\alpha, \beta, \chi, \delta, \epsilon, \phi, \gamma\}$.

These game trees represent sequences or orders of moving. Nodes (positions) can be reached along the branches of the tree from left to right.

[1]It is common practice to denote the last element of the player set by the same symbol as the set itself where this does not cause confusion.

Note that in the left diagram each position has a unique predecessor, whereas in the right diagram node e has two predecessors, namely b and c. ■

In games like chess it is not uncommon that different sequences of moves lead to the same position of the figures. For example, in the right tree of example 1.2, action δ, moving down from b, and action ϵ, moving up from c, lead to the same node e. It is therefore not possible to see how the player reached node e. For some games this may make sense, but it is not appropriate in general. Even in chess, the way in which a certain position has been reached may provide information about the behavior of the opponent that is not captured by the position of the figures alone. Therefore two identical positions (nodes), reached in different sequences of moves, will be distinguished, as in the left tree of example 1.2.

Remark 1.1 (infinite sets of nodes and actions). In this exposition of the extensive form, examples will always involve finite sets of actions A and finite sets of nodes N. This restriction is necessary for a diagrammatic representation of game trees. In the formal definitions, however, such a restriction is not necessary. It is therefore possible to apply the following definitions to infinite sets of nodes and actions.

Economic applications often require infinite sets of nodes and actions. In oligopoly theory, for example, one considers arbitrary output levels as possible actions of a firm. This means that the set of actions for this firm is the set of positive real numbers, which is a continuum and therefore an infinite set. Such examples will appear in Chapter 3.

Another example is a monopoly under permanent threat of entry. *Permanent* means here that there is an infinite sequence of potential moves to enter the market of the monopolist. This implies an infinite set of actions ("to enter") and an infinite set of nodes. This example will be studied in Chapter 8. ■

Denote by N the set of nodes of the game tree and by o ("origin") the initial node. Let σ: $N \to N$ be the function that associates with each node other than the origin o, its predecessor, and, for the origin, define $\sigma(o) = o$. For any node $n \in N$ and any positive integer k, $\sigma^k(n)$ indicates the kth iteration of the function σ, that is,

$$\sigma^k(n) \equiv \underbrace{\sigma(\sigma(\sigma(\ldots\sigma(n)\ldots)))}_{k\text{-times}}.$$

Definition 1.1. A *game tree* is a set of nodes N and a function σ: $N \to N$, $\sigma(o) = o$, such that, for all nodes $n \in N$, $\sigma^k(n) = o$ for some positive integer k holds.

The condition $\sigma^k(n) = o$ for all n is necessary to ensure that all nodes are connected to the origin, that is, that the nodes form a tree. Example 1.2 can be used to illustrate the definition.

EXAMPLE 1.2 (*resumed*). The set of nodes is $N = \{a, b, c, d, e, f, g, h\}$ and the origin is node a, $o = a$. In the left tree one has

$$\sigma(h) = \sigma(g) = \sigma(f) = c, \qquad \sigma(e) = \sigma(d) = b, \qquad \sigma(b) = \sigma(c) = a.$$

The condition that all nodes must be connected with the origin is satisfied, since one has

$$\sigma^2(h) = \sigma(\sigma(h)) = a, \qquad \sigma^2(g) = \sigma(\sigma(g)) = a,$$
$$\sigma^2(f) = \sigma(\sigma(f)) = a, \qquad \sigma^2(e) = \sigma(\sigma(e)) = a,$$
$$\sigma^2(d) = \sigma(\sigma(d)) = a.$$

On the other hand, the right tree does not satisfy the definition since e has two predecessors and the function σ would not be defined for node e. ■

According to the definition of the game tree, nodes are connected to other nodes by the predecessor node function σ. Players choose actions to move from one situation to another. Thus, one has to specify how actions lead from one node to another. This is accomplished by a predecessor action function α: $N \setminus \{o\} \to A$, which associates with each node n (except the origin o) the action $\alpha(n)$ leading from the predecessor node $\sigma(n)$ to node n. Example 1.2 may be used again to illustrate the concept of the predecessor action function.

EXAMPLE 1.2 (*resumed*). Reconsider the left tree of example 1.2 repeated here.

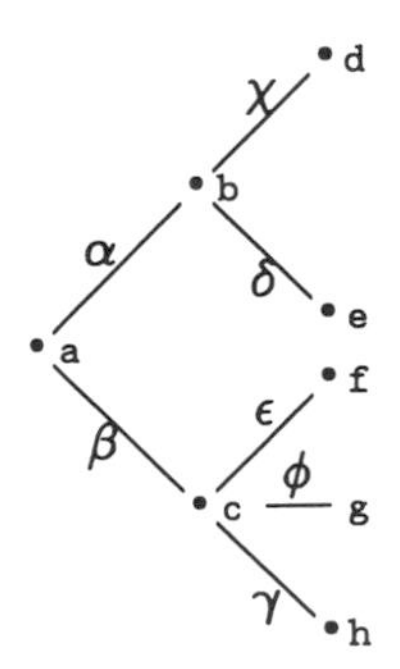

In this example, the set of nodes is $N = \{a, b, c, d, e, f, g, h\}$ and the set of actions is $A = \{\alpha, \beta, \chi, \delta, \epsilon, \phi, \gamma\}$. The function α can be written explicitly in this case:

$$\alpha(b) = \alpha, \quad \alpha(c) = \beta, \quad \alpha(d) = \chi, \quad \alpha(e) = \delta,$$
$$\alpha(f) = \epsilon, \quad \alpha(g) = \phi, \quad \alpha(h) = \gamma.$$

Note that the action leading from $\sigma(n)$ to n is uniquely defined for all nodes in N except the initial node a. By definition, there is no action leading to the initial node. ■

Situations of social interaction (nodes) can be classified in various ways: one may want to distinguish situations in which one agent has to act from those in which another agent is due to move, or indicate situations in which the game ends. In this way one distinguishes

- terminal nodes
- decision nodes
- decision nodes of a player (player partition)

First, there are situations in which some agent has to move, and situations in which the game ends and no further moves take place. Accordingly, one can distinguish *decision nodes* and *terminal nodes*. Let $\sigma^{-1}(n)$ be the inverse image of the function σ, that is, the set of nodes that have node n as a predecessor. A node is called a *terminal node* if it is the predecessor of no other node, that is, if $\sigma^{-1}(n) = \varnothing$ holds. A *decision node* is a node that is not a terminal node. Hence, one can write the set of terminal nodes as

$$\mathscr{T}(N) = \{n \in N | \sigma^{-1}(n) = \varnothing\} \qquad \text{(set of terminal nodes)},$$

and the set of decision nodes as

$$\mathscr{D}(N) = N \setminus \mathscr{T}(N) \qquad \text{(set of decision nodes)}.$$

Notice there may be games with no terminal nodes. In those games all nodes are decision nodes. Examples of such games will appear in Chapter 8. Example 1.2 may be used to illustrate these concepts.

EXAMPLE 1.2 (*resumed*). Consider the left game tree. In this case, the set of terminal nodes is $\mathscr{T}(N) = \{d, e, f, g, h\}$, and the set of decision nodes is $\mathscr{D}(N) = \{a, b, c\}$. ■

Observe that nodes can be unambiguously classified as decision or terminal nodes. Each node belongs either to the set of terminal nodes $\mathscr{T}(N)$ or to the set of decision nodes $\mathscr{D}(N)$, but never to both. Such a classification is called *partition* of the set of nodes. Formally, one can express the fact that decision nodes and terminal nodes partition the set of nodes by the following two conditions: $\mathscr{D}(N) \cup \mathscr{T}(N) = N$ and $\mathscr{D}(N) \cap \mathscr{T}(N) = \varnothing$.

With the predecessor action function α, one can identify the actions available at any decision node of the game tree. For any node $n \in \mathscr{D}(N)$, denote by $A(n)$ the set of actions $\alpha(m)$ from nodes m that have n as their predecessor node; formally, $A(n) = \{\alpha(m) | \sigma(m) = n\}$. Notice that the set of actions at terminal node n, $A(n)$, is not well defined. This justifies the terminology of calling nodes in $\mathscr{D}(N)$ decision nodes.

Secondly, a classification of decision nodes according to the player who makes the decision there has to be introduced. Denote by N_i the set of decision nodes at which player $i \in I$ has to choose an action. In a particular situation (node) one and only one agent is assumed to move.[2] Hence, the set of decision nodes, $\mathscr{D}(N)$, can be partitioned into mutually exclusive subsets N_i, $i \in I$.

DEFINITION 1.2. A list of mutually exclusive sets of decision nodes for each player, $(N_i)_{i \in I}$, is called *player partition* of $\mathscr{D}(N)$.

A player partition assigns nodes to players who have to make a decision there. Therefore, one can define the set of all possible actions of some player i, A_i, as the union of all sets $A(n)$ over nodes n in N_i. Example 1.2 shows a player partition and the way it is usually indicated in the diagram of a game tree.

EXAMPLE 1.2 (*resumed*). There are three decision nodes in this example: a, b, and c. Suppose the description of the social interaction indicates there are two players, player 1 and player 2 respectively, and that player 1 moves first at node a, and that player 2 moves second at node b or c, depending on the move player 1 made previously. Then the set of decision nodes $\mathscr{D}(N)$ can be split up further into decision nodes of player 1, $N_1 = \{a\}$, and of player 2, $N_2 = \{b, c\}$. (N_1, N_2) is the player partition of this game.

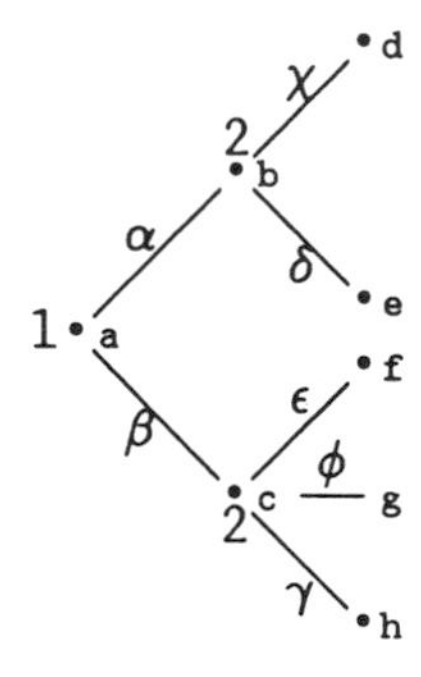

[2]At this stage the possibility of simultaneous moves of several players is disregarded. The possibility of representing such a case in a game tree will be discussed later.

Notice that decision nodes are now labeled with 1 or 2 depending on who is to move. In addition, one has the following sets of actions at the decision nodes $\mathscr{D}(N) = \{a, b, c\}$:

$$A(a) = \{\alpha, \beta\}, \qquad A(b) = \{\chi, \delta\}, \qquad A(c) = \{\epsilon, \phi, \gamma\}.$$

The set of actions for player 1, A_1, is $\{\alpha, \beta\}$, and the set of actions for player 2, A_2, is $\{\chi, \delta, \epsilon, \phi, \gamma\}$. ■

It remains to specify the payoffs of the players at the end of the game. For finite games,[3] it is sufficient to associate the payoffs of the players with the terminal nodes. The payoff function associates with each node $n \in \mathscr{T}(N)$ a payoff vector $(r_1(n), r_2(n), \ldots, r_I(n))$ that specifies the payoff for each player at this terminal node. For each player $i \in I$, $r_i(n)$ is a number indicating the payoff to player i if the terminal node n is reached.

Definition 1.3. The payoff function r: $\mathscr{T}(N) \to \mathbb{R}^I$ associates with each terminal node a vector of real numbers, the payoff to each of the players for each terminal node.

The following example will illustrate how payoffs are indicated in the game tree.

Example 1.2 (*resumed*). Payoffs will be indicated in diagrams of game trees as rows or columns of numbers at each terminal node. The interpretation is simply that the first number indicates the payoff of player 1 at that terminal node, the second number gives the payoff for player 2 at that node, and so on.

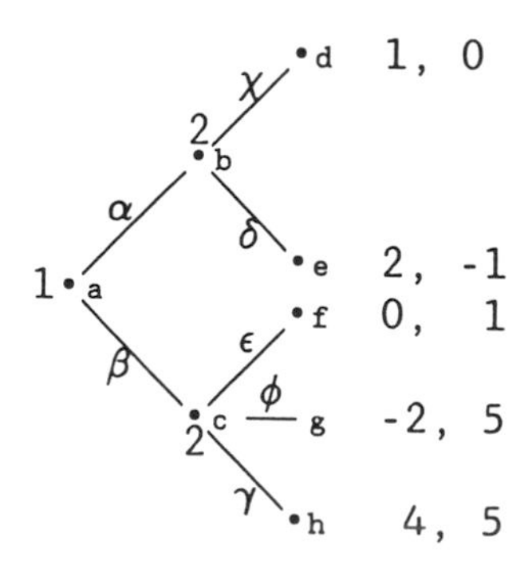

[3]For games with an infinite number of nodes, one has to associate payoffs with sequences of actions directly. This procedure will be introduced in Chapter 8, the only place it is needed in this book.

Hence, one has the following payoff function for player 1:

$$r_1(d) = 1, \quad r_1(e) = 2, \quad r_1(f) = 0, \quad r_1(g) = -2, \quad \text{and} \quad r_1(h) = 4,$$

and for player 2:

$$r_2(d) = 0, \quad r_2(e) = -1, \quad r_2(f) = 1, \quad r_2(g) = 5, \quad \text{and} \quad r_2(h) = 5.$$

■

1.1.2 A First Result: Chess and Zermelo's Theorem

No behavioral assumptions have been made so far. The structure of a game could be described without reference to how the game should or would be played. Of course, what one would be interested to know is whether, for a particular game, there is an unbeatable strategy. If a player can win no matter what the opponent does, then the outcome is a foregone conclusion.

In fact, for many simple board games, one knows that a player can play the game with a guaranteed win. For more complex games like chess, however, even with supercomputers, the search for an optimal strategy has not yet been successful. This raises the question of whether the quest for an optimal strategy might not have been doomed from the beginning because none exists. The following theorem, probably one of the earliest in game theory, provides an answer to this question.

THEOREM 1.1 *(Zermelo, 1913). In chess, either white can force a win, or black can force a win, or both sides can force a draw.*

Before proving this theorem, note that it claims that chess is a game with a determinate result. The theorem, however, gives no hint as to which player will win, let alone what a winning strategy looks like. But it indicates that the search for an optimal strategy may be successful. The following example illustrates the idea of the proof.

EXAMPLE 1.1 (*resumed*). Chess is a game with a finite tree since its rules usually declare the game a draw if some position is repeated a predetermined number of times. Hence, if it were not too complex to be drawn, its game tree would be of the type shown. In the diagram, W denotes the white player, B the black player, and ω, β, and δ denote the outcomes "white wins," "black wins," or "draw." It is assumed that each player prefers winning to drawing and drawing to losing. The game ends in maximal four moves.

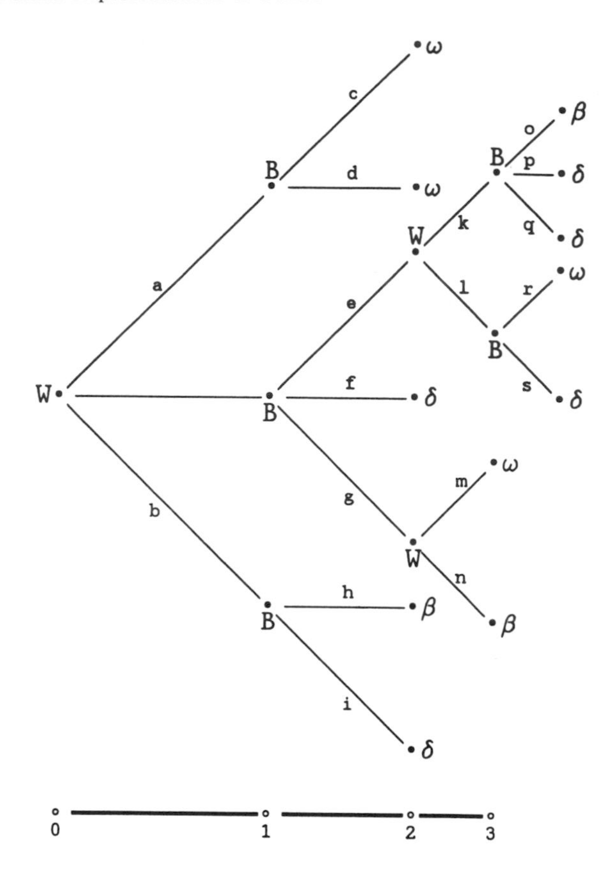

Consider the last stage first. If stage 3 is reached, black has to move. If black is at the upper node she will choose the action o; at the lower node she will choose the action s. Hence, in stage 2 when white comes to move, provided the game does not end earlier, white will choose move l, which yields a draw if she is at the upper node, or m, the winning move, if she is at the lower node. Finally, in stage 1, black will choose the winning action h at the lower node, an action leading to a draw (either e or f) at the middle node, but cannot avoid a white win in the upper node. This allows the white player to force a win by moving to the upper node in stage 0. ■

This procedure for deriving a game's solution by working backward in the way described in the example is called *backward induction*. It can be applied to finite games in which players can observe all moves and in which no ties in the payoffs of some player occur. A formal and more complete description of the backward induction procedure will follow in Chapter 6. The proof of Zermelo's theorem concludes this subsection.

Proof. Note that chess is a finite game, that is, there is a finite number of moves T such that no game lasts longer than T stages. Consider a class of games called chess(t), where t is any natural number. The game chess(t) has the same rules as normal chess with the exception that the game is declared a draw if it does not end with a win until the tth move. Since regular chess ends in T moves, it coincides with chess(T). It suffices, therefore, to show that the theorem holds for an arbitrary game chess(t).

The proof that, in chess(t), for any t either white can force a win or black can force a win or both can force a draw proceeds by induction.

For $t = 1$, if black moves first,

- either black has a move that wins, that is, black can force a win
- or for any move of black, white will win, that is, white can force a win
- or both can force a draw, since the game chess(1) is declared a draw after one move that does not result in a win

If white moves first, the same argument holds. Hence, the theorem is true for chess(1).

Now suppose the theorem is true for chess($t - 1$). Clearly, if chess($t - 1$) allows black to force a win, chess(t) will have the same outcome and our theorem holds for chess(t). Similarly, if chess($t - 1$) allows white to force a win, then chess(t) has the same outcome and the theorem is true. Therefore, suppose that white and black can force a draw in chess($t - 1$).

If white moves next

- either white has a move that wins, that is, white can force a win
- or for any move of white, black will win, that is, black can force a win
- or both can force a draw since the game chess(t) is declared a draw after t moves have brought no win

A similar argument holds if black comes to move in the last stage of chess(t). Thus, the theorem holds for chess(t). ■

1.1.3 Information Sets

The description of the extensive form has not yet accounted for the possibility that a player may be unable to observe the moves of opponents. Since actions lead from nodes to other nodes, a player who cannot observe the action of another player will not know at which decision node she finds herself. To understand the information problem consider the following example.

EXAMPLE 1.3 (*Matching pennies*). Two players, player 1 and player 2, each put a coin on the table but keep their moves hidden from each other.

Player 1 puts her coin down first, then player 2 does the same. Finally, they reveal to each other the sides of the coins lying face up on the table. If the sides match, player 1 wins a dollar from player 2; if the sides do not match, player 2 wins a dollar.

Denote by H ("heads") and T ("tails") the choices of player 1 and by h and t the choices of player 2.

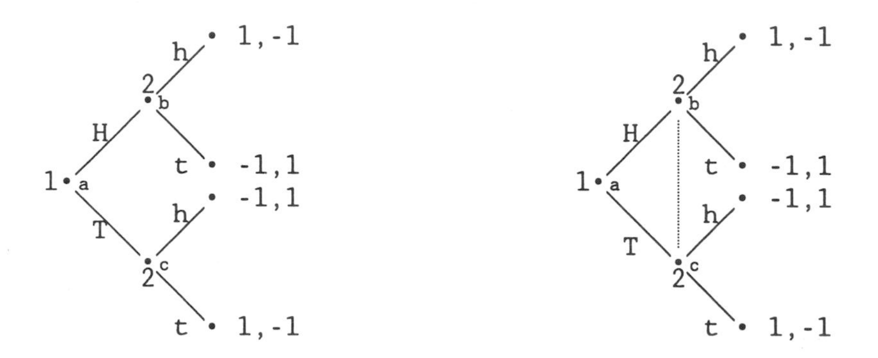

For each player, it has been assumed that an upward move indicates a choice of H (h), whereas a downward move indicates the choice T (t). The left tree, however, does not correspond to the verbal description of the situation, since it does not indicate that player 2 cannot observe player 1's move and, therefore, does not know whether she is at node b or node c when she makes her choice. To show that a player may not be able to distinguish nodes, one joins the decision nodes that are indistinguishable in a set, called an *information set*. In diagramming game trees, information sets are indicated by joining the particular nodes with a dotted line (-----). ■

Example 1.3 illustrates the importance of informational assumptions. If player 2 knew whether she was at node b or c, she could choose t at b and h at c, which would guarantee a win. Without knowing player 1's move, player 2 would not be in such a fortunate situation.

To model situations in which players have to make a decision but do not know at which decision node they find themselves because they are unable to observe some moves which led to these nodes, one splits the set of decision nodes of each player into sets of nodes a player cannot distinguish, which are called information sets. An information set, denoted by u, is a set of decision nodes for the same player, for example, $u = \{b, c\}$ for player 2 in example 1.3. It is possible that an information set contains a single node only. The information set $u' = \{a\}$ of player 1, in example 1.3, is such a case. In this way, an information set also covers the case in which players know exactly their situation.

Several conditions must hold for nodes that are joined in an information set for this concept to reflect the idea that a player is ignorant about the precise node of the information set in which she finds herself:

(i) Information sets of player $i \in I$ contain only decision nodes of player i.
(ii) Each decision node of player i is contained in one and only one information set of this player.
(iii) The same choices must be available to a player at all nodes of an information set.

The first two conditions say that information sets partition the set of decision nodes of a player, N_i. Nodes in the same information set are indistinguishable for this player. For each player $i \in I$, U_i denotes the set of all information sets of this player.

To see that condition (iii) is necessary, imagine two nodes at which the player has two moves available at one, and three moves at the other, for example, nodes b and c in example 1.2. To know that she has three moves available (instead of two) would reveal to the player at which node she makes the choice. Recall that $A(n) = \{\alpha(m) | \sigma(m) = n\}$ denotes the set of actions (moves, choices) available at node n. Condition (iii) requires that, for any information set u, $A(x) = A(y)$ for all $x, y \in u$ holds. The set of choices at u, $A(u)$, is, therefore, unambiguously given by the set of actions available at some node of this information set, $A(u) = A(x)$ for some $x \in u$. Example 1.3 illustrates these concepts.

Example 1.3 (*resumed*). Consider the right-hand tree: Each player has one information set, $U_1 = \{\{a\}\}$ and $U_2 = \{\{b, c\}\}$. The set of actions at the three decision nodes a, b, and c are

$$A(a) = \{H, T\}, \qquad A(b) = \{h, t\}, \qquad A(c) = \{h, t\}.$$

Since the information set of player 1, $\{a\}$, contains only a single node, the set of choices at $\{a\}$, $A(\{a\}) = A(a) = \{H, T\}$. For player 2, however, the information set contains two nodes and the actions at these two nodes must be the same, $A(\{b, c\}) = A(b) = A(c) = \{h, t\}$. ■

The following definition summarizes these properties.

Definition 1.4. U_i is the set of information sets of player $i \in I$. For any $u \in U_i$, the set of choices at u is denoted by $A(u)$. The list of all these sets $(U_i)_{i=1}^{I}$ is called *information partition*.

It is now possible to see how simultaneous moves can be included in the formal description of a game. In example 1.3, player 1 moved first by making a choice of H or T, which player 2 could not observe. Then player 2 made a decision on which action to take, h or t. Therefore, both players do not know what the other player's move is when they have to decide what to do. In the case of player 1 this is so because the move of player 2 has not taken place yet, whereas in the case of player 2 the move of player 1 could not be observed. Hence, when they make their moves, the state of information about the behavior of the other player is the same for both of them.

What matters in the case of simultaneous moves is not that the choices literally take place at the same moment, but that no player knows the opponent's decision when she makes her own move. Therefore, the same formal representation as in the right-hand diagram of example 1.3 could have been used to describe a simultaneous move of both players. The sequence of moves is immaterial in a simultaneous move game. What is important is the information that players have when they make their choices. Therefore, there are usually many equivalent ways to formally represent a simultaneous move game in the extensive form. The following modification of example 1.3 to a simultaneous move game illustrates this point.

EXAMPLE 1.3 (*modified*). Consider the game of example 1.3 for the case where both players put down their coins simultaneously. Everything else remains unchanged.

The following two game trees are equivalent representations of this game.

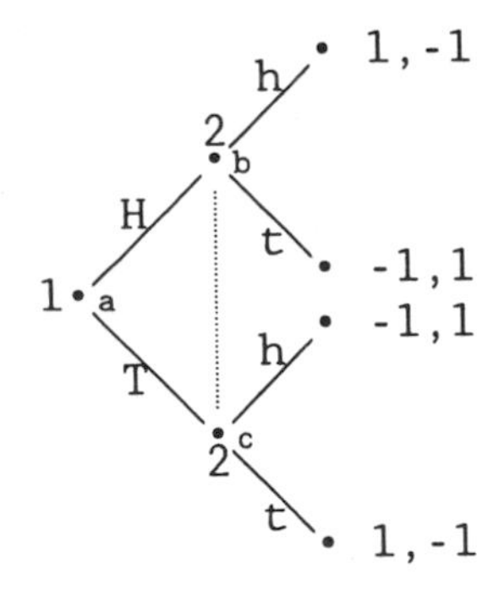

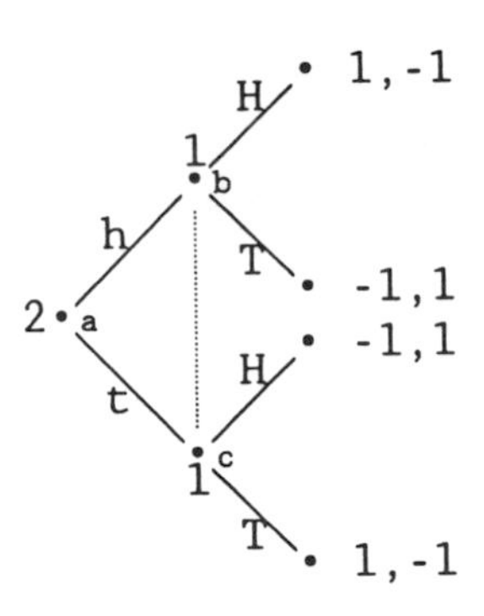

The fact that player 1 moves first in the left-hand tree while player 2 moves first in the right-hand tree is without consequence for the choice situation in which the agents find themselves. ■

This completes the description of a game in extensive form. The following list summarizes the concepts introduced so far.

- the set of players I
- a set of situations N and the sequence of their appearance σ, the game tree (N, σ)

- the set of actions and a function that relates actions to nodes (A, α)
- who comes to move at the decision nodes, the player partition $(N_i)_{i \in I}$
- which decision nodes each player can distinguish, the set of information sets U
- what choices a player has at each information set, the set of choices at information sets $(A(u))_{u \in U}$
- what the payoffs to the players are when the game ends, the payoff function r.

DEFINITION 1.5. A game in *extensive form* is completely specified by the following list Γ:

$$\Gamma = \big(I, (N, \sigma),\ (A, \alpha),\ (N_i)_{i \in I},\ U,\ (A(u))_{u \in U}, r\big).$$

This subsection closes with an example illustrating these concepts.

EXAMPLE 1.4. Consider the following game in extensive form.

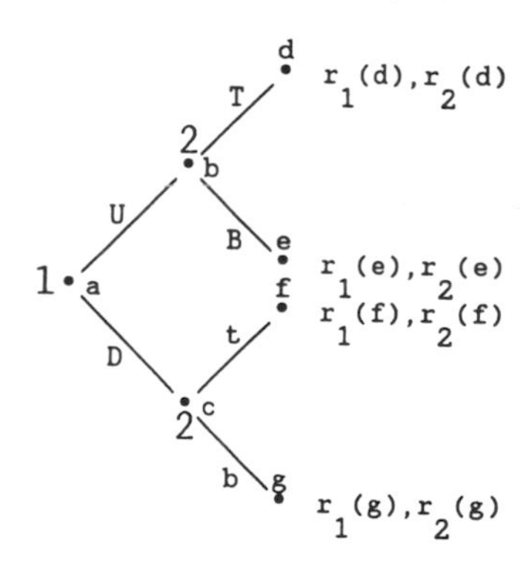

This game tree can be described by the following sets and functions:

$$I = \{1, 2\},$$

$$N = \{a, b, c, d, e, f, g\},$$

$$\sigma(a) = a, \qquad \sigma(b) = a, \qquad \sigma(c) = a, \qquad \sigma(d) = b,$$

$$\sigma(e) = b, \qquad \sigma(f) = c, \qquad \sigma(g) = c,$$

$$A = \{U, D, T, B, t, b\},$$

$$\alpha(b) = U, \qquad \alpha(c) = D, \qquad \alpha(d) = T, \qquad \alpha(e) = B,$$

$$\alpha(f) = t, \qquad \alpha(g) = b,$$

$$N_1 = \{a\}, \qquad N_2 = \{b, c\},$$

$$U_1 = \{\{a\}\}, \qquad U_2 = \{\{b\}, \{c\}\},$$

$$A(\{a\}) = \{U, D\}, \qquad A(\{b\}) = \{T, B\}, \qquad A(\{c\}) = \{t, b\}. \qquad \blacksquare$$

1.1.4 What Do Players Know about a Game?

Given definition 1.5, one can distinguish games in extensive form according to the information that players have about the actions chosen by the other players. Games in which each player knows exactly what has happened in previous moves are called *games with perfect information*. Games in which there is some uncertainty about previous moves are called *games with imperfect information*. Formally stated:

DEFINITION 1.6. The game $\Gamma = (I, (N, \sigma), (A, \alpha), (N_i)_{i \in I}, U, A(u))_{u \in U}, r)$ is a game with perfect information if each information set $u \in U$ contains only a single node. Otherwise it is called a game of imperfect information.

Chess (example 1.1) is a game with perfect information because each player can observe what has happened in previous moves. This implies for the game tree that no agent is ever uncertain as to the node at which she is located. Hence, all information sets contain just one element and each move is made with perfect information about the situation. Matching pennies, on the other hand, is a game with imperfect information because player 2 does not know what player 1 chose in the first stage of the game. This is reflected in the fact that the only information set of player 2, $u = \{b, c\}$, has two elements.

The notion of perfect and imperfect information refers to a property of the extensive form of a game and has to be distinguished from the question of whether agents are informed about the extensive form of the game. Even if an outside observer is able to describe a situation of social interaction in the detail required for the extensive form of the game, this in itself does not indicate whether each player knows all the elements of this description. For example, there may be games in which one agent does not know another player's payoff from a certain course of play. Or, a player may not know whether her opponents can observe her move.

At the beginning of this chapter, rational behavior was introduced as payoff-maximizing behavior of players given their information about the environment. To analyze the behavior of a rational player, it is therefore important to describe what each player knows about the structure of the game. Furthermore, if the optimal behavior of a player depends on the opponent's action, then the player needs to know what this opponent knows about the game and her behavior. But if this opponent's behavior is affected by the first player's action, then the first player would want to know if the opponent knows that she is fully informed about the game.[4]

The concept of *common knowledge* states the assumption that, to any degree of mutual understanding, all players of a game are completely

[4]Rubinstein (1989) provides a good example of how ignorance about whether a message has been received may influence the optimal behavior of agents.

informed about some aspect of the game. A piece of information is common knowledge if all players know it, and know that all other players know it, and know that all other players know that all other players know it, up to any degree of mutual knowledge.[5] The notion of common knowledge allows one to distinguish games according to how much information players have about the game.

DEFINITION 1.7. The game $\Gamma = (I, (N, \sigma), (A, \alpha), (N_i)_{i \in I}, U, (A(u))_{u \in U}, r)$ is a game with complete information if each element of Γ is common knowledge. Otherwise it is a game with incomplete information.

Notice the difference between perfect and imperfect information and complete and incomplete information. The former is a structural property of a game, whereas the latter refers to an informational characteristic of players. Chapter 5 will show how one can extend the description of a game to capture incompleteness of information. Up to this stage only games with complete information will be considered.

1.2 Concepts of Strategies

The basic approach for determining or predicting how the interaction between players in a given game will proceed consists of two elements:

(i) Descriptions of strategies of players
(ii) Principles for selection of particular strategies

Thus, the concept of a strategy is fundamental to game-theoretic analysis.

A *strategy* is a complete plan for how to play the game. "Complete" means in this context that for any contingency the plan must specify what the player would do. Thus, a strategy can be viewed as an instruction for a referee of how the player will move when the play reaches a node at which she has to move.

1.2.1 Pure Strategies

The most immediate concept of a strategy is a pure strategy. A *pure strategy* describes for each information set of a player a unique action that will be taken if this information set is reached during the course of play.

DEFINITION 1.8. A *pure strategy* of player $i \in I$ is a function $s_i: U_i \rightarrow A$ with $s_i(u) \in A(u)$ for all $u \in U_i$. A *pure strategy combination* $s = (s_1, s_2, \ldots, s_I)$ is a list of pure strategies s_i, for each agent $i \in I$.

Recall that A is the set of all possible actions that an agent may choose at some information set. But players are constrained to choose

[5]Common knowledge is a subtle concept of game theory. Binmore (1992, Chapter 10) provides a good introduction to this concept.

from a particular subset of actions if they move at an information set u, $A(u)$, the set of choices at u. Therefore, according to definition 1.8, a pure strategy for a player specifies for each information set at which she must move, that is, for each $u \in U_i$, which action $s_i(u)$ from those available at that information set $A(u)$ this player will choose.

Denote by S_i the set of pure strategies of player i. Notice that, for games with finite sets of nodes, only a finite number of pure strategies exist and that S_i is a finite set in these cases. Given a pure strategy combination s, that is, a pure strategy $s_i \in S_i$ for each agent $i \in I$, a course of play for the game is completely determined. From playing these strategies the payoffs of the players can be associated, therefore, with a strategy combination $s = (s_1, s_2, \ldots, s_I)$. Let $p_i(s)$ be the payoff that player i obtains if the strategy combination s is played. For games in extensive form with a finite set of nodes, any pure strategy combination will lead to a terminal node. Denote by $n(s)$ the terminal node that is reached if the strategy combination s is played, then $p_i(s) \equiv r_i(n(s))$ holds for all players $i \in I$. The following example illustrates these concepts.

EXAMPLE 1.5. Consider the game in extensive form given by the following game tree.

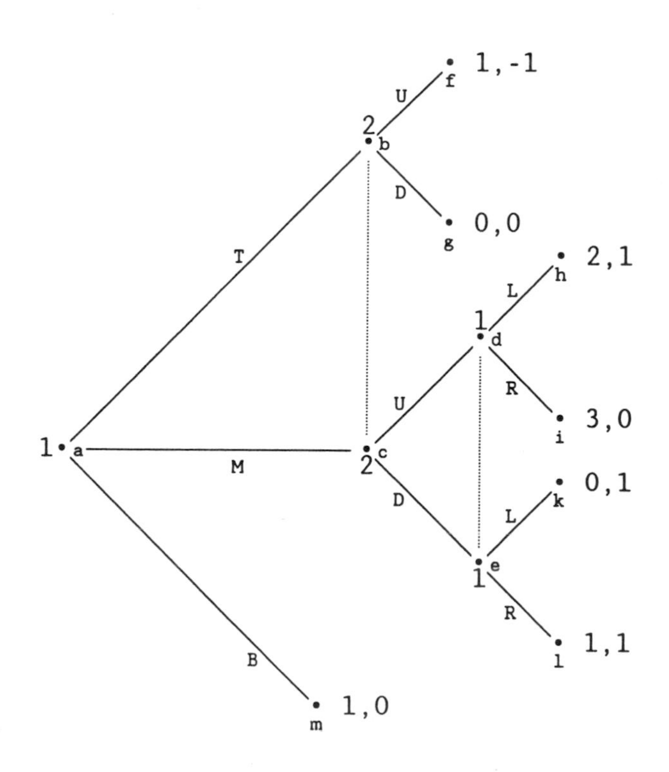

In this case one can write the set of strategies for each player explicitly.

$$S_1 = \{s_1^1, s_1^2, s_1^3, s_1^4, s_1^5, s_1^6\},$$

with

$$s_1^1 \equiv (s_1^1(\{a\}), s_1^1(\{d,e\})) = (T, L), \quad s_1^2 \equiv (s_1^2(\{a\}), s_1^2(\{d,e\})) = (T, R),$$
$$s_1^3 \equiv (s_1^3(\{a\}), s_1^3(\{d,e\})) = (M, L), \quad s_1^4 \equiv (s_1^4(\{a\}), s_1^4(\{d,e\})) = (M, R),$$
$$s_1^5 \equiv (s_1^5(\{a\}), s_1^5(\{d,e\})) = (B, L), \quad s_1^6 \equiv (s_1^6(\{a\}), s_1^6(\{d,e\})) = (B, R),$$

and

$$S_2 = \{s_2^1, s_2^2\},$$

with

$$s_2^1 \equiv (s_2^1(\{b,c\})) = (U), \qquad s_2^2 \equiv (s_2^2(\{b,c\})) = (D).$$

The payoffs for player 1, $p_1(s)$, from the twelve strategy combinations $s = (s_1, s_2)$ are given below.

$$p_1(s_1^1, s_2^1) = r_1(f) = 1, \quad p_1(s_1^2, s_2^1) = r_1(f) = 1, \quad p_1(s_1^3, s_2^1) = r_1(h) = 2,$$
$$p_1(s_1^4, s_2^1) = r_1(i) = 3, \quad p_1(s_1^5, s_2^1) = r_1(m) = 1, \quad p_1(s_1^5, s_2^1) = r_1(m) = 1,$$
$$p_1(s_1^1, s_2^2) = r_1(g) = 0, \quad p_1(s_1^2, s_2^2) = r_1(g) = 0, \quad p_1(s_1^3, s_2^2) = r_1(k) = 0,$$
$$p_1(s_1^4, s_2^2) = r_1(1) = 1, \quad p_1(s_1^5, s_2^2) = r_1(m) = 1, \quad p_1(s_1^6, s_2^2) = r_1(m) = 1.$$

Similarly, one can derive the payoff of player 2 from these strategy combinations. ■

A strategy specifies what a player will do even in situations that will not occur during the course of the game. In example 1.5, player 1 will never reach the information set $\{d, e\}$ if she chooses action T at information set $\{a\}$ and player 2 chooses D, or if she chooses action B. Nevertheless, a strategy has to specify what she would do at the information set $\{d, e\}$. Similarly, player 2 has to indicate what she would do at her information set even if she does not have to move, because player 1 chose a strategy which has B as the action at the information set $\{a\}$. Thus, the concept of a strategy requires considerable redundancy. This is unavoidable because a player may want to choose a strategy that avoids an information set of her opponent precisely because of the action the other player would choose at that information set.

1.2.2 Mixed Strategies

Choosing a pure strategy that specifies what the agent will do whenever she has to make a move, that is, at all information sets, may appear the obvious way to determine a course of action. But consider player 2 in the matching pennies game. Since she does not know the choice of player 1, it would not be unreasonable to toss the coin instead of putting it down deliberately, leaving to chance which side will lie face up. But tossing a coin means choosing with a certain probability the pure strategy to play H and with the complementary probability the pure strategy T. Such a behavior is called mixing pure strategies and the resulting strategy is called a *mixed strategy*.

Notice that the set of mixed strategies always includes all pure strategies because one can consider a pure strategy as the special case of a mixed strategy in which the respective pure strategy is played with probability 1 and any other pure strategy with probability 0. Hence, mixed strategies can be viewed as a natural generalization of pure strategies.

Remark 1.2. Generalizing the strategy concept is not the only reason for considering mixed strategies. In Chapters 2 through 4 it will be shown that there are games in which mutually consistent rational behavior is impossible if only pure strategies are allowed. Furthermore, in Chapter 5 it will be argued that, in games with incomplete information, one can interpret mixed strategies as the probability of meeting a particular opponent playing a pure strategy if the player is uncertain which opponent she faces. ■

Choosing a mixed strategy simply means that a player chooses a random device for selecting the pure strategy to be played. The type of random device chosen determines the probabilities with which the different pure strategies will be selected. Therefore, such a random device represents a probability distribution on the set of pure strategies. Players are assumed to choose their random device independently.[6]

Let S_i be the set of pure strategies of player i, and denote by M_i the set of all probability distributions on the set S_i. In general, the set of pure strategies S_i may be a finite set, that is, contain just a finite number of strategies, or it may be infinite. In this book, however, only mixed strategies on finite sets S_i are considered.

DEFINITION 1.9. A *mixed strategy* m_i is a probability distribution on the set of pure strategies, that is, $m_i \in M_i$. A list of mixed strategies $m = (m_1, m_2, \ldots, m_I)$ is called a *mixed strategy combination*.

If the set of pure strategies is finite, then M_i, the set of probability distributions on the set S_i, has a particularly simple form. Suppose there are n pure strategies in S_i, then the set of probability distributions on S_i is

[6] If players jointly choose a random device and condition their strategy choices on the outcome of this joint randomization, then they would play a correlated strategy; Fudenberg and Tirole, (1991, pp. 52–59).

the unit simplex of dimension n. This simplex will be denoted by $\Delta(S_i)$, or Δ^n if the number of strategies in S_i is known to be n. Formally,

$$\Delta^n = \left\{ (p^1, p^2, \ldots, p^n) \in \mathbb{R}^n_+ \,\middle|\, \sum_{\hbar=1}^{n} p^\hbar = 1 \right\}.$$

In a diagram, the 2-dimensional simplex can be represented as the points on the line segment joining the points (0, 1) and (1, 0), and the 3-dimensional simplex as the triangle determined by the points (1, 0, 0), (0, 1, 0), and (0, 0, 1).

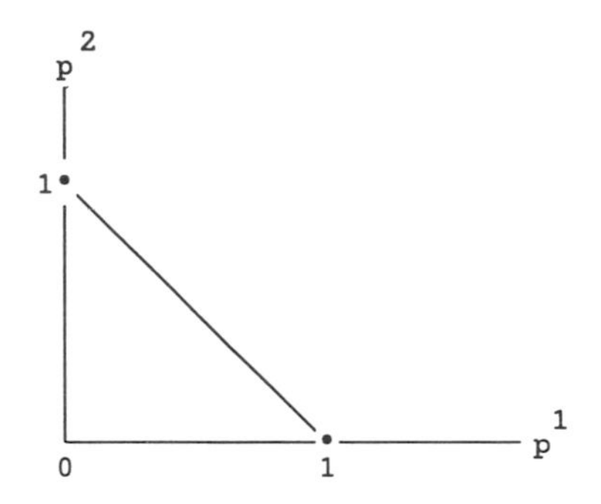

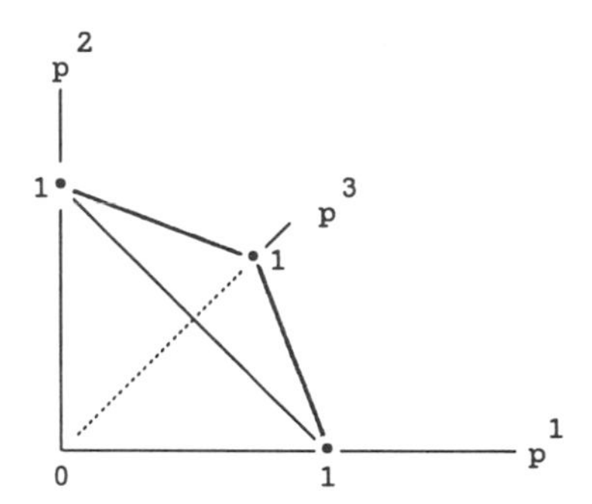

Since only mixed strategies on finite pure strategy sets will be considered, $M_i = \Delta(S_i)$ holds throughout this book. Let $m_i \in M_i$ be such a probability distribution and assume that the set of pure strategies contains n elements, $S_i = \{s_i^1, s_i^2, \ldots, s_i^n\}$, then $m_i(s_i^k)$ denotes the probability that the pure strategy s_i^k will be played according to the probability distribution m_i.

If players choose mixed strategies, that is, choose their pure strategies randomly, then the resulting play and outcome of the game become random variables and it is not obvious what payoff should be assigned for a mixed strategy combination. Allowing for mixed strategies makes the agents' choices a problem of decision making under uncertainty. It is therefore not surprising that von Neumann and Morgenstern (1947) provided one of the earliest axiomatic foundations for decision making under uncertainty in their book *The Theory of Games and Economic Behavior*.[7] Since this seminal contribution, the most common solution to this problem is to assume that agents rank the random outcomes of different mixed strategy combinations according to the expected payoff obtained.

Definition 1.10. The *expected payoff* of player $i \in I$ from a mixed strategy combination m, $P_i(m)$, is the expected value of the payoffs of

[7]Compare Machina (1987) for a modern exposition of this theory, a critical evaluation, and references to more recent developments.

player i generated by the mixed strategy combination m,

$$P_i(m) = \sum_{s \in S} p_i(s_1, s_2, \ldots, s_I) \cdot [m_1(s_1) \cdot m_2(s_2) \cdot \ldots \cdot m_I(s_I)].$$

Note that the summation takes place over the set of all possible pure strategy combinations $S \equiv S_1 \times S_2 \times \cdots \times S_I$.

The following example will illustrate the concept of expected payoff from a mixed strategy combination.

EXAMPLE 1.5 (*resumed*). Suppose that player 1 and player 2 choose the mixed strategies

$$\begin{aligned} m_1 &= (m_1(s_1^1), m_1(s_1^2), m_1(s_1^3), m_1(s_1^4), m_1(s_1^5), m_1(s_1^6)) \\ &= (0, .2, .2, .3, .1, .2) \end{aligned}$$

and

$$m_2 = (m_2(s_2^1), m_2(s_2^2)) = (.5, .5)$$

respectively. Then the following expected payoff for player 1 results:

$$\begin{aligned} P_1(m_1, m_2) = {} & p_1(s_1^1, s_2^1) \cdot [m_1(s_1^1) \cdot m_2(s_2^1)] + p_1(s_1^1, s_2^2) \cdot [m_1(s_1^1) \cdot m_2(s_2^2)] \\ & + p_1(s_1^2, s_2^1) \cdot [m_1(s_1^2) \cdot m_2(s_2^1)] + p_1(s_1^2, s_2^2) \cdot [m_1(s_1^2) \cdot m_2(s_2^2)] \\ & + p_1(s_1^3, s_2^1) \cdot [m_1(s_1^3) \cdot m_2(s_2^1)] + p_1(s_1^3, s_2^2) \cdot [m_1(s_1^3) \cdot m_2(s_2^2)] \\ & + p_1(s_1^4, s_2^1) \cdot [m_1(s_1^4) \cdot m_2(s_2^1)] + p_1(s_1^4, s_2^2) \cdot [m_1(s_1^4) \cdot m_2(s_2^2)] \\ & + p_1(s_1^5, s_2^1) \cdot [m_1(s_1^5) \cdot m_2(s_2^1)] + p_1(s_1^5, s_2^2) \cdot [m_1(s_1^5) \cdot m_2(s_2^2)] \\ & + p_1(s_1^6, s_2^1) \cdot [m_1(s_1^6) \cdot m_2(s_2^1)] + p_1(s_1^6, s_2^2) \cdot [m_1(s_1^6) \cdot m_2(s_2^2)] \\ = {} & (1) \cdot [0 \cdot .5] + (0) \cdot [0 \cdot .5] + (1) \cdot [.2 \cdot .5] + (0) \cdot [.2 \cdot .5] \\ & + (2) \cdot [.2 \cdot .5] + (0) \cdot [.2 \cdot .5] + (3) \cdot [.3 \cdot .5] + (1) \cdot [.3 \cdot .5] \\ & + (1) \cdot [.1 \cdot .5] + (1) \cdot [.1 \cdot .5] + (1) \cdot [.2 \cdot .5] + (1) \cdot [.2 \cdot .5] = 1.2. \end{aligned}$$

Similarly, one obtains the expected payoff of player 2. ■

In economics, one often works with expected utility in the case of decision problems involving uncertainty. In particular, the curvature of the expected utility function is taken to represent the agent's attitudes towards risk. In game theory, it is usually assumed that payoffs represent the subjective evaluation of the outcome of a play by the agent. This means that the payoffs are assumed to reflect a player's risk attitudes. Consequently, attitudes to risk are rarely explicitly considered in game-theoretic analysis.

1.2.3 Behavior Strategies

In moving from pure to mixed strategies the concept of a strategy was generalized by allowing agents to randomize over pure strategies. However, under a mixed strategy an agent still makes a deterministic choice at each information set at which she has to move. Another concept of strategy, *behavior strategy*, allows agents to randomize at the level of action choices. That means that agents are allowed random choices at each information set at which they have to make a move.

To formalize the notion of an agent choosing to randomize over possible choices at an information set u, denote by $B(u)$ the set of all probability distributions on actions at u, $A(u)$. Clearly, if $A(u)$ is a finite set (as it usually is in this book), then $B(u) = \Delta(A(u))$ holds, that is, the set of probability distributions on $A(u)$ is the simplex of a dimension equal to the number of choices at u. Finally, let $B = \bigcup_{u \in U} B(u)$ be the set of all possible probability distributions on choice sets.

DEFINITION 1.11. A *behavior strategy* of player i, $i \in I$, is a function b_i: $U_i \to B$ with $b_i^u \in B(u)$ for all $u \in U_i$. A *behavior strategy combination* b is a list of behavior strategies for all players, $b = (b_1, b_2, \ldots, b_I)$.

Analogous to the notation used in the case of mixed strategies, $b_i^u(a)$ will denote the probability that action $a \in A(u)$ will be chosen by player i according to the behavior strategy b_i^u.

In contrast to a pure strategy combination, a behavior strategy combination will not determine a unique path of play in the game, but rather will induce a probability distribution on the set of terminal nodes $\mathscr{T}(N)$ and, therefore, a probability distribution on agents' payoffs.

Denote by $q_b(n)$ the probability that terminal node n is reached if the behavior strategy combination b is played. This probability $q_b(n)$ can be easily calculated by multiplying all the probabilities the behavior strategy b assigns to the actions along the path to the terminal node n. If, for any node m, one denotes by $j(m)$ the agent who has to move at node m and by $u(m)$ the information set which contains the node m, then one can write formally:

$$q_b(n) \equiv \left[b_{j(\sigma(n))}^{u(\sigma(n))}(\alpha(n)) \cdot b_{j(\sigma(\sigma(n)))}^{u(\sigma(\sigma(n)))}(\alpha(\sigma(n)) \cdot \ldots \right].$$

The probability distributions on payoffs induced by different behavior strategy combinations can be evaluated according to the expected payoff criterion as with mixed strategies.

DEFINITION 1.12. The *expected payoff* of player i, $i \in I$, from a behavior strategy combination $b = (b_1, b_2, \ldots, b_I)$, $R_i(b)$, is the expected value of the payoffs given the probability distribution on terminal nodes induced by b, that is,

$$R_i(b) = \sum_{n \in \mathscr{T}(N)} r_i(n) \cdot q_b(n).$$

The following example helps to understand these concepts.

EXAMPLE 1.5 (*resumed*). Recall that player 1 has two information sets, $U_1 = \{\{a\}, \{d, e\}\}$, and player 2 has just one information set, $U_2 =$

$\{\{b, c\}\}$. Consequently, their behavior strategies have the following form:

$$b_1 = \left(b_1^{\{a\}}(T), b_1^{\{a\}}(M), b_1^{\{a\}}(B); b_1^{\{d,e\}}(L), b_1^{\{d,e\}}(R)\right)$$

and

$$b_2 = \left(b_2^{\{b,c\}}(U), b_2^{\{b,c\}}(D)\right).$$

Of course, $\Sigma_{a \in A(u)} b_i^u(a) = 1$ must hold for all players $i \in I$ and all information sets $u \in U$.

Given a behavior strategy combination b, the probabilities of reaching the seven terminal nodes $\{f, g, h, i, k, l, m\}$ can now be easily computed:

$$q_b(f) = b_2^{\{b,c\}}(U) \cdot b_1^{\{a\}}(T), \quad q_b(g) = b_2^{\{b,c\}}(D) \cdot b_1^{\{a\}}(T), \quad q_b(m) = b_1^{\{a\}}(B),$$
$$q_b(h) = b_1^{\{d,e\}}(L) \cdot b_2^{\{b,c\}}(U) \cdot b_1^{\{a\}}(M), \quad q_b(i) = b_1^{\{d,e\}}(R) \cdot b_2^{\{b,c\}}(U) \cdot b_1^{\{a\}}(M)$$
$$q_b(k) = b_1^{\{d,e\}}(L) \cdot b_2^{\{b,c\}}(D) \cdot b_1^{\{a\}}(M), \quad q_b(l) = b_1^{\{d,e\}}(R) \cdot b_2^{\{b,c\}}(D) \cdot b_1^{\{a\}}(M)$$

It is easy to check that

$$q_b(f) + q_b(g) + q_b(h) + q_b(i) + q_b(k) + q_b(l) + q_b(m) = 1$$

holds. Thus, q_b is indeed a probability distribution on terminal nodes induced by the behavior strategy b. ■

Once again pure strategies can be viewed as special cases of behavior strategies in which a player makes a choice with probability one at each information set. Of course, the probability of reaching some of the terminal nodes will be zero in such a case.

1.2.4 Behavior Strategies versus Mixed Strategies

Since behavior strategies and mixed strategies are generalizations of pure strategies, which are obtained by allowing for some kind of randomization, the question arises whether the two concepts, behavior strategies and mixed strategies, are not "essentially the same." "Essentially the same" means in this context that the set of all mixed strategy combinations will create exactly the same probability distributions over outcomes as the set of all behavior strategy combinations. If this were the case, then assuming that agents can play mixed strategies would be the same as assuming that they play behavior strategies, because their opportunity sets would be the same. The choice of one or the other of these assumptions would become a matter of pure convenience without any behavioral content.

To answer this question, note firstly that for every behavior strategy combination b there is a mixed strategy combination m which induces the same distribution over payoffs. To see this, note that the pure strategy s_i of player i specifies a move $s_i(u)$ at every information set of player i. Since the behavior strategy b_i assigns probability $b_i^u(s_i(u))$ to player i choosing action $s_i(u)$ at the information set u, one can define a mixed strategy as the product of the probabilities that player i assigns in her behavior strategy b_i to the actions determined by the pure strategy s_i; formally, $m_i(s_i) \equiv \Pi_{u \in U_i} b_i^u(s_i(u))$. It remains to check that this constructed mixed strategy combination induces the same probability distribution over terminal nodes. In general, several pure strategies may lead to the same terminal node, for example, s_1^5, s_1^6 to node m in example 1.5. The probability of reaching a particular terminal node n according to a given mixed strategy combination equals the sum of the probabilities assigned by the mixed strategies to all those pure strategy combinations that lead to this terminal node. Let $S(n)$ be the subset of pure strategy combinations leading to terminal node n, then one can express the probability of reaching terminal node n with mixed strategy combination $m, q_m(n)$ as follows:

$$q_m(n) = \sum_{s \in S(n)} [m_1(s_1) \cdot m_2(s_2) \cdot \ldots \cdot m_I(s_I)].$$

To prove in all generality that $q_m(n) = q_b(n)$ holds is quite tedious and needs additional definitions. It is, however, easy to check the claim in the examples given in this chapter.

Unfortunately, it is not true in general that there is a behavior strategy combination b for every mixed strategy combination m that yields the same probability distribution on the terminal nodes. The following example illustrates this point.

EXAMPLE 1.6. Consider the following game tree.

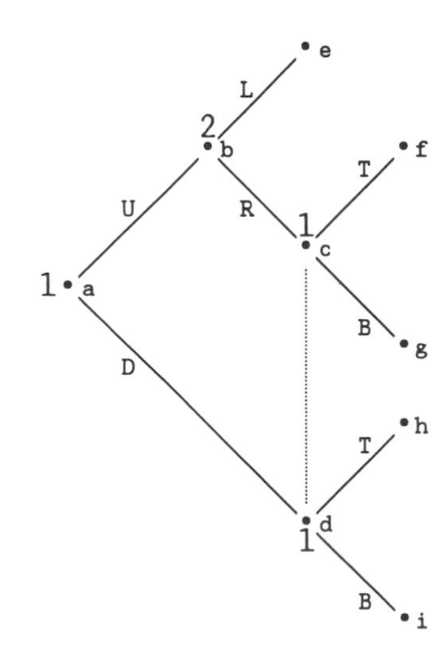

The following mixed strategy combination produces a probability distribution on the terminal nodes that cannot be achieved by any behavior strategy combination. Suppose player 1 plays each of the following two pure strategies, s_1^1, s_1^2 with probability .5, and any other pure strategy with probability zero:

$$s_1^1 \equiv \left(s_1^1(\{a\}), s_1^1(\{c,d\})\right) = (U,T)$$

and

$$s_1^2 \equiv \left(s_1^2(\{a\}), s_1^2(\{c,d\})\right) = (D,B).$$

Player 2 is assumed to play the pure strategy $s_2^1 \equiv (s_2^1(\{b\})) = (R)$ with probability one. This mixed strategy combination induces the following probability distribution on terminal nodes:

$$q_m(f) = 1/2, \qquad q_m(i) = 1/2, \qquad \text{and} \qquad q_m\{n\} = 0$$

for all other terminal nodes. To see that there is no behavior strategy combination b yielding this probability distribution on terminal nodes, note that any behavior strategy combination b that would induce the same probability distribution as the given mixed strategy must satisfy the following conditions:

$$q_b(f) = b_1^{\{c,d\}}(T) \cdot b_2^{\{b\}}(R) \cdot b_1^{\{a\}}(U) = 1/2$$

and

$$q_b(i) = b_1^{\{c,d\}}(B) \cdot b_1^{\{a\}}(D) = 1/2.$$

This implies, however, that

$$b_1^{\{a\}}(U) > 0, \qquad b_1^{\{c,d\}}(T) > 0, \qquad b_1^{\{a\}}(D) > 0,$$
$$b_1^{\{c,d\}}(B) > 0, \quad \text{and} \quad b_2^{\{b\}}(R) > 0$$

holds. Therefore,

$$q_b(h) = b_1^{\{a\}}(D) \cdot b_1^{\{c,d\}}(T) > 0$$

must hold as well. This contradicts the fact that the mixed strategy induces a probability of zero for node h. Thus, there can be no behavior strategy in this example that yields the same probability distribution on terminal nodes as the mixed strategy given earlier. ■

This question of whether there is a class of games for which the two strategy concepts are equivalent has been addressed and answered in the early stages of game-theoretic analysis. The answer depends on a special property of a game called *perfect recall*. Loosely speaking a game has perfect recall if each player remembers all her past moves.

DEFINITION 1.13. A game Γ is a *game with perfect recall* if for any player $i \in I$ and any two information sets $u, v \in U_i$ it is true that $n \in v$ and

$\sigma^k(n) \in u$ for some k imply that, for all $m \in v$, there is some k' such that $\alpha(\sigma^{k'-1}(m)) = \alpha(\sigma^{k-1}(n))$ holds.

The definition simply requires that if it happens that a player has made a move $\alpha(\sigma^{k-1}(n))$ k steps before node n in information set v, then the same move out of the information set u must lead to all the nodes in the information set v (maybe in a different number of steps k'). Example 1.6 shows a game tree for a game without perfect recall.

EXAMPLE 1.6 (*resumed*). Obviously, at the information set $\{c, d\}$ player 1 must have forgotten her choice at the information set $\{a\}$. The game clearly does not satisfy definition 1.13, since $\alpha(\sigma^0(d)) = D \neq U = \alpha(b) = \alpha(\sigma^1(c))$ in spite of the fact that c and d lie in the same information set. Hence, the game is without prefect recall. ■

The fundamental theorem on the relationship between behavior strategies and mixed strategies can now be stated.[8]

THEOREM 1.2 *(Kuhn, 1953). In games with perfect recall, for any mixed strategy combination m, there is a behavior strategy combination b that induces the same probability distribution on the set of terminal nodes* $\mathcal{T}(N)$.

According to this theorem, mixed strategies and behavior strategies are equivalent for games with prefect recall. Example 1.6 illustrates that by contrast, in games without perfect recall a certain mixed strategy combination induces a probability distribution on the terminal nodes that cannot be obtained by a behavior strategy combination.

1.3 Games in Strategic Form

The previous sections provided precise language to describe the details of an interactive situation. This required considerable notation, in particular to model the sequence of moves and the information available to each agent. On the other hand, to analyze the behavior of the agents, all that is needed is a plan for each agent indicating what she would do in each situation when she has to make a move. Thus, to analyze behavior it seems sufficient to list all plans of action or strategies. Since each strategy combination, that is, each list of individual strategies, determines a unique outcome to the game, payoffs can be associated with strategy combinations.

Therefore, a description of all the strategic options of each agent by indicating the pure strategy sets S_i for each agent $i \in I$, and the payoffs that can be achieved by all the possible strategy combinations $(s_1, \ldots, s_I)$, that is, $p_i(s_1, \ldots, s_I)$ for all $i \in I$, seems sufficient to capture all the

[8]Theorem 1.2 will not be proved here. For a proof see Kuhn (1953), theorem 4.

elements of strategic importance for a game. A game is in *strategic* (or *normal*) *form* if only the set of players I, the set of strategies S_i, and the payoff functions $p_i(s)$ for each player $i \in I$ are given.

DEFINITION 1.14. A game Γ is in *strategic* (*normal*) *form* if $(I, (S_i)_{i \in I}, (p_i)_{i \in I})$ is specified.

It is clear that the strategic form of a game is a much simpler description than the extensive form. This allows two-player games with finite strategy sets to be represented by their *payoff matrix*. Thus, for a game with two players, $I = \{1, 2\}$, in which player 1 has m pure strategies and player 2 has n pure strategies, $S_1 = \{s_1^1, s_1^2, \ldots, s_1^m\}$, $S_2 = \{s_2^1, s_2^2, \ldots, s_2^n\}$, one can display all the information of the strategic form in the following matrix.

	s_2^1	s_2^2	$\cdots$	s_2^n
s_1^1	$p_1(s_1^1, s_2^1), p_2(s_1^1, s_2^1)$	$p_1(s_1^1, s_2^2), p_2(s_1^1, s_2^2)$	$\cdots$	$p_1(s_1^1, s_2^n), p_2(s_1^1, s_2^n)$
s_1^2	$p_1(s_1^2, s_2^1), p_2(s_1^2, s_2^1)$	$p_1(s_1^2, s_2^2), p_2(s_1^2, s_2^2)$	$\cdots$	$p_1(s_1^2, s_2^n), p_2(s_1^2, s_2^n)$
$\vdots$	$\vdots$	$\vdots$		$\vdots$
s_1^m	$p_1(s_1^m, s_2^1), p_2(s_1^m, s_2^1)$	$p_1(s_1^m, s_2^2), p_2(s_1^m, s_2^2)$	$\cdots$	$p_1(s_1^m, s_2^n), p_2(s_1^m, s_2^n)$

The matching pennies game illustrates how this simplifies the description of a game.

EXAMPLE 1.3 (*resumed*). In this game, each player has just two information sets and just two choices. Player 1 moves first and can decide whether to put down H or T. Player 2 moves second and has to decide, without knowing what player 1 did, whether to choose h or t to lie face up. Hence, a strategy of each player can be identified with an action of a player.[9] This game can be represented by the following payoff matrix.

		Player 2	
		h	t
Player 1	H	$1, -1$	$-1, 1$
	T	$-1, 1$	$1, -1$

One can read easily from this matrix that the strategy set of player 1 is $S_1 = \{H, T\}$ and the strategy set of player 2 is $S_2 = \{h, t\}$. In addition, if strategy combination (H, h) is played, the payoff to player 1 will be $p_1(H, h) = 1$ and to player 2 $p_2(H, h) = -1$. Similarly, one deduces from this matrix that player 2 wins a dollar from player 1 if (H, t) or (T, h) is played, that is, if the sides of the coins do not match. ■

[9]Note that one can always identify a strategy with an action of a player if there is just one information set at which this player has to move.

This greatly simplified structure of games in strategic form makes analysis of the behavior of agents much easier. The reason for this simplification is that the strategic form of a game abstracts from the dynamic and informational details of a game in extensive form. This raises the question of whether something essential is lost by abstracting from these aspects of a game.

Analyzing just the choice of a strategy by an agent is sufficient only if one can be certain that the agent can commit herself to the chosen strategy under all circumstances. Commitment possibilities of agents are, therefore, important for answering the question of whether the strategic form of a game is sufficient to analyze the behavior of players. Assume that agents are allowed to choose a plan of action, a strategy, to submit to another agent who is instructed to carry through those plans. If this agent would not accept any "corrections" or "changes" once she has received her first instruction from each player, then it would be certainly adequate to consider only the strategic form of a game.

On the other hand, if commitment possibilities do not exist, then a player may very well choose a strategy which would require her to act against her interests at some stage of the game. These problems are central to concepts of equilibrium refinements and will be discussed extensively in Chapter 6. For now, it will be assumed that commitment is possible in order to allow us to concentrate on the strategic choice behavior.

The fact that the strategic form of a game abstracts from certain aspects of the extensive form is evident, since

- each game in extensive form has a strategic form representation
- but, in general, there are several extensive form representations for each game in strategic form

The following example shows this point.

EXAMPLE 1.7. Consider the following game trees.

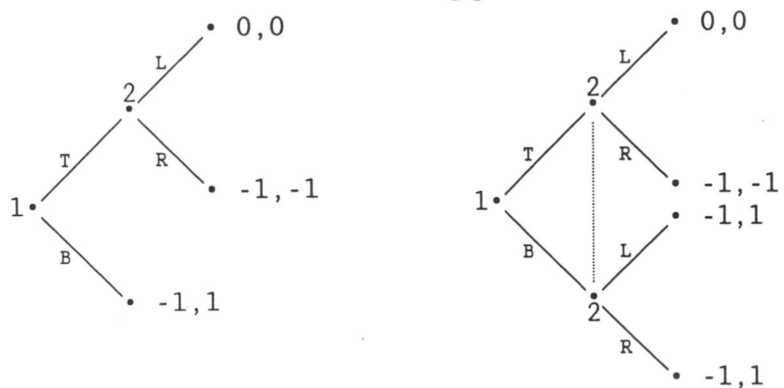

Both games can be described in the following way. Player 1 moves first and can choose either T ("top") or B ("bottom"). Player 2 moves second and can choose either L ("left") of R ("right"). In the left diagram, player 1 can end the game by choosing B and player 2 knows that player 1 chose T if she comes to move. In the right tree, player 2 comes to move no matter what player 1 did, but she is uninformed about player 1's action. Nevertheless, because the payoff to both players is identical if player 1 moves B in the right tree, one can argue that both games are strategically equivalent. In any case, both extensive forms have the same normal form given by the following payoff matrix.

		Player 2	
		L	R
Player 1	T	$0, 0$	$-1, -1$
	B	$-1, 1$	$-1, 1$

Note that one can identify strategies and actions in this case, since there is just one information set for each player. ■

This section concludes with another example in which actions and strategies cannot be identified.

Example 1.8 (*matching pennies with perfect information*). The rules of this game are exactly the same as in matching pennies (example 1.3) except that player 2 is allowed to observe the move of player 1.

The extensive form of this game can be represented by the following game tree.

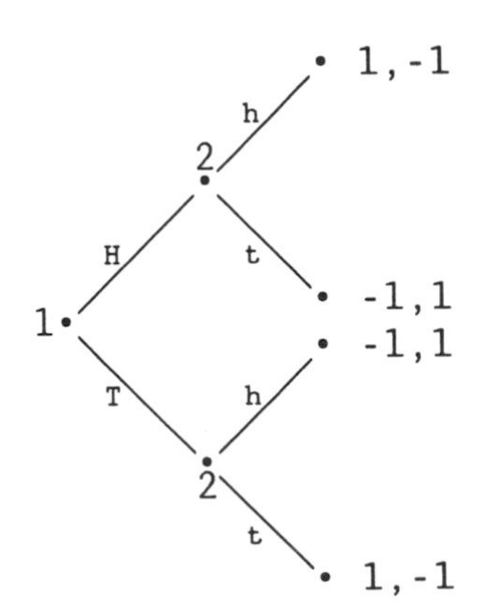

From the extensive form of the game, one can immediately see that player 2 has two information sets and, therefore, that there are two situations in which this player may have to choose an action. Since player 2 has two choices (h and t) in both situations there are four possible strategies available for her, $S_2 = \{s_2^1, s_2^2, s_2^3, s_2^4\}$, which can be described as follows:

	s_2^1	s_2^2	s_2^3	s_2^4
if H	h	t	h	t
if T	t	h	h	t

Player 1 has only one information set and strategies and actions can be identified. Therefore, the following strategic form of this game emerges.

$$I = \{1,2\}$$

$$S_1 = \{H,T\}, \qquad S_2 = \{s_2^1, s_2^2, s_2^3, s_2^4\},$$

$$\begin{array}{llll} p_1(H,s_2^1) = 1, & p_1(H,s_2^2) = -1, & p_1(H,s_2^3) = 1, & p_1(H,s_2^4) = -1, \\ p_1(T,s_2^1) = 1, & p_1(T,s_2^2) = -1, & p_1(T,s_2^3) = -1, & p_1(T,s_2^4) = 1, \\ p_2(H,s_2^1) = -1, & p_2(H,s_2^2) = 1, & p_2(H,s_2^3) = -1, & p_2(H,s_2^4) = 1, \\ p_2(T,s_2^1) = -1, & p_2(T,s_2^2) = 1, & p_2(T,s_2^3) = 1, & p_2(T,s_2^4) = -1. \end{array}$$

Or in the more compact form of the payoff matrix:

		Player 2			
		s_2^1	s_2^2	s_2^3	s_2^4
Player 1	H	1, −1	−1, 1	1, −1	−1, 1
	T	1, −1	−1, 1	−1, 1	1, −1

■

1.4 Cooperative and Noncooperative Games

Since von Neumann und Morgenstern's (1947) fundamental work on game theory, it has become a tradition to distinguish between cooperative and noncooperative game theory. Indeed, in the 1960s and early 1970s it could be said that cooperative game theory had become the dominant part of game-theoretic research. The main difference between these two branches lies in the type of questions they try to answer.

Cooperative game theory is concerned with the kind of coalitions a group of players will form if different coalitions produce different outcomes and if these joint outcomes then have to be shared among the members. In contrast, *noncooperative game theory* focuses on strategies players will choose.

The description of a game or social interaction as presented in the extensive and strategic form is designed precisely to address the latter question. To find an answer to the question of what coalitions of players are likely to form and which will be stable eventually, one has to specify

(i) What different groups of players can achieve jointly
(ii) To what extent the joint results can be shared among participating players

These questions will be addressed in this section. The following example may help to understand these issues.

EXAMPLE 1.9 (*duopoly*). Consider two firms which produce the same product. Given their technologies, firm 1 can produce either 10 units of output or nothing, while firm 2 can produce either 16 units of output or nothing. For simplicity, there are no costs of production. It is common knowledge that the inverse demand function for the product has the following form:

$$p(x) = 15 - .5 \cdot x,$$

where x denotes the total quantity of output put on the market.

If neither of the firms can observe the action of the opponent, then they have the following strategy sets: $S_1 = \{10, 0\}$ and $S_2 = \{16, 0\}$. Assuming that both firms are interested in profit, one derives the following payoff for firm 1 if the strategy combination $(10, 16)$ is chosen:

$$p_1(10, 16) \equiv (15 - .5 \cdot 26) \cdot 10 = 20.$$

Similarly, one derives the payoffs for the other strategy combinations and for player 2. The complete strategic form of this game can be summarized in a payoff matrix.

		Firm 2	
		16	0
Firm 1	10	20, 32	100, 0
	0	0, 112	0, 0

This payoff matrix shows clearly that the highest total profit is achieved if firm 1 stays out of the market. On the other hand, this is unlikely to happen, because staying out of the market yields zero profit for firm 1, whereas entering the market yields at least 20 units of profit. ■

It seems natural to ask whether firm 2 might not offer some of the profit it would earn if firm 1 stayed out of the market to compensate firm 1 for doing so. This type of cooperation between the two firms cannot be expected however, if firm 2 could break its promise to pay compensation after it has made the profit.

Example 1.9 highlights the problems in cooperative analysis: If there are gains from cooperation, that is, if there are strategy combinations that yield a better payoff than each player could obtain individually, but if the distribution of these gains does not provide enough incentives for both players to conform to the cooperative strategy combination, then there must be an exogenous institution to enforce a cooperative agreement. Otherwise, one of the cooperating agents should be expected to renege on the agreement. Again the possibility of commitment is crucial.

For this reason, the basic assumption of cooperative game theory is the existence of some mechanism to guarantee cooperative agreements. If binding agreements are possible among players, then one can simply assume that any joint payoff a coalition can achieve through a particular play of its members can be actually realized by an agreement to play that way. *Ex post* deviations from the agreement need not be considered.

The second question mentioned concerns the scope for agents to redistribute payoffs. Since payoffs are subjective utilities, they cannot be redistributed directly. What can be transferred between players are monetary payments or private goods. In general, a coalition can obtain a set of payoff vectors by redistributing the joint outcome or by making monetary side payments. If the preferences of the players can be represented by quasi-linear utility functions,[10] however, then there exists at least one good (money) in which transfers of utilities can be made at a constant rate. In this case, utility is transferable and it is possible to express the value of the joint outcome in terms of this transfer good.

Games in which players have quasi-linear utility functions are called games with *transferable utility* or games with *side payments*. Note that transferable utility represents a strong restriction on the preferences of the players. Since the assumption of transferable payoffs is problematic, cooperative game theory has been developed for nontransferable payoffs as well. The presentation of cooperative game theory in this book will be restricted to the transferable utility case since all concepts covered in this book have analogues in the theory with nontransferable utility and since games with transferable utility have a simpler representation.

[10]Compare Varian (1984, pp. 278–283) for a formal definition and results on quasi-linear utility functions.

1.5 Games in Coalitional Form

Throughout this section it will be assumed that

- binding agreements among players are possible
- payoffs can be transferred among players

Under the first of these assumptions, any group of players, or coalition, can choose the strategy combination that yields the highest possible joint payoff to this coalition. The highest possible payoff a coalition may achieve is called *worth of the coalition*. The second assumption makes any distribution of the worth of the coalition among members possible as part of a binding agreement among members of that coalition.

The definition of the worth of a coalition poses a problem regarding the behavior of those players who are not members of the coalition. Since their behavior, in general, influences the outcome for the coalition, the worth of that coalition will typically also depend on the choices of players who are not members of the coalition. The most common interpretation is that the worth of a coalition is the best outcome it can guarantee itself against the other players.[11]

The description of a cooperative game now becomes straightforward. A game in characteristic function form consists of

- a set of players I
- a function v that associates with each possible coalition the maximum joint payoff this coalition can achieve

Let $\mathscr{P}(I)$ denote the set of all subsets of I, that is, the set of all coalitions that could be formed from the player set I, then one can give the following definition.

DEFINITION 1.15. The characteristic function v associates with each coalition in $\mathscr{P}(I)$ the maximum joint payoff of this coalition, $v\colon \mathscr{P}(I) \to \mathbb{R}$. A game Γ is in *coalitional form* if (I, v) is specified.

Note that the power set of I, $\mathscr{P}(I)$, contains the empty set $\varnothing$, the total set I, and all sets with a single agent only. Thus, a "single player coalition" is considered a special case of a coalition as well as the "grand coalition" that comprises all players. By convention, the worth of the empty coalition will be zero, that is, $v(\varnothing) = 0$ is assumed for all games in coalitional form. For obvious reasons, games in coalitional form are

[11]Von Neumann and Morgenstern (1947), working mainly with zero-sum games, suggested taking the maximin value of a coalition as the worth of the coalition. For the case of nonzero-sum games, however, using the maxmin value as the worth of the coalition assigns a very conservative payoff to this coalition.

sometimes called *games in characteristic function form* as well. The following two examples may illustrate this definition.

EXAMPLE 1.9 (*resumed*). For the duopoly game discussed earlier, one obtains the following characteristic function:

$$v(\{1,2\}) = 112, \qquad v(\{1\}) = 20, \qquad v(\{2\}) = 32,$$

where it has been assumed that single players can achieve at best the payoff $(20, 32)$ if they cannot agree on playing $(0, 16)$ which yields the best joint outcome.

Thus, $I = \{1,2\}$ and v, defined by $v(I) = 112$, $v(\{1\}) = 20$, $v(\{2\}) = 32$, $v(\emptyset) = 0$, describe a game in coalitional form. ■

Notice that in example 1.9 it is not explained how players acting on their own would achieve the outcome $(20, 32)$. In fact, in many cooperative games the precise manner in which a coalition will realize its worth remains unexplained. Mostly the characteristic function is given directly without describing how it is derived from an underlying game in strategic or extensive form, as in the following example which concludes this section.

EXAMPLE 1.10 (*majority voting*). Suppose that three players vote on projects yielding some monetary payoff. The project for which the majority votes will be implemented and the entire payoff is received by those in favor of the project. Without loss of generality, one can assume that winning the ballot will yield a payoff of one, and losing it will result in a payoff of zero.

This leads to the following characteristic function:

$$v(I) = v(\{1,2\}) = v(\{1,3\}) = v(\{2,3\}) = 1,$$
$$v(\emptyset) = v(\{1\}) = v(\{2\}) = v(\{3\}) = 0.$$

Hence, given the player set $I = \{1,2,3\}$, (I, v) is a game in coalitional form. ■

1.6 Summary

This chapter introduced three different forms of representation for games: the extensive form, the strategic form, and the coalitional form. These are distinguished by the degree of abstraction, ranging from the most detailed description of the extensive form, through the rather abstract strategic form, to the coalitional form that neglects even strategic details by the assumption of commitment possibilities.

During this exposition, three strategy concepts were presented:

- *pure strategies* give a complete plan of action for the game in which a player chooses an action at every stage the player has to move.
- *mixed strategies* are random choices of pure strategies where the player controls the randomization.
- *behavior strategies* are a complete plan of action for the game in which a player may choose to randomize over actions at every stage she has to move.

In addition, it could be demonstrated that, for games with perfect recall, mixed and behavior strategies are equivalent in the sense that they produce the same probability distributions on outcomes of the game.

The following diagram summarizes the classification of games described in this chapter.

GAME THEORY

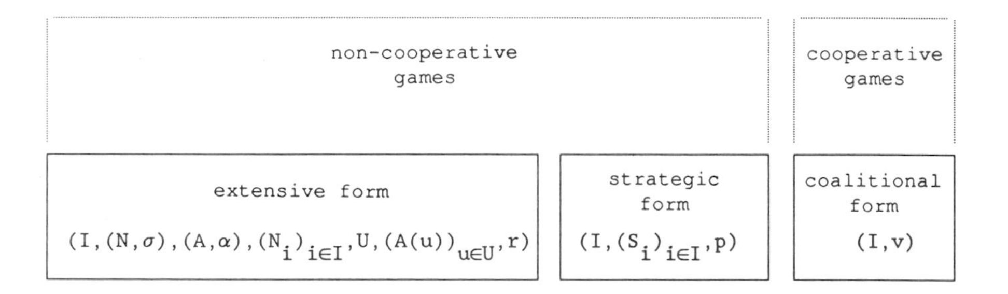

1.7 Remarks on the Literature

There are few references for the formal description of extensive form games. Most of the literature introduces the extensive form informally with a few examples of game trees. Section 2 (pp. 865–869) of Kreps and Wilson (1982) is still one of the best expositions of the notation involved in the extensive form. The presentation of Zermelo's theorem in this chapter is inspired by Aumann's (1975a) lecture notes. The relationship between mixed and behavior strategies is discussed by Binmore (1992, pp. 459–462), and Myerson (1991, pp. 154–161) provides a proof of Kuhn's theorem. Binmore (1992, pp. 467–488) offers an excellent introduction to the notion of common knowledge and demonstrates the importance of this assumption. Rubinstein's (1989) "Electronic-Mail" example shows the consequences of a failure of the common-knowledge assumption.

Exercises

1.1 *Consider the following graphs.*

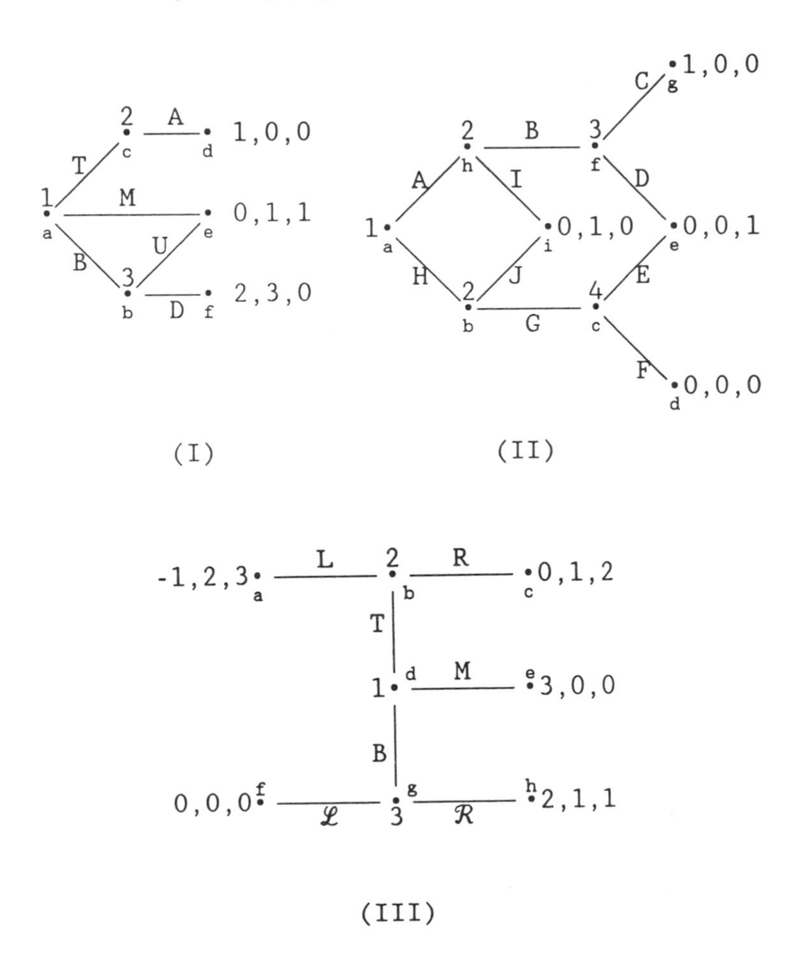

(a) For each graph, write down the player set, the set of nodes N, the set of actions A, the predecessor function σ, and the action function α.

(b) Which of these graphs are game trees?

1.2 *Consider the following game trees*:

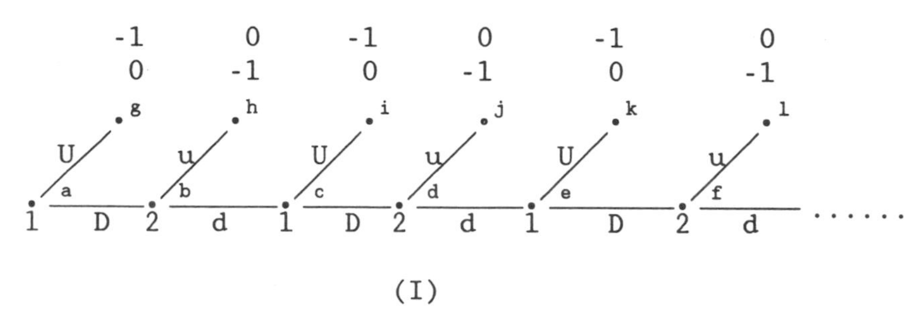

(I)

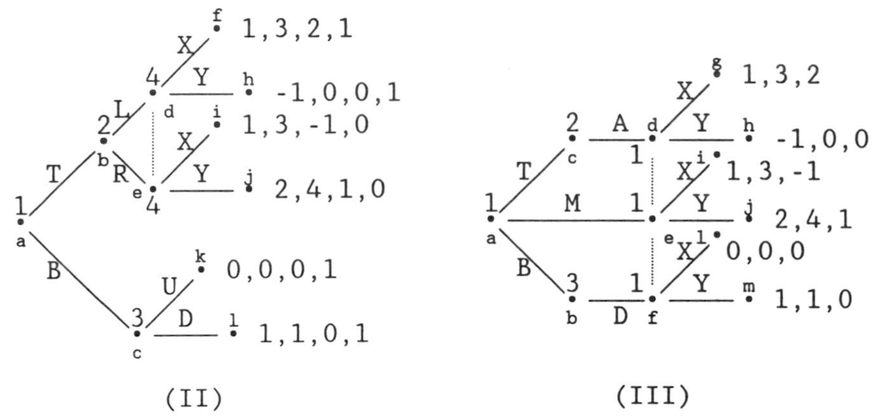

(II) (III)

(a) For each graph, write down the player set, the set of nodes N, the set of actions A, the predecessor function σ, and the action function α.

(b) Write down the information sets of the players and the set of choices available at these information sets.

1.3 *Consider two parties that have invested in a joint venture. The value of the investment in any period t will be t. In each period, each of the parties may withdraw from the project. If a party interrupts the joint venture the other party loses one unit of its return. The party that withdraws second in each period will gain an extra unit of return from withdrawing. If the venture remains uninterrupted until period T, then the game ends with the payoff of the party that would have had the opportunity to withdraw next.*

For $T = 2$, give a formal description of this situation and draw the game tree. Analyze the game and make a prediction about its outcome. Does your prediction depend on the length of the game T?

1.4 *Consider the following game in extensive form.*

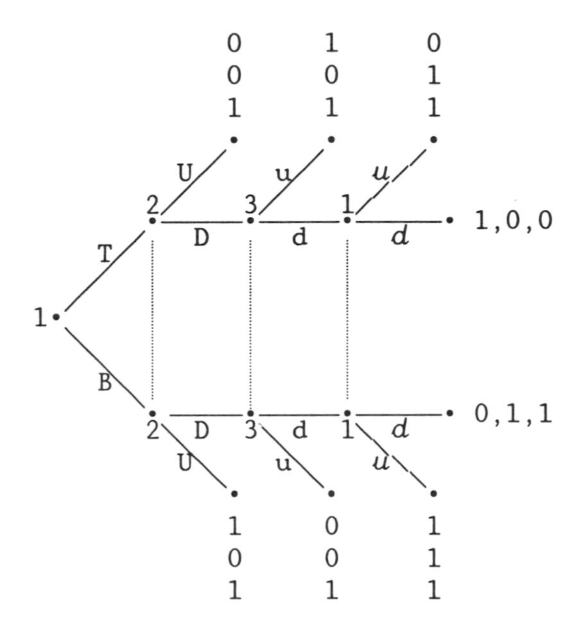

(a) Check whether this is a correct extensive form.
(b) Give an example of a pure strategy for each player and an example of a behavior strategy for each player.
(c) Is there a mixed strategy for each player that yields the same payoff distribution to the players as the behavior strategies suggested under (b)?

1.5 *Consider the game in exercise 1.4.*
(a) Check whether this is a game with perfect recall.
(b) Give an example of a mixed strategy combination that yields a payoff distribution that cannot be achieved by a behavior strategy combination.

1.6 *Consider two firms which produce the same product and sell it in a market with the following demand function*:

$$d(p) = \max\{0, 12 - p\}.$$

Suppose that, for technological reasons, firm 1 can produce either 4 *units of the good at a cost of* 10 *or* 6 *units at a cost of* 15. *Similarly, firm 2 can produce either* 3 *units at a cost of* 8 *or* 4 *units at a cost of* 10.

(a) Derive the payoff function of this game.

(b) Write down this game in the extensive form and in the strategic form for the case in which both firms make their supply decision simultaneously and for the case in which firm 1 makes its decision first.

1.7 *Consider the case in which player 1 has the choice to play matching pennies with player 2 or not to do so. The game, if it is played, proceeds as described in example 1.3. For the payoffs, assume that player 1 wins a dollar from player 2 if the sides of the coins match, and player 2 wins a dollar from player 1 if they do not match. If the no-play option is taken, the payoff is zero for both players.*

Draw the game tree of this game. Write down this game in the extensive form and in the strategic form.

1.8 *Reconsider exercise 1.7 with the following modification. Player 1 is asked first whether she wants to participate in the game. If she accepts, player 2 is asked for her consent. If she agrees as well the game is played.*

Draw the game tree for this case. How many pure strategies has each player? Write down the sets of pure strategies for both players.

1.9 *Consider the following three games*:

- *Three members of a committee vote to select a president out of a group of four candidates. The voting procedure is as follows: in order of seniority, each player is allowed to veto one candidate. The remaining candidate is elected.*
- *Three companies consider a joint venture that will yield a certain outcome of* \$100,000. *For successful completion of the project, the participation of at least two of the companies is required.*
- *A committee of three has to decide on a course of action. Three proposals are under consideration. The decision rule is a secret ballot in which each member of the committee suggests one of the proposals. If a proposal gets a majority of votes, it will be implemented. If there is no majority, the project suggested by the chairperson will be chosen.*

(a) Which of these situations are cooperative and which are noncooperative games?

(b) For the noncooperative games, draw the game tree and indicate the information sets. If possible, write down the strategic form of the game.

(c) For the cooperative games, write down the coalitional form of the game.

2

Two-Player Zero-Sum Games

In Chapter 1, different ways to describe a game and game strategies were introduced. The analysis of games, however, was deliberately postponed. The following chapters will focus on what constitutes an appropriate way to analyze a game. While introducing those equilibrium and solution concepts that are most commonly used in economics, these chapters will show that no generally accepted way to analyze a game has been established yet. In fact, there is a growing perception that no universally acceptable solution concept may exist. Chapters 3 and 4 will be concerned with equilibrium concepts for general games in strategic form and Chapters 6 and 7 will extend and adapt these concepts in the context of the extensive form. In Chapter 5, a method to analyze games with incomplete information will be presented.

Before that, however, it is useful to consider a special class of games in strategic form. *Two-player zero-sum games* provide the opportunity to introduce different ways of reasoning about the outcome of the interaction modeled by the game. In these games, it is possible to predict the outcome because all modes of reasoning lead to the same result. In addition, one can relate the problem of existence of an equilibrium to the problem of existence of a solution to an optimization problem. In this sense, two-player zero-sum games provide a good introduction to the more general theory of games in strategic form.

Zero-sum games are strictly competitive in that what one player gains the other loses. A typical example in this category of games is matching pennies (example 1.3). Player 1 wins a dollar from player 2 if both choose a matching side of a coin. Hence, player 1 wins exactly the amount that player 2 loses, and vice versa.

DEFINITION 2.1. A *two-player zero-sum game* is a game in strategic form such that

$$p_1(s_1, s_2) + p_2(s_1, s_2) = 0$$

for all $s_1 \in S_1$ and all $s_2 \in S_2$ holds.

From the definition of a zero-sum game, $p_1(s_1, s_2) = -p_2(s_1, s_2)$ follows. Thus, in a payoff matrix, one needs to note the payoff of player 1 only because the payoff of player 2 is the negative of this number.

EXAMPLE 2.1 (*matching pennies*; *example* 1.3 *resumed*). The payoff matrix of this game is usually written in the following way.

		Player 2	
		H	T
Player 1	H	1	-1
	T	-1	1

Player 2's payoff is simply the negative of player 1's payoff. ■

2.1 What Is a Solution to a Game?

In Chapter 1, Zermelo's theorem demonstrated that in chess there is a strategy that allows one player to force a win or at least a draw. Hence, the theorem suggests there is an optimal way to play chess. Obvious as it may appear for games such as chess, some explanation of what "optimal" means in games in which a player's payoff depends on the actions of the opponent as well as on his own behavior is required. As this chapter will show, for zero-sum games there is a natural notion of an "optimal" strategy. From the perspective of agents, two lines of reasoning for how to play a game will be presented. Both provide a possible answer to the question: What is a solution of a game? Fortunately, for zero-sum games, both arguments suggest the same strategies as a solution. The following example illustrates the two approaches to the question of "how to play the game."

EXAMPLE 2.2. Consider the following payoff matrix of an arbitrary two-person zero-sum game. Player 1 has two strategies, $S_1 = \{s_1^1, s_1^2\}$, whereas player 2 has four strategies, $S_2 = \{s_2^1, s_2^2, s_2^3, s_2^4\}$.

		Player 2			
		s_2^1	s_2^2	s_2^3	s_2^4
Player 1	s_1^1	-3	2	-1	4
	s_1^2	5	1	0	3

■

2.1.1 The Value of a Game

Considering a player in the game of example 2.2, the following reasoning might provide some insight on how to play this game.

Argument 1.

For player 1: "If I play my strategy s_1^1, then the worst that can happen is that player 2 plays his first strategy s_2^1. Therefore, I get at least -3 by playing s_1^1. On the other hand, if I play my second strategy s_1^2, then the worst that can happen is player 2 plays s_2^3. In this case, I guarantee myself an outcome of 0. Therefore, strategy s_1^2 is better for me."

For player 2: "If I play s_2^1, I guarantee myself -5; if I play s_2^2, I get at least -2; for s_2^3 the worst that can happen is 0, and for s_2^4 the worst is -4. Of these alternatives, obviously s_2^3 is the best."

If both players play according to this reasoning, then player 1 plays s_1^2 and player 2 plays s_2^3, yielding an outcome of 0 for both players.

The logic of this approach is in considering the worst case. Each player considers the worst that could result from each of his strategies and then simply chooses the strategy that yields the best "worst outcome." By this procedure, each player can guarantee himself a minimal payoff level. This guaranteed payoff level is called *security level* or *maximin value*.

Denote by

$$\tilde{v}_1(s_1) \equiv \left[\min_{s_2 \in S_2} p_1(s_1, s_2)\right] \quad \text{and} \quad \tilde{v}_2(s_2) \equiv \left[\min_{s_1 \in S_1} p_2(s_1, s_2)\right]$$

the worst outcome for player 1 and player 2, respectively, for a given strategy choice of their own.

DEFINITION 2.2. The *maximin value* (or *security level*) of player $i \in \{1, 2\}$ is the maximum value $v_i \equiv \max_{s_i \in S_i} \tilde{v}_i(s_i)$. A *maximin strategy* (or *prudent strategy*) is a maximizing strategy $\bar{s}_i \in \operatorname{argmax}_{s_i \in S_i} \tilde{v}_i(s_i)$.

An important feature of a maximin strategy is the independence of a player's choice from any information about the opponent's payoff. All a player needs to know to determine his maximin strategy is his own payoff and the strategies available to the opponent. No speculation about the opponent's motivation, his payoff function, is necessary. The security level of a player can be derived without a belief about the opponent's play.

Since in zero-sum games with two players $p_1(s_1, s_2) = -p_2(s_1, s_2)$ holds, the following equivalent definition of the security level of player 2 is possible:[1]

$$v_2 \equiv \max_{s_2 \in S_2} \tilde{v}_2(s_2) = -\min_{s_2 \in S_2} [-\tilde{v}_2(s_2)] = -\min_{s_2 \in S_2} \left[-\left(\min_{s_1 \in S_1} p_2(s_1, s_2) \right) \right]$$
$$= -\min_{s_2 \in S_2} \left[\max_{s_1 \in S_1} -p_2(s_1, s_2) \right] = -\min_{s_2 \in S_2} \left[\max_{s_1 \in S_1} p_1(s_1, s_2) \right].$$

Hence, the maximin value of player 2, v_2, can be written as the negative of the minimax value of player 1. This is an immediate consequence of the fact that the payoff of player 2 is simply the negative of the payoff of player 1 in a zero-sum game.

This alternative representation of the maximin value of player 2 makes it easy to prove the following relationship between the maximin values of both players.

LEMMA 2.1. *For any two-player zero-sum game, $-v_2 \geq v_1$ holds.*

Proof. Consider the following three steps:

(i) Let $\bar{s}_1$ be a strategy in S_1 which maximizes $\tilde{v}_1(s_1)$, then $v_1 = \tilde{v}_1(\bar{s}_1)$ holds. By the definition of $\tilde{v}_1(s_1)$, for any $s_1 \in S_1$,

$$\tilde{v}_1(s_1) \equiv \left[\min_{s_2 \in S_2} p_1(s_1, s_2) \right] \leq p_1(s_1, s_2) \quad \text{for all} \quad s_2 \in S_2.$$

Hence, in particular,

$$v_1 = \tilde{v}_1(\bar{s}_1) \leq p_1(\bar{s}_1, s_2) \quad \text{for all} \quad s_2 \in S_2.$$

(ii) Define $\bar{v}(s_2) \equiv [\max_{s_1 \in S_1} p_1(s_1, s_2)]$. This is the function that yields the maximum value of p_1 for each strategy of player 2, s_2. Furthermore, let $\bar{s}_2$ be a strategy of player 2 that minimizes $\bar{v}(s_2)$. Note that

$$-v_2 = \min_{s_2 \in S_2} \left[\max_{s_1 \in S_1} p_1(s_1, s_2) \right] = \bar{v}(\bar{s}_2)$$

[1]Note that for any function $f(x)$, $\max f(x) = -\min[-f(x)]$ and $\min f(x) = -\max[-f(x)]$ holds.

holds. By the definition of $\bar{v}(s_2)$, for any $s_2 \in S_2$,

$$\bar{v}(s_2) \equiv \left[\max_{s_1 \in S_1} p_1(s_1, s_2)\right] \geq p_1(s_1, s_2) \quad \text{for all} \quad s_1 \in S_1.$$

Hence, in particular

$$-v_2 = \bar{v}(\bar{s}_2) \geq p_1(s_1, \bar{s}_2) \quad \text{for all} \quad s_1 \in S_1.$$

(iii) From (i), $v_1 \leq p_1(\bar{s}_1, \bar{s}_2)$ and, from (ii), $-v_2 \geq p_1(\bar{s}_1, \bar{s}_2)$ follow. Hence, $-v_2 \geq p_1(\bar{s}_1, \bar{s}_2) \geq v_1$ holds, as was claimed. ■

Lemma 2.1 establishes that the negative value of player 2's security level v_2 is always greater than or equal to the value of player 1's security level v_1. Note, however, that this fact does not imply anything about the relative position of the players; neither player 1 nor player 2 can be said to be better or worse off because of the result $-v_2 \geq v_1$. The following example helps to clarify this.

Example 2.1 (*resumed*). In the matching pennies game, one can easily check that the worst that can happen to player 1 is to lose a dollar for each of his strategies, "heads" or "tails." Hence, in this case, both strategies are maximin strategies and the maximin value is $v_1 = -1$. Similarly, for player 2 both strategies are maximin strategies and $v_2 = -1$ holds. Therefore, $-v_2 = 1 > -1 = v_1$ follows, confirming the result of lemma 2.1. ■

Example 2.1 demonstrates that $-v_2$ may be strictly greater than v_1. But equality can occur as well as one can see from example 2.2.

Example 2.2 (*resumed*). Checking the worst outcome for each strategy, player 1 is guaranteed -3 by playing strategy s_1^1 and 0 by playing strategy s_1^2. Hence, s_1^2 is player 1's maximin strategy yielding a security level of 0. Similar considerations show that s_2^3 is player 2's prudent strategy and 0 his maximin value. In this case, one has $-v_2 = 0 = v_1$. ■

These examples show that $-v_2$ may be strictly greater than v_1 (example 2.1) or equal to v_1 (example 2.2). If these two values coincide, this common security level is called *value of the game*.

Definition 2.3. The value of a two-person zero-sum game is the number v that satisfies

$$v \equiv \max_{s_1 \in S_1}\left[\min_{s_2 \in S_2} p_1(s_1, s_2)\right] = \min_{s_2 \in S_2}\left[\max_{s_1 \in S_1} p_1(s_1, s_2)\right],$$

or equivalently $v \equiv v_1 = -v_2$.

Note that example 2.1 shows that not every game has a value. Indeed, games that have a value are sometimes called *inessential*, because the only outcome is the security level for each of the players. In games with a value there is a natural solution to the game. Each player plays his maximin strategy and realizes his maximin value. The following subsection provides another argument for the solution of a game.

2.1.2 Equilibrium of a Game

The previous approach for finding a solution to a game was based on worst case considerations. It recommended a "safety first" type of strategy, and one may wonder whether this approach is not overly pessimistic. The following reasoning provides an alternative argument for how to play the game.

Argument 2. Returning to the game in example 2.2, one can argue as follows.

For player 1: "If player 2 plays s_2^1, the best I can do is play s_1^2 to obtain 5, but if player 2 plays s_2^2 or s_2^4, I should play s_1^1. If he plays s_2^3, however, s_1^2 would be the best choice for me."

For player 2: "If player 1 plays s_1^1, the best for me is to play s_2^1, and if he plays s_1^2, I should play s_2^3."

In this argument each player works out what his best response to each strategy choice of the opponent would be. This in itself does not tell which strategy each agent should or would choose. If one assumes, however, that a solution of a game has to be a situation in which each player plays a best response to the other player's actual strategy choice, then the strategy pair (s_1^2, s_2^3) will be the solution of the game.

There are two elements to this argument:

(i) Each player should choose a best response to the expected strategy choice of the opponent.
(ii) Beliefs about the opponent's strategy choice should be "rational" in the sense that, in equilibrium, these expectations are fulfilled.

The first of these requirements assumes that a player tries to predict the opponent's strategy choice, whereas the second imposes consistency of expectations about the opponent's play with those strategies that are actually played.

In contrast to the maximin argument discussed, this approach does not provide an immediate suggestion of how to play; rather it postulates that any solution for a game should have the properties of optimal behavior of all agents and rational expectations. Even if players were to agree to these principles, no immediate suggested course of action emerges. In particular, an agent would have to know the payoff of his opponent to determine what would be his best response. Thus, common knowledge of the payoff structure is a necessary requirement for this approach. But, even given this information, there may be more than one strategy combination that satisfies this criterion of optimality and consistency. In such a case, even with common knowledge of the payoff functions, no recommended behavior can be deduced by a player. The following definition formalizes the two principles of optimality and consistency suggested.[2]

DEFINITION 2.4. An *equilibrium* of a game is a strategy combination (s_1^*, s_2^*) such that

$$p_1(s_1^*, s_2^*) \geq p_1(s_1, s_2^*) \quad \text{for all} \quad s_1 \in S_1 \quad \text{and}$$

$$p_2(s_1^*, s_2^*) \geq p_2(s_1^*, s_2) \quad \text{for all} \quad s_2 \in S_2 \quad \text{hold.}$$

Note that any strategy combination that satisfies these conditions is an equilibrium and, therefore, qualifies as a solution. The first line of the definition requires that s_1^* be a best response to s_2^* and the second condition requires s_2^* to be a best response to s_1^*. Thus, optimality and mutual consistency are satisfied for any such strategy combination (s_1^*, s_2^*).

The following examples will provide some insight into the characteristics of an equilibrium.

EXAMPLE 2.1 (*resumed*). Reconsider matching pennies. For player 1, one has the following best response function:

- If player 2 plays H, then playing H is the best response.
- If player 2 plays T, then playing T is the best response.

Similarly, for player 2, one obtains the following best responses:

- If player 1 plays H, then playing T is the best response.
- If player 1 plays T, then playing H is the best response.

It is not hard to see that there is no mutually consistent best response for both players in this case. If player 1 plays H, then player 2 should play T;

[2]Note that this is the definition of a Nash equilibrium for a two-player game. A general definition will be given in Chapter 4.

but if player 2 plays T, player 1 will find it optimal to play T as well; yet for player 1 playing T, playing H would be the best response of player 2; and player 1 would want to play H in response to player 2 playing H. This creates a cycle as the following diagram shows.

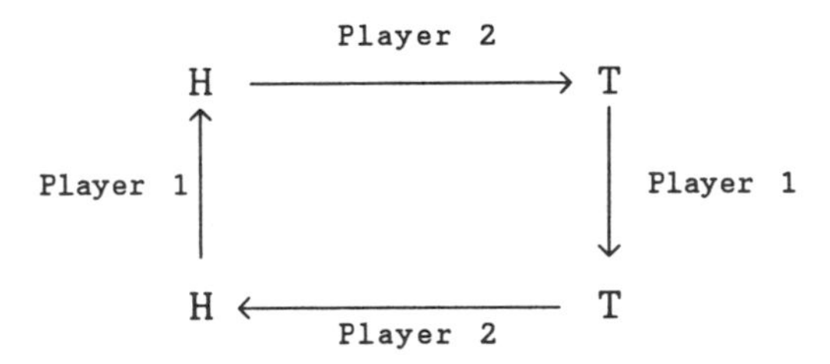

Hence, an equilibrium does not exist in matching pennies. ■

Note that there were maximin strategies for matching pennies. Actually, the recommendation from argument 1 in the previous section was that both strategies were maximin strategies and would yield at least -1. Hence, a player could choose either of his two strategies. The message from argument 2 for matching pennies is that there is no mutually consistent optimal play. In contrast, for example 2.2, it was demonstrated that there is an equilibrium, namely (s_1^2, s_2^3). In addition, the equilibrium strategies were exactly the maximin strategies for this game.

The following example will illustrate that there may be more than one equilibrium in a two-player zero-sum game.

EXAMPLE 2.3. Consider the following zero-sum game.

		Player 2		
		s_2^1	s_2^2	s_2^3
Player 1	s_1^1	1	2	1
	s_1^2	0	3	-1

Obviously, it is best for player 1 to play his first strategy in response to strategy s_2^1 and s_2^3 of player 2, and his second strategy against s_2^2. For player 2 both strategies s_2^1 and s_2^3 are best responses to player 1's first strategy, whereas s_2^3 is his best response to s_1^2. Thus, (s_1^1, s_2^1) and (s_1^1, s_2^3) are equilibrium strategy combinations. The following diagram illustrates the best response relationships.

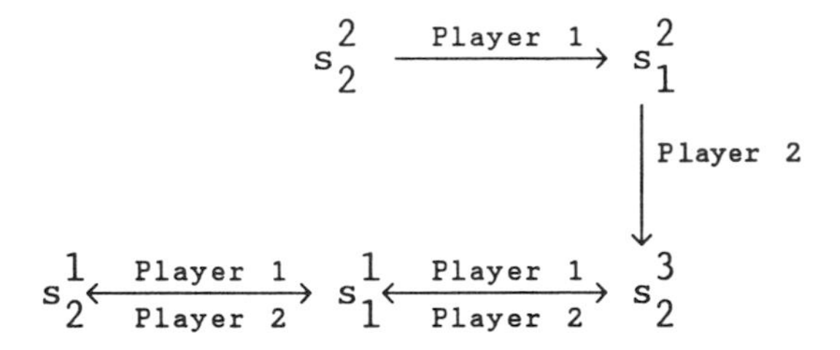

The two equilibria are visible as double arrows in the diagram. Note, however, that the equilibrium payoff for both players is the same in both equilibria, namely 1 for player 1 and -1 for player 2. Moreover, it is easy to check that s_1^1 is a maximin strategy for player 1, and that s_2^1 and s_2^3 are maximin strategies for player 2. The value of this game is $v = 1$. ■

Example 2.3 illustrates a property of multiple equilibria that holds for zero-sum games in general. If a game has more than one equilibrium, then equilibrium strategies are exchangeable. That means that one can combine any strategy of player 1 that occurs in some equilibrium strategy combination and any strategy of player 2 that is a component of an equilibrium strategy combination to form a new equilibrium strategy combination yielding the same payoff to the players.

2.2 Relationship of Equilibrium and Value

The last section presented two concepts for a solution of a game. The first was based on a prescription for each player to choose a strategy that would minimize losses. The second suggested choosing a best response to the opponent's choice. If the maximin strategies are consistent in the sense that $v_1 + v_2 = 0$ holds, the game was said to have a value. If there is a strategy combination that is a mutual best response, then the game has an equilibrium. In addition, several examples provided the following insights:

(i) There are zero-sum games that have no equilibrium, one equilibrium, or several equilibria.
(ii) A zero-sum game may or may not have a value.
(iii) If a game has a value it has an equilibrium as well, and the strategy combinations of the equilibrium are maximin strategies of the players.

The results in this section show that these observations can be generalized.

THEOREM 2.1. *A two-person zero-sum game has an equilibrium if and only if it has a value.*

The following proof will show that

(i) Every equilibrium strategy combination (s_1^*, s_2^*) is a maximin strategy combination with a common security level, the value of the game.
(ii) Every maximin strategy combination $(\bar{s}_1, \bar{s}_2)$ of a game with a value is an equilibrium.

Proof.

(i) First it will be shown that an equilibrium implies

$$v_1 \geq p_1(s_1^*, s_2^*) \geq -v_2.$$

A game has an equilibrium if there is a strategy combination (s_1^*, s_2^*) such that $p_1(s_1^*, s_2^*) \geq p_1(s_1, s_2^*)$, for all $s_1 \in S_1$, and $p_2(s_1^*, s_2^*) \geq p_2(s_1^* s_2)$, for all $s_2 \in S_2$ holds.

Recall that

$$\bar{v}(s_2) \equiv \max_{s_1 \in S_1} p_1(s_1, s_2)$$

is the maximum value of $p_1(s_1, s_2)$ on S_1 for a given s_2. Hence, $\bar{v}(s_2^*) = p_1(s_1^*, s_2^*)$ follows from $p_1(s_1^*, s_2^*) \geq p_1(s_1, s_2^*)$, for all $s_1 \in S_1$. Since the minimum value of $\bar{v}(s_2)$ must be smaller than $\bar{v}(s_2)$ for any other s_2,

$$-v_2 \equiv \min_{s_2 \in S_2} \bar{v}(s_2) \leq \bar{v}(s_2^*) = p_1(s_1^*, s_2^*) \tag{a}$$

follows.

Multiplying the second equilibrium condition, $p_2(s_1^*, s_2^*) \geq p_2(s_1^*, s_2)$, for all $s_2 \in S_2$, by (-1) and using the fact that $p_1(s_1, s_2) = -p_2(s_1, s_2)$ holds for zero-sum games, one can write this equilibrium condition equivalently as $p_1(s_1^*, s_2) \leq p_1(s_1^*, s_2^*)$ for all $s_2 \in S_2$. Thus, one can view the second equilibrium condition as a requirement that $p_1(s_1, s_2)$ be minimized for a given s_1^*.

Recall the minimum value function

$$\tilde{v}(s_1) \equiv \min_{s_2 \in S_2} p_1(s_1, s_2).$$

Clearly, $p_1(s_1^*, s_2^*) = \tilde{v}(s_1^*)$ must be true for the equilibrium strategy combination (s_1^*, s_2^*). Since the maximum value of $\tilde{v}(s_1)$ must be larger than the value of $\tilde{v}(s_1)$ for any other s_1,

$$v_1 \equiv \max_{s_1 \in S_1} \tilde{v}(s_1) \geq \tilde{v}(s_1^*) = p_1(s_1^*, s_2^*) \tag{b}$$

follows.

Inequalities (a) and (b) together imply $v_1 \geq p_1(s_1^*, s_2^*) \geq -v_2$. By lemma 2.1, one has $-v_2 \geq v_1$ and, therefore, $v_1 = -v_2 = p_1(s_1^*, s_2^*)$.

(ii) Using the same definition for $\bar{v}(s_2)$ and $\tilde{v}(s_1)$ as before, let $\bar{s}_1$ be a maximizer of $\tilde{v}(s_1)$ and $\bar{s}_2$ a minimizer of $\bar{v}(s_2)$. Clearly, $v_1 = \tilde{v}(\bar{s}_1)$ and $-v_2 = \bar{v}(\bar{s}_2)$ holds. Furthermore, note that, for any s_2,

$$\bar{v}(s_2) \equiv \max_{s_1 \in S_1} p_1(s_1, s_2) \geq p_1(s_1, s_2) \quad \text{for all} \quad s_1 \in S_1, \tag{c}$$

and, for any s_1,

$$\tilde{v}(s_1) \equiv \min_{s_2 \in S_2} p_1(s_1, s_2) \leq p_1(s_1, s_2) \quad \text{for all} \quad s_2 \in S_2 \tag{d}$$

holds. In particular, for $(\bar{s}_1, \bar{s}_2)$,

$$\bar{v}(\bar{s}_2) \equiv \max_{s_1 \in S_1} p_1(s_1, \bar{s}_2) \geq p_1(\bar{s}_1, \bar{s}_2) \geq \min_{s_2 \in S_2} p_1(\bar{s}_1, s_2) \equiv \tilde{v}(\bar{s}_1)$$

must hold. Hence, if the game has a value, then $v_1 = -v_2$ implies

$$-v_2 = \bar{v}(\bar{s}_2) = p_1(\bar{s}_1, \bar{s}_2) = \tilde{v}(\bar{s}_1) = v_1.$$

From (c) and (d),

$$p_1(\bar{s}_1, s_2) \geq \tilde{v}(\bar{s}_1) = p_1(\bar{s}_1, \bar{s}_2) = \bar{v}(\bar{s}_2) \geq p_1(s_1, \bar{s}_2)$$

for all $s_1 \in S_1$ and all $s_2 \in S_2$ follows. Hence, $(\bar{s}_1, \bar{s}_2)$ is an equilibrium of the game. ■

This theorem shows that in two-person zero-sum games maximin strategies support a value of the game if and only if they are equilibrium strategies. However, it does not indicate the solution of games such as matching pennies that have no equilibrium point. This will be considered in the next section.

2.3 Mixed Strategies

Existence of a solution means that the respective solution concept is applicable to the problem under consideration. In matching pennies, for example, neither an equilibrium nor a value exists. That means that both concepts cannot be applied to this example because no strategy pair (s_1, s_2) exists that would satisfy the requirements of definition 2.3. Similarly, there is no common value for the security levels of the players in this game and definition 2.2 does not apply. In this sense, neither of the solution concepts discussed so far can be used in general. From theorem 2.1 however, it follows that both the equilibrium concept and the value concept will always be jointly applicable or jointly fail.

Nonexistence of solution concepts is closely related to characteristics of the strategy sets and the payoff functions. In the case of the equilibrium and the value, it is sufficient to expand the strategy space to include mixed strategies[3] to guarantee existence of an equilibrium for all two-person zero-sum games with finite strategy sets. The following definition extends the equilibrium concept to games with mixed strategies.

DEFINITION 2.5. An *equilibrium in mixed strategies* is a mixed strategy combination (m_1^*, m_2^*) such that

$$P_1(m_1^*, m_2^*) \geq P_1(m_1, m_2^*) \quad \text{for all} \quad m_1 \in M_1$$
$$P_2(m_1^*, m_2^*) \geq P_2(m_1^*, m_2) \quad \text{for all} \quad m_2 \in M_2.$$

The definition of an equilibrium in mixed strategies requires that the equilibrium mixed strategy combination (m_1^*, m_2^*) be a best response for each player, exactly as in the case of pure strategies. Note, however, that expected payoffs P_i have to be considered because of the randomization introduced by the mixed strategies.

It is equally straightforward to adapt the definition of the value to included mixed strategies.

DEFINITION 2.6. *The* mixed-strategy value *of a game is the number v^m that satisfies*

$$v^m \equiv \max_{m_1 \in M_1} \left[\min_{m_2 \in M_2} P_1(m_1, m_2) \right] = \min_{m_2 \in S_2} \left[\max_{m_1 \in S_1} P_1(m_1, m_2) \right].$$

The security level of a player using mixed strategies is denoted by v_i^m.

The extension of the equilibrium and value concept to games in which players can choose mixed strategies requires only a change of the strategy sets, from S_i to M_i for each player, and an appropriate redefinition of their payoff functions. Thus, it is not difficult to check that theorem 2.1 remains true for games with mixed strategies. Hence, a game has a value in mixed strategies if and only if it has an equilibrium in mixed strategies. Extending a game to allow for mixed strategies is, however, sufficient to make the equilibrium and value concept applicable to all two-person zero-sum games with finitely many pure strategies.

THEOREM 2.2 *(von Neumann, 1928). A game with finitely many strategies has an equilibrium in mixed strategies.*

This theorem is proved for general games in strategic form (not just for two-player zero-sum games) in Chapter 4. A reader not interested in this proof may proceed directly to the end of the proof. The following

[3]See section 1.2.2 for a general description of the concept of a mixed strategy and its associated payoff.

proof of this theorem for two-player zero-sum games is of special interest since it can be related to linear optimization techniques familiar to most economists. Furthermore, it provides a useful insight into the relationship between existence of an equilibrium and existence of a solution to an optimization problem. The proof proceeds in two steps. First, it is shown that, for two-player zero-sum games an equilibrium forms a *saddle point* of an optimization problem. Then it is demonstrated that the optimization problem has a solution.

Proof. Using the zero-sum property,

$$-P_1(m_1^*, m_2^*) = P_2(m_1^*, m_2^*) \geq P_2(m_1^*, m_2) = -P_1(m_1^*, m_2) \quad \text{for all} \quad m_2 \in M_2$$

follows from the definition of an equilibrium (2.4). Multiplying this inequality by -1 and using the first condition of definition 2.4, the definition of an equilibrium can be written equivalently as a saddle point,[4]

$$P_1(m_1^*, m_2) \geq P_1(m_1^*, m_2^*) \geq P_1(m_1, m_2^*)$$

for all $m_1 \in M_1$ and all $m_2 \in M_2$. A game, therefore, has an equilibrium if and only if it has a saddle point. The following argument will show that a game with finite strategy sets has a saddle point in mixed strategies.

For any set of numbers $a(s_1), b(s_2), s_1 \in S_1,\ s_2 \in S_2$, consider the following two optimization problems:

$$\text{Choose} \quad m_1 \in M_1 \quad \text{to maximize} \quad \sum_{s_1 \in S_1} a(s_1) \cdot m_1(s_1) \quad \text{subject to}$$

$$\sum_{s_1 \in S_1} p_1(s_1, s_2) \cdot m_1(s_1) \leq b(s_2) \quad \text{for all} \quad s_2 \in S_2. \tag{A}$$

$$\text{Choose} \quad m_2 \in M_2 \quad \text{to minimize} \quad \sum_{s_2 \in S_2} b(s_2) \cdot m_2(s_2) \quad \text{subject to}$$

$$\sum_{s_2 \in S_2} p_1(s_1, s_2) \cdot m_2(s_2) \geq a(s_1) \quad \text{for all} \quad s_1 \in S_1. \tag{B}$$

The duality theorem for linear programming problems (see Takayama, 1988, p. 156) states that (i) the maximization problem A has a solution m_1^* if and only if the minimization problem B has a solution m_2^*, and that (ii)

[4]For a formal definition of a saddle point, see Takayama, (1988 p. 66).

the following equalities hold:

$$m_2^*(s_2)\cdot\left[b(s_2)-\sum_{s_1\in S_2}p(s_1,s_2)\cdot m_1^*(s_1)\right]$$
$$=m_1^*(s_1)\cdot\left[\sum_{s_2\in S_2}p(s_1,s_2)\cdot m_2^*(s_2)-a(s_1)\right]=0 \qquad (*)$$
$$\sum_{s_1\in S_1}a(s_1)\cdot m_1^*(s_1)=\sum_{s_2\in S_2}b(s_2)\cdot m_2^*(s_2). \qquad (**)$$

For a finite set of pure strategies S_1, the mixed strategy set M_1 is compact and the objective function $P_1(m_1, m_2)$ is continuous. By the maximum theorem (see Varian, (1984, p. 318), the maximization problem A has a solution m_1^*, and therefore, by the duality theorem, the minimization problem B has a solution m_2^* as well. From (*),

$$a(s_1)\cdot m_1^*(s_1)=\sum_{s_2\in S_2}p(s_1,s_2)\cdot m_1^*(s_1)\cdot m_2^*(s_2) \quad \text{for all} \quad s_1\in S_1,$$
$$b(s_2)\cdot m_2^*(s_2)=\sum_{s_1\in S_1}p(s_1,s_2)\cdot m_1^*(s_1)\cdot m_2^*(s_2) \quad \text{for all} \quad s_2\in S_2$$

follows. Summing the first equations over s_1, one obtains

$$P_1(m_1^*,m_2^*)\equiv\sum_{s_1\in S_1}\sum_{s_2\in S_2}p_1(s_1,s_2)\cdot m_1^*(s_1)\cdot m_2^*(s_2)$$
$$=\sum_{s_1\in S_1}a(s_1)\cdot m_1^*(s_1)=\sum_{s_2\in S_2}b(s_2)\cdot m_2^*(s_2) \qquad [\text{by } (**)]$$
$$\geq\sum_{s_1\in S_1}\sum_{s_2\in S_2}p_1(s_1,s_2)\cdot m_1(s_1)\cdot m_2^*(s_2)$$
$$\equiv P_1(m_1,m_2^*) \qquad \text{for all } m_1\in M_1.$$

The last inequality follows from summing the constraints of problem A over s_2. Thus, m_1^* is a maximizer of $P_1(m_1, m_2^*)$. Similarly, using the constraints of B, one can conclude:

$$P_1(m_1^*,m_2^*)\equiv\sum_{s_1\in S_1}\sum_{s_2\in S_2}p_1(s_1,s_2)\cdot m_1^*(s_1)\cdot m_2^*(s_2)$$
$$=\sum_{s_1\in S_1}a(s_1)\cdot m_1^*(s_1)=\sum_{s_2\in S_2}b(s_2)\cdot m_2^*(s_2) \qquad [\text{by } (**)]$$
$$\leq\sum_{s_1\in S_1}\sum_{s_2\in S_2}p_1(s_1,s_2)\cdot m_1^*(s_1)\cdot m_2(s_2)$$
$$\equiv P_1(m_1^*,m_2) \qquad \text{for all } m_2\in M_2.$$

This shows that m_2^* is a minimizer of $P_1(m_1^*, m_2)$ and completes the proof. ∎

Since a two-player zero-sum game has an equilibrium in mixed strategies if and only if it has a value in mixed strategies, it follows from theorem 2.2 that a two-person zero-sum game with finitely many strategies has a value if mixed strategies are allowed. In fact, one can show that the mixed-strategy value lies between the pure-strategy security levels of the two players. The following lemma shows this result.

LEMMA 2.2. *The maximin value over mixed strategies, v_i^m, satisfies $v_i^m \geq v_i$ for both players* 1 *and* 2.

Proof. Without loss of generality, the argument will be conducted for player 1. Recalling the definition of the maximin value,

$$v_1^m = \max_{m_1 \in M_1} \min_{m_2 \in M_2} P_1(m_1, m_2)$$

must hold.

Denote by $\tilde{m}_i^{sk}$ the following mixed strategy of player i:

$$\tilde{m}_i^{sk}(s_i^l) = \begin{cases} 1 & \text{if } l = k \\ 0 & \text{if } l \neq k, \end{cases}$$

that is, the mixed strategy that plays the pure strategy s_i^k with probability one and any other pure strategy of player i with probability zero. Hence, playing the mixed strategy $\tilde{m}_i^{sk}$ is the same as playing the pure strategy s_i^k.

Note that, for any s_1',

$$\min_{s_2 \in S_2} p_1(s_1', s_2) \leq P_1(\tilde{m}_1^{s'}, m_2) \quad \text{for all} \quad m_2 \in M_2 \quad \text{holds,}$$

because a weighted average of any set of numbers is always at least as high as the smallest number in the set. In particular, for any $s_1' \in S_1$,

$$\min_{s_2 \in S_2} p_1(s_1', s_2) \leq \min_{m_2 \in M_2} P_1(\tilde{m}_1^{s'}, m_2)$$

must hold.

Suppose that s_1' is a strategy that maximizes

$$\tilde{v}(s_1) \equiv \min_{s_2 \in S_2} p_1(s_1, s_2),$$

then

$$\begin{aligned} v_1 \equiv \max_{s_1 \in S_1} \left[\min_{s_2 \in S_2} p_1(s_1, s_2) \right] &= \min_{s_2 \in S_2} p_1(s_1', s_2) \\ &\leq \min_{m_2 \in M_2} P_1(\tilde{m}_1^{s'}, m_2) \\ &\leq \max_{m_1 \in M_1} \min_{m_2 \in M_2} P_1(m_1, m_2) \equiv v_1^m \end{aligned}$$

holds, where the last inequality follows because the maximum of the expected payoff function with respect to m_1 is always greater than the value of this function for a particular m_1, say $\tilde{m}_1^{s'}$. ■

By lemma 2.2, it follows that enlarging the strategy set of a player by allowing him to use mixed strategies cannot be harmful to that player. As an immediate consequence, one can see that the maximin value over mixed strategies v^m (which always exists in two-player zero-sum games with finite strategy sets) must lie between v_1 and $-v_2$, that is, $v_1 \leq v^m \leq -v_2$. This follows because $v_2 \leq v_2^m$ implies $-v_2 \geq -v_2^m$ and, by definition of the value, $v^m \equiv v_1^m = -v_2^m$ holds.

Indeed, whenever a game has no value in pure strategies, that is, $v_1 < -v_2$ holds, then the mixed strategy maximin value v^m will lie strictly between these two numbers. If a game has a value in pure strategies, however, then the pure strategy value will coincide with the mixed strategy value, since in this case $v_1 = -v_2$ and $v_1 \leq v^m \leq -v_2$ must hold simultaneously.

The following example illustrating the existence of an equilibrium and value in mixed strategies for matching pennies concludes this chapter.

EXAMPLE 2.1 (*resumed*). As shown, the game matching pennies has no equilibrium in pure strategies and, therefore, no value. Allowing for mixed strategies is, however, particularly attractive in this case, because intuition would suggest that players make a random choice whenever there is no clearly defined "best strategy."

If mixed strategies are considered, the strategy sets take the following form:

$$M_1 = \{(m_1(H), m_1(T)) \in \mathbb{R}_+^2 \| m_1(H) + m_1(T) = 1\} \equiv \Delta^2,$$
$$M_2 = \{(m_2(H), m_2(T)) \in \mathbb{R}_+^2 \| m_2(H) + m_2(T) = 1\} \equiv \Delta^2.$$

These strategy sets are simply the sets of all probability distributions on the sets of pure strategies $S_i = \{H, T\}$. Since there are just two pure strategies, the mixed strategy sets are 2-dimensional simplexes, that is, sets the elements of which are pairs of positive numbers that add to one. Given a mixed strategy for each agent, $m_1 \equiv (m_1(H), m_1(T))$ and $m_2 = (m_2(H), m_2(T))$, the expected payoffs of players 1 and 2 are easily derived as

$$P_1(m_1, m_2) = m_1(H) \cdot [m_2(H) - m_2(T)] + m_1(T) \cdot [m_2(T) - m_2(H)],$$
$$P_2(m_1, m_2) = m_2(H) \cdot [m_1(T) - m_1(H)] + m_2(T) \cdot [m_1(H) - m_1(T)].$$

To find the equilibrium mixed strategy combination (m_1^*, m_2^*), consider the best response strategies of player 1. If player 2 plays H with a probability $m_2(H) > .5$, then $m_2(T) = 1 - m_2(H) < .5$ holds, and, therefore,

$$[m_2(H) - m_2(T)] > 0 \quad \text{and} \quad [m_2(T) - m_2(H)] < 0$$

follows. Hence, making $m_1(H)$ as large as possible and $m_1(T)$ as small as possible yields the highest payoff for player 1. Given that probabilities of mutually exclusive events must sum to one, $m_1(H) = 1$ and $m_1(T) = 0$ is the best response for player 1 to any mixed strategy of player 2 that gives H a probability greater than .5.

A similar argument shows that for $m_2(H) < .5$, the best response of player 1 is to play T with probability one, that is, $(m_1(H), m_1(T)) = (0, 1)$. Finally, if player 2 plays H and T with probability .5 each, then $P_1(m_1, (.5, .5)) = 0$ would hold for any mixed strategy player 1 chooses. Therefore, he would be indifferent about which strategy to play. In summary, the best response of player 1 to any mixed strategy of player 2 has the following form:

$$m_1(H) = \begin{cases} 1 & \text{if } m_2(H) > .5 \\ \text{any probability between 0 and 1} & \text{if } m_2(H) = .5 \\ 0 & \text{if } m_2(H) < .5 \end{cases}$$

where only the probability of playing $H, m_i(H)$, is noted for each player since $m_i(T) = 1 - m_i(H)$ holds.

In a similar way one derives the best response for player 2 to any mixed strategy choice of player 1,

$$m_2(H) = \begin{cases} 1 & \text{if } m_1(H) < .5 \\ \text{any probability between 0 and 1} & \text{if } m_1(H) = .5 \\ 0 & \text{if } m(H) > .5 \end{cases}$$

The following diagram shows these two best response mappings.

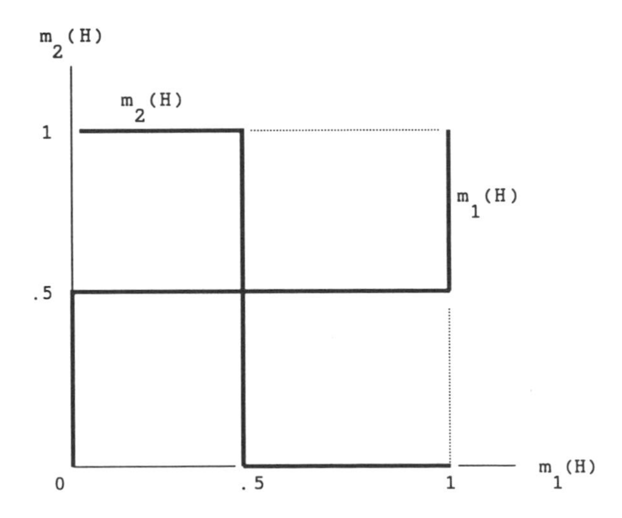

The solid lines indicate the best response mappings. The best response for player 2 runs from 0 to .5 at the level of 1 and from .5 to 1 at level 0 parallel to the horizontal axis. The best response of player 1 runs parallel to the vertical axis first at 0 and then at 1.

Since an equilibrium is a mixed strategy combination (m_1^*, m_2^*) that is a mutually best response, an equilibrium mixed strategy combination must lie on both best response mappings. Hence, one can see from the diagram that $m_1(H) = .5$ and $m_2(H) = .5$ is the only mixed strategy combination satisfying this condition. Thus,

$$(m_1^*, m_2^*) \equiv ((m_1^*(H), m_1^*(T)), (m_2^*(H), m_2^*(T))) = ((.5, .5), (.5, .5))$$

is the unique equilibrium of matching pennies in mixed strategies. Notice that both players are indifferent as to their strategy choice in this equilibrium.

It is obvious that matching pennies has a value in terms of mixed strategies as well. By playing the mixed strategy $m_i = (.5, .5)$, each player can guarantee himself an expected payoff of zero. Hence, the maximin value over mixed strategies is $v_i^m = 0$ for each player, and $v^m = 0$ is the value in mixed strategies. ■

2.4 Summary

In this chapter a special class of games was analyzed. These games are characterized by the fact that one player's win is another player's loss. In particular, two-player zero-sum games were analyzed with the question in mind of what could constitute a solution to the game. Two types of reasoning were introduced:

(i) A security oriented choice of strategy was proposed, in which each player was to consider the worst payoff for each of his strategies and then to play the best of them. If these maximin strategies were consistent in the sense that their payoffs would sum to zero, the game was said to have a value.

(ii) The second line of argument stressed the need for a strategy to be a best response to the behavior of the opponent. In this case consistency required that only strategy combinations that are best response strategies for each player be considered as solutions to a game.

Fortunately, in two-player zero-sum games, the two concepts coincide. Hence, a solution has a "natural" meaning in this context. The problem of nonexistence of such a solution can be overcome by considering mixed strategies.

As Chapter 4 will show, the equilibrium concept developed in this chapter and the existence result of theorem 2.2 carry over to general games in strategic form. The notion of the value and the equivalence theorem 2.1, however, have no analogues in general n-person games.

2.5 Remarks on the Literature

The results presented here date back to von Neumann and Morgenstern (1947) and earlier literature. Though the study of zero-sum games has been important in shaping ideas about what could be a solution for a game, a special discussion of zero-sum games is often omitted from modern books on game theory. The exposition of the two arguments for a solution of a game follows Shubik (1982, Chapter 8). Aumann (1975a) contains the equivalence theorem and Moulin (1986) covers all the results of this chapter. Moulin (1986) in particular presents further results, including results on zero-sum games in extensive form and with perfect information. In addition, Moulin (1986) and Shubik (1982) are excellent references for further examples of zero-sum games. Binmore (1992) discusses zero-sum games in Chapter 6.

Exercises

2.1 *Consider the following game in strategic form:*

		Player 2			
		s_2^1	s_2^2	s_2^3	s_2^4
	s_1^1	-3	2	-1	-3
Player 1	s_1^2	-5	1	0	-4
	s_1^3	-4	0	1	-3

Derive the maximin strategies of the two players and show that this game has a value. Does this game have a unique equilibrium?

2.2 *(Shubik, 1982, p. 226) Colonel Blotto has three divisions to defend two mountain passes. He will defend successfully against equal or smaller strength but lose against superior forces. The enemy has two divisions. The battle is lost if either pass is captured. Neither side has advance information on the disposition of the opponent's divisions. Assume that the worth of overall victory is* 1 *and of defeat is* -1.

(a) Write down the payoff matrix of this game.

(b) Is there an equilibrium in pure strategies?

(c) Derive the equilibrium strategies of the mixed extension of this game.

Hint: Compute the mixed strategies that make the opponent indifferent about those pure strategies he uses with positive probability. Why is such a mixed strategy combination an equilibrium?

2.3 *Consider a firm and a trade union bargaining over their shares in a productivity gain* $\bar{x}$. *The firm and the trade union make their claims* x_f *and* x_u *respectively. If the claims are compatible, that is,* $x_f + x_u = \bar{x}$, *then they are implemented. If they are incompatible however,* $x_f + x_u \neq \bar{x}$, *then they go to an arbitration commission known to allocate them a share of* $\bar{x}$ *proportional to their claims.*

(a) Write down the strategic form of this game.

(b) Derive a maximin strategy combination and show that the security levels sum to $\bar{x}$.

(c) Derive an equilibrium strategy combination or show that the game has no equilibrium in pure strategies. *Hint:* Draw the best response functions for the firm and the trade union.

Remark: Notice that this is not a zero-sum game. The sum of the two players' payoffs is, however, equal to $\bar{x}$ for any strategy combination.

2.4 *Consider a population that has to vote on a road development project that half favor and half reject. Denote by* $q = 1$ *the decision to carry out the project and by* $q = 0$ *the decision not to. Let* $v_i(q)$, $i = 1, 2$, *be the payoff of a player from group* i, *and assume that* $v_1(q) = q$ *and* $v_2(q) = -q$ *represent the preferences of the players. Assume further that one can treat players of the same group as identical.*

Write down the payoff matrix for this game and derive the equilibrium strategy combination for the following two voting rules:

(*i*) *a strict majority is needed to accept the project*

(*ii*) *a strict majority is needed to reject the project.*

2.5 *(Adapted from Moulin, 1986, p. 15) Consider a government and an opposition planning the publication dates of their election programs. Represent the time from this decision to the election date by the interval* $[0, 1]$. *Denote by* $x(t)$ *the government's probability of gaining a majority of votes in the election if the program is published at time* t. *Similarly,* $y(t)$ *denotes the opposition's probability of obtaining a majority if it makes its program public in period* t. *To capture the notion that the probability of success increases as the publication date is chosen closer to the election date at* $t = 1$, $x(t)$ *and* $y(t)$ *are assumed to be continu-*

ous and nondecreasing functions with $x(0) = y(0) = 0$ *and* $x(1) = y(1) = 1$.

The party that moves first determines the probability of winning. If both parties move simultaneously, then the probability of winning the election is the probability of gaining a majority and the other party not gaining a majority. Finally, assume that, for either party, the payoff from winning is 1 and from losing -1.

(a) Determine the strategy sets and payoff functions of the players.

(b) Draw a diagram of the payoff functions. Is this a zero-sum game?

(c) Does this game have a value. If yes, derive the value; if no, show that it has no value.

3

Equilibrium Concepts I: Dominance Arguments

In the last chapter, the special class of two-player zero-sum games was used to motivate two different approaches to the question of what may be considered a solution of a game. The discussion of zero-sum games presented arguments that players may advance for how to play the game together with assumptions on mutual consistency of their behavior. Mutual consistency of players' maximin strategies led to the concept of the *value*, and mutual consistency of best responses defined an *equilibrium*. By requiring that the players' maximin values should sum to zero, the value makes direct reference to the zero-sum property of the game, and hence the concept of a value for a game will not carry over to games in general. On the other hand, the definition of an equilibrium makes no reference to specific properties of zero-sum games, so that the equilibrium concept is more suitable as a solution concept for general games.

This chapter focuses on equilibrium concepts based on a different kind of argument. If a player has a strategy better than any other, independent of the strategic choices of opponents, then it appears reasonable to assume a player will choose this strategy. Such dominance arguments are intuitive and often applied in economics. The dominance notion is formalized in this chapter and related to the maximin behavior analyzed in Chapter 2. The discussion will be conducted in terms of pure strategies wherever possible. It should be clear, however that, if one substitutes the expected payoff P_i for the payoff function p_i, all definitions and results in this chapter carry over directly to mixed strategies.

Remark 3.1 (on notation). The following notational conventions will be useful for reasoning in games with an arbitrary number of players and arbitrary numbers of strategies.

In many cases, it is necessary to distinguish between the strategies of a particular player $i \in I$ and the strategies of all other players. For that reason, it is sometimes useful to write a strategy combination $s = (s_1, s_2, \ldots, s_I)$ as

$$s = (s_i, s_1, s_2, \ldots, s_{i-1}, s_{i+1}, \ldots, s_I) = (s_i, s_{-i}).$$

Obviously, s_{-i} stands for the strategy combination of all players except player i,

$$s_{-i} = (s_1, s_2, \ldots, s_{i-1}, s_{i+1}, \ldots, s_I).$$

The same notational convention can be used for mixed and behavior strategies, $m = (m_i, m_{-i})$ or $b = (b_i, b_{-i})$.

It also simplifies notation to denote the set of all pure strategy combinations by S. S is the Cartesian product of the individual strategy sets S_i, $S \equiv S_1 \times S_2 \times \cdots \times S_I$. The notation $s \in S$ means that s is a strategy combination where the strategy of the first player comes from S_1, the strategy of the second player from S_2, etcetera; that is,

$$\overbrace{(s_1, s_2, \ldots, s_I)}^{s} \in \overbrace{S_1 \times S_2 \times \cdots \times S_I}^{S}$$

where $s_1 \in S_1$, $s_2 \in S_2, \ldots, s_I \in S_I$ holds. S_{-i} denotes the set of all strategy combinations of players other than player i,

$$S_{-i} = S_1 \times \cdots \times S_{i-1} \times S_{i+1} \times \cdots \times S_I.$$

Thus, $s \in S$ is a well-defined notation for a list of strategies for all players, s, from the set of all such lists S, and $s_{-i} \in S_{-i}$ means a list of strategies for all players except player i, s_{-i}, from the set of all such lists S_{-i}.

In an analogous manner, $m \in M$, $m_{-i} \in M_{-i}$ and $b \in B$, $b_{-i} \in B_{-i}$ will be used for mixed and behavior strategies where appropriate. ■

With the help of this notation, it is possible to generalize the definition of a maximin strategy as follows. Denote by $\tilde{v}_i(s_i)$ the minimal payoff player i could obtain by playing strategy s_i; formally,

$$\tilde{v}_i(s_i) \equiv \min_{s_{-i} \in S_{-i}} p_i(s_i, s_{-i}).$$

DEFINITION 3.1. The *maximin value* (*security level*) of player $i \in I, v_i$, is

$$v_i = \max_{s_i \in S_i} \tilde{v}_i(s_i).$$

A *maximin strategy* (or *prudent strategy*) of player $i \in I$ is a strategy $\bar{s}_i$ that maximizes $\tilde{v}_i(s_i)$, that is, for which

$$\bar{s}_i \in \operatorname{argmax}_{s_i \in S_i} \tilde{v}_i(s_i)$$

holds.

Note that the notion of a maximin strategy and a maximin value that a player can guarantee herself remain well defined concepts, even if the value concept cannot be defined in general games.

3.1 Dominant Strategy Equilibrium

One of the attractive features of the value is that agents do not need any information about their opponents' payoff functions to determine their maximin strategies. All that is required for finding a maximin strategy is knowledge of their own payoff function and of the opponents' strategy sets.

The notion of a best response, on the other hand, generally requires some knowledge of the behavior of opponents, and this in turn cannot be determined without knowledge of their payoff functions. However there are games in which it is possible to find a best response to the opponents' behavior without this knowledge, simply because there may be a strategy that is a best response to any possible strategy combination of the other players. The following example illustrates this possibility.

EXAMPLE 3.1 (*prisoners' dilemma*). Two criminals have been caught by the police. Because of lack of evidence the prosecution needs a confession to convict. If no confession ensues, they will be charged and convicted for a minor offense earning them one year less than a conviction for the main crime.

The prosecutor offers each prisoner the following deal. If she confesses, and the other does not, she will get three years off her sentence whereas the other prisoner will get an extra year (-1) in prison. If both confess, they will be punished according to the law (no reductions).

This story easily translates into the following game in strategic form. There are two players, $I = \{1, 2\}$, and each has a strategy set with two strategies, "to confess" (C) or "not to confess" (N), that is, $S_1 \equiv \{C, N\}$ and $S_2 \equiv \{C, N\}$. The following matrix shows the payoffs from these strategies.

		Player 2	
		N	C
Player 1	N	1, 1	−1, 3
	C	3, −1	0, 0

Actually, there is no dilemma for the prisoners in terms of how they should behave. If prisoner 2 confesses, prisoner 1 is better off cooperating with the prosecution because otherwise she will serve an extra year in prison; and if prisoner 2 does not confess, it is best for prisoner 1 to confess, since she will get three years off her term. Thus, no matter what prisoner 2 does, it is best for prisoner 1 to cooperate with the prosecution.

A similar argument for prisoner 2 shows that prisoner 2 will confess in any case. This leads one to predict (C, C) as the outcome of the game. The dilemma lies in the fact that this individually rational behavior precludes the best outcome for the prisoners, namely not to confess (N, N) and to get a lighter sentence.

Notice that the source of the dilemma is not lack of communication between the prisoners but incentives. Even if they could talk and agree not to confess, each has an incentive to break the agreement at the cost of the one who sticks to it. Thus, an agreement to not confess would be incredible. ■

As example 3.1 shows, in some games there are strategies that are a best response to any strategy combination of the other players. In these cases, players can be expected to play such a dominant strategy. The following definition makes precise the concept of a dominant strategy.

Definition 3.2. A strategy $\tilde{s}_i \in S_i$ is called *dominant strategy* if

$$p_i(\tilde{s}_i, s_{-i}) \geq p_i(s_i, s_{-i}) \quad \text{for all} \quad s_i \in S_i \text{ and all } s_{-i} \in S_{-i}$$

holds. A strategy combination $\tilde{s} = (\tilde{s}_1, \tilde{s}_2, \ldots, \tilde{s}_I)$ with a dominant strategy for each player is called a *dominant strategy equilibrium*.

In example 3.1, the strategy combination (C, C) is a dominant strategy equilibrium. The following example shows that there may be more than one dominant strategy equilibrium in a game.

Example 3.2 (*Moulin*, 1986, *p*. 65). Consider the following two-player game in which player 1 can choose an action called "top" (T) or an action called "bottom" (B), and player 2 may choose between "left" (L) and "right" (R). The payoffs are given by the following table.

		Player 2	
		L	R
Player 1	T	1, 1	0, 1
	B	1, 0	0, 0

In this case, each player is indifferent as to what strategy to choose regardless of the opponent's strategy. Hence, for both players, each strategy is a dominant strategy, and any strategy combination is a dominant strategy equilibrium. ■

Notice that (T, L) would be the best strategy combination for the players, and, in contrast to example 3.1, it is a dominant strategy equilibrium. However, any other strategy combination is a dominant strategy equilibrium as well. Once again, this example demonstrates that individually rational behavior may result in suboptimal outcomes.

The argument that players will choose a dominant strategy if they have one available is quite powerful. If a strategy is optimal under all circumstances no matter what the opponents do, then there is a strong incentive to use this strategy. In particular, a player with a dominant strategy need not speculate about the behavior of opponents, that is, she need not know their payoff functions. Thus, it is difficult to imagine any other solution to a game with a dominant strategy equilibrium. Unfortunately, as the following example demonstrates, most games do not possess a dominant strategy equilibrium.

EXAMPLE 3.3. Consider the two-player game given by the following payoff matrix.

		Player 2			
		s_2^1	s_2^2	s_2^3	s_2^4
	s_1^1	4, 2	3, 3	1, 2	7, 2
Player 1	s_1^2	3, 8	2, 4	0, 2	5, 5
	s_1^3	4, 1	4, 2	0, 1	5, 0

Player 1 has no dominant strategy because she would want to play her third strategy, s_1^3, if player 2 plays s_2^2, and her first strategy, s_1^1, if player 2 plays s_2^3. Hence, there is no strategy for player 1 that is a best response to all strategies of player 2.

This suffices to show that there is no dominant strategy equilibrium in this example. ■

These examples demonstrate that dominant strategy equilibria may or may not exist and may or may not be unique. Furthermore, though dominant strategies have a strong appeal because every player has good reason to use a dominant strategy if she has one, dominant strategy equilibria may fail to be socially optimal.

The following examples show two economic applications of the dominant strategy equilibrium concept.

EXAMPLE 3.4 (*price-setting duopoly*). Two firms sell two goods that are close substitutes. Hence, a price change for one good has a major impact on the demand for the other. Demand functions are given as

$$d_1(p_1, p_2) = 10 - p_1 + .5 \cdot \frac{p_2}{p_1} \quad \text{and} \quad d_2(p_1, p_2) = 20 - 2 \cdot p_2 + \frac{p_1}{p_2},$$

where p_1 and p_2 denote the prices of firm 1 and firm 2 respectively. Suppose, for computational simplicity, that both firms have zero costs of production and maximize profits through price setting.

This description can be translated easily into the following game in strategic form:

- player set: $I = \{1, 2\}$
- strategy sets: $S_1 = [0, 10]$ and $S_2 = [0, 10]$
- payoff functions: $\pi_1(p_1, p_2) = 10 \cdot p_1 - p_1^2 + .5 \cdot p_2$
  $\pi_2(p_1, p_2) = 20 \cdot p_2 - 2 \cdot p_2^2 + p_1$

Note that in this case the strategy sets are infinite. Without loss of generality prices have been assumed to be between zero and ten, because for prices larger than ten the demand for both goods could become zero. The payoff functions are simply the profit functions obtained by multiplying the demand functions by the respective price. Recall that costs of production equal zero.

For any price of good 2, p_2, the optimal price response of firm 1 is obtained by maximizing $\pi_1(p_1, p_2)$ with respect to p_1. Note that the profit function $\pi_1(p_1, p_2)$ is differentiable and strictly concave in p_1. Hence, the first order condition

$$\frac{\partial \pi_1(p_1, p_2)}{\partial p_1} = 0$$

is necessary and sufficient for a maximum of the profit function with respect to price p_1. It is easy to calculate that $p_1 = 5$ solves the first order condition for all p_2. This means that firm 1 has a best response of $p_1 = 5$ to any price p_2 that firm 2 might choose. Hence, $p_1 = 5$ is a dominant strategy for firm 1. Similarly, one deduces that firm 2 has a dominant strategy of $p_2 = 5$. Therefore, this duopoly problem has a unique dominant strategy equilibrium $(p_1, p_2) = (5, 5)$. The following diagram shows the best response functions for both firms.

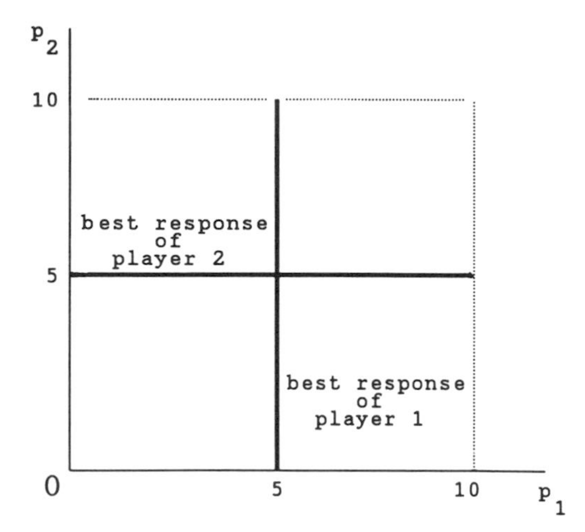

■

A second field of economics in which the dominant strategy equilibrium has found widespread application is social choice theory. One of the problems in this field concerns the possibility of extracting the "true" information necessary for a correct evaluation of public choice projects from self-interested agents. The question of whether one can develop schemes or mechanisms that make telling the truth a dominant strategy for each agent has become a focus of research. The following example shows one type of mechanism that induces agents to report truthfully.

EXAMPLE 3.5 (*Clarke–Groves mechanism for allocation of a public good*). Consider a government planning to upgrade a street in an area where 10 firms are potential users. The options are to upgrade the street, (U) or not to do so, (N). To assess the need for its road improvement program, the government sends a questionnaire to the firms in this area asking them to state their expected gain or loss from the upgrading of the street.

Suppose it is common knowledge that the government will undertake the project if and only if

$$\sum_{i=1}^{10} \pi_i(U) \geq \sum_{i=1}^{10} \pi_i(N)$$

holds, where $\pi_i(U)$ and $\pi_i(N)$ denote the reported gains from upgrading or not upgrading the street respectively.

This situation can be formalized as the following game in strategic form. There are ten players, $I = \{1, \ldots, 10\}$. The strategy of each player is a pair of numbers $(\pi_i(U), \pi_i(N))$ indicating the report about its gains from the street upgrading decision that firm i sends to the government. Since gains or losses are possible, any real number pair may be chosen as a message to the government. Thus, the strategy space for each agent is the space of all possible messages, $\mathbb{R}^2$.

Knowing the decision-making process of government, each player can try to manipulate the outcome of this process in its favor by over- or understating its gains or losses. Clearly, if the outcome a^* is unfavorable to firm i, the firm can manipulate its message $\pi_i = (\pi_i(U), \pi_i(N))$ to achieve a more favorable outcome. To avoid such misrepresentation of the true gains or losses, the following scheme is proposed.

Assume the government makes its decision about the road project according to the same principle as before, namely to choose the action $a \in \{U, N\}$ in order to maximize the sum of the payoffs $\Sigma_i \pi_i(a)$, but announces that, in addition, it will collect a tax from each firm based on their reports $(\pi_1, \pi_2, \ldots, \pi_{10})$:

$$t_i(\pi_1, \pi_2, \ldots, \pi_{10}) = \sum_{j \neq i} \pi_j(a^*(\pi_1, \ldots, \pi_{10})) - \max_{a \in \{U, N\}} \sum_{j \neq i} \pi_j(a),$$

where $a^*(\pi_1, \ldots, \pi_{10})$ denotes the maximizing decision of the government,

$$a^*(\pi_1, \ldots, \pi_{10}) = \begin{cases} U & \text{for } \sum_{i=1}^{10} \pi_i(U) \geq \sum_{i=1}^{10} \pi_i(N) \\ N & \text{otherwise.} \end{cases}$$

This tax is designed to reduce the incentives of the firms to misrepresent their true valuation. Notice that, by construction of this tax schedule, $t_i(\cdot) \leq 0$ holds.

If one denotes by $\hat{\pi}_i = (\hat{\pi}_i(U), \hat{\pi}_i(N))$ the "true" gain or loss of firm i, then one obtains the following payoff function:

$$p_i(\pi_1, \ldots, \pi_{10}) = \hat{\pi}_i(a^*(\pi_1, \ldots, \pi_{10})) + t_i(\pi_1, \ldots, \pi_{10}).$$

Notice that player i can report a valuation $\pi_i \neq \hat{\pi}_i$ to influence the choice of government, $a^*(\pi_1, \ldots, \pi_{10})$. Player i's evaluation of this outcome, however, will take place according to the true valuation $\hat{\pi}_i$. The tax schedule $t_i(\pi_1, \ldots, \pi_{10})$ affects the payoff of player i and, thus, has the potential to change the player's incentives in regard to her report π_i. Interestingly, this payment schedule $t_i(\cdot)$ makes it a dominant strategy for each firm to announce the true gains.

CLAIM. *A truthful report (i.e., $\pi_i = \hat{\pi}_i$) is a dominant strategy for each firm i.*

Proof. Note that firm i cannot influence the part $K(\pi_{-i})$ of its payoff,

$$K(\pi_{-i}) \equiv \max_{a \in A} \sum_{j \neq i} \pi_j(a).$$

Consider an arbitrary message combination

$$\pi_{-i} = (\pi_1, \ldots, \pi_{i-1}\, \pi_{i+1}, \ldots, \pi_{10})$$

for the other firms. Given firm i's report π_i, the government will choose $a \in \{U, N\}$ to maximize $\pi_i(a) + \Sigma_{j \neq i}\pi_j(a)$. If firm i could determine the action a, it would be chosen to maximize $p_i(\pi_i, \pi_{-i}) \equiv \hat{\pi}_i(a) + \Sigma_{j \neq i}\pi_j(a) - K(\pi_{-i})$, given the transfer scheme $t_i(\cdot)$. Since $K(\pi_{-i})$ cannot be influenced by her choice, her payoff is maximized if $\hat{\pi}_i(a) + \Sigma_{j \neq i}\pi_j(a)$ is maximized. This is, however, the government's objective function provided player i truthfully reports her valuation. Thus, by misrepresenting her gains from the project, player i can only lose because the government might make a decision that does not maximize her payoff. Formally, since $a^*(\hat{\pi}_i, \pi_{-i})$ maximizes $\hat{\pi}_i(a) + \Sigma_{j \neq i}\pi_j(a)$,

$$\begin{aligned} p_i(\hat{\pi}_i, \pi_{-i}) &\equiv \hat{\pi}_i(a^*(\hat{\pi}_i, \pi_{-i})) + \sum_{j \neq i} \pi_j(a^*(\hat{\pi}_i, \pi_{-i})) - K(\pi_{-i}) \\ &\geq \hat{\pi}_i(a^*(\pi)) + \sum_{j \neq i} \pi_j(a^*(\pi)) - K(\pi_{-i}) \equiv p_i(\pi_i, \pi_{-i}) \end{aligned}$$

must hold for all π_i that firm i could choose regardless of the other firms' reports π_{-i}. This proves that telling the truth is a dominant strategy for firm i.

However, in general, these taxes create a surplus, $\Sigma t_i(\cdot) \neq 0$. The allocative inefficiency of this surplus can be viewed as the cost of inducing truthful reports from firms. ■

It is interesting to note that dominant strategies are prudent strategies. This section concludes with a lemma showing the relationship between dominant strategies and maximin strategies.

LEMMA 3.1. *Every dominant strategy is a maximin strategy.*

Proof. For any $s_i \in S_i$, denote by $\tilde{s}_{-i}(s_i)$ the strategy combination of the other players that minimizes the payoff of player i and thus satisfies

$$p_i(s_i, \tilde{s}_{-i}(s_i)) = \min_{s_{-i} \in S_{-i}} p_i(s_i, s_{-i}).$$

By definition, for a dominant strategy $\tilde{s}_i$, $p_i(\tilde{s}_i, s_{-i}) \geq p_i(s_i, s_{-i})$ must hold for all $s_i \in S_i$ and all $s_{-i} \in S_{-i}$. Hence, in particular,

$$p_i(\tilde{s}_i, \tilde{s}_{-i}(\tilde{s}_i)) \geq p_i(s_i, \tilde{s}_{-i}(\tilde{s}_i)) \quad \text{for all} \quad s_i \in S_i$$

follows. But, since $\tilde{s}_{-i}(s_i)$ is a minimizer of $p_i(s_i, s_{-i})$ for any $s_i \in S_i$,

$$p_i(s_i, \tilde{s}_{-i}(\tilde{s}_i)) \geq p_i(s_i, \tilde{s}_{-i}(s_i)) \quad \text{for all} \quad s_i \in S_i$$

must be true. Combining these two inequalities that hold for all $s_i \in S_i$, one obtains

$$p_i(\tilde{s}_i, \tilde{s}_{-i}(\tilde{s}_i)) \geq p_i(s_i, \tilde{s}_{-i}(s_i)) \quad \text{for all} \quad s_i \in S_i.$$

This is, however, the definition of a maximin strategy. Hence, $\tilde{s}_i$ is a maximin strategy. ■

This lemma implies that every dominant strategy equilibrium is a maximin strategy combination. In games with dominant strategy equilibria, though these games need not have a value, the maximin strategy combinations form dominant strategy equilibria. Example 3.2 illustrates, however, that these maximin strategy combinations no longer have identical payoffs. It is easy to see from the matching pennies game that in general the reverse statement of the lemma is not true: Maximin strategies need not be dominant strategies.

3.2 Iterated Dominance Equilibrium

In the previous section, it was argued that a player will choose a dominant strategy, one that is a best response to any behavior of the other players, if such a strategy is available. The same logic leads one to assume that no player would want to play a dominated strategy.

DEFINITION 3.3. A strategy $\tilde{s}_i$ *dominates* strategy s_i, denoted $\tilde{s}_i \succ s_i$, if for all $s_{-i} \in S_{-i}$

$$p_i(\tilde{s}_i, s_{-i}) \geq p_i(s_i, s_{-i})$$

holds, with a strict inequality for at least one $s_{-i} \in S_{-i}$.

If a player never considers playing a dominated strategy, then one can eliminate this strategy and restrict attention to the subset of strategies of a player that are not dominated.

Denote by $D_i(p_i;\ S_1, S_2, \ldots, S_I)$ the subset of strategies of player $i \in I$ that are not dominated by any other strategy in the set. The list of arguments $(p_i;\ S_1, S_2, \ldots, S_I)$ lists all those elements of a game in strategic form that an agent must know to determine whether any strategy is dominated. Observe that it is not necessary to know the payoff functions of the other players.

The set of *undominated* strategies $D_i(p_i;\ S_1, S_2, \ldots, S_I)$ is a subset of S_i but may be equal to the latter, that is, $D_i(p_i;\ S_1, S_2, \ldots, S_I) \subseteq S_i$. Consider the game matching pennies (example 2.1) in which no player has a dominated strategy. Therefore, the set of undominated strategies equals the set of all strategies, that is, $D_i(p_i;\ S_1, S_2, \ldots, S_I) = S_i$, in this case. On the other hand, the game prisoners' dilemma has N as a dominated strategy for each player. Hence, in this case, $D_i(p_i;\ S_1, S_2, \ldots, S_I) \subset S_i$ holds. The following example is less trivial.

EXAMPLE 3.3 (*resumed*). For convenience, the payoff matrix of this game is repeated here.

		Player 2			
		s_2^1	s_2^2	s_2^3	s_2^4
	s_1^1	4, 2	3, 3	1, 2	7, 2
Player 1	s_1^2	3, 8	2, 4	0, 2	5, 5
	s_1^3	4, 1	4, 2	0, 1	5, 0

It is easy to check that for player 1, strategy s_1^2 is dominated by s_1^1 and by s_1^3. Hence, $D_1(p_1;\ S_1, S_2) = \{s_1^1, s_1^3\} \subset S_1$ holds. Similarly, for player 2 one notes $s_2^2 \succ s_2^3$ and $s_2^1 \succ s_2^4$. Therefore, the set of undominated strategies of player 2 is a strict subset of S_2: $D_2(p_2;\ S_1, S_2) = \{s_2^1, s_2^2\} \subset S_2$.

If the possibility can be ruled out, however, that any player will ever play a dominated strategy, then one may be tempted to consider the following reduced game as the "true" strategic situation.

		Player 2	
		s_2^1	s_2^2
Player 1	s_1^1	4, 2	3, 3
	s_1^3	4, 1	4, 2

Note that omitting dominated strategies and considering the resulting strategy sets as the relevant ones presumes that each player knows the payoffs of all other players. This is because, without information about player j's payoff, another player i cannot determine which strategies of player j are dominated. To argue therefore that

$$(I, (D_1(p_1;\ S_1, S_2), D_2(p_2;\ S_1, S_2)), (p_1, p_2))$$

is the relevant strategic form of a game implies a much stronger assumption about the information that the players have about their opponents than was required to determine maximin strategies or dominant strategies.

Obviously, this "new" derived game again has dominated strategies. Using the same argument that no dominated strategy will be chosen a second time, one can eliminate strategies s_1^1 and s_2^1 ($s_1^3 \succ s_1^1$ and $s_2^2 \succ s_2^1$). This reduces the game to the strategy combination (s_1^3, s_2^2).

Thus, if one assumes that no dominated strategies will ever be played by any player, and that every player knows that every player knows this, then one has to conclude that (s_1^3, s_2^2), the remaining strategy combination, is in fact the solution of this game. ■

This example suggests that the following two informational assumptions may be sufficient to determine a solution of the game:

(i) The complete strategic form of a game is common knowledge, in particular, all payoff functions.
(ii) It is common knowledge that no player will ever play a dominated strategy.

With these assumptions, it is possible to eliminate iteratively dominated strategies until only undominated strategies remain. Note that for games with finite strategy sets, this process must end in finite time.[1] At the end of this iterated elimination of dominated strategies one is left with a game with smaller strategy sets for most players. If this final game has only strategy combinations left in which each player gets the same payoff from all her remaining strategies, then one calls this game *dominance solvable* and any remaining strategy combination an *iterated dominance equilibrium* (or *sophisticated equilibrium*). Of course, in many cases, the remaining game after iterated elimination of dominated strategies does not have this property of an equal payoff for each player from all remaining strategies. In this case, an iterated dominance equilibrium does not exist.

Define the following sequence of strategy sets:

$$\text{for all} \quad i \in I,\ S_i^0 \equiv S_i, \quad \text{and} \quad S_i^{t+1} \equiv D_i\left(p_i;\ S_1^t, \dots, S_I^t\right) \quad \text{for all} \quad t \geq 0.$$

This sequence of strategy sets satisfies

$$S_i^{t+1} \subseteq S_i^t \subseteq \cdots \subseteq S_i^0 \equiv S_i$$

because each strategy sets S_i^{t+1} is derived from the previous one, S_i^t, by eliminating some strategies if this is possible.

DEFINITION 3.4. A game is called *dominance solvable* if there exists an integer T such that for all $i \in I$ and for all $s_{-i} \in S_{-i}^T$

$$s_i, s_i' \in S_i^T \text{ implies } p_i(s_i, s_{-i}) = p_i(s_i', s_{-i}).$$

Any strategy combination $\tilde{s} \in S_1^T \times \cdots \times S_I^T$ is called *iterated dominance equilibrium* (or *sophisticated equilibrium*).

In many cases, the iterated elimination of dominated strategies leads to a unique iterated dominance equilibrium, as in example 3.3, where (s_1^3, s_2^2) is the unique sophisticated equilibrium. If the process of iterated elimination of dominated strategies ends with a unique strategy combina-

[1] For an extension to games with infinite strategy sets, see Fudenberg and Tirole, (1991, p. 45).

tion, then this strategy combination trivially satisfies the equal payoff condition and, therefore, represents an iterated dominance equilibrium. The following example adapted from Moulin (1986, pp. 73–75) has such a unique outcome of the iterated elimination of dominated strategies.

EXAMPLE 3.6 (*voting with a president*). Three persons have to select a project from the set $\{a, b, c\}$. If the majority votes for a particular project, then it will be carried out. If, however, there is no majority for any project, then the project proposed by player 1, the president, is selected. Suppose that the utilities from the projects, if they are implemented, are as follows:

$$u_1(a) = 3, \quad u_2(a) = 2, \quad u_3(a) = 1$$
$$u_1(b) = 2, \quad u_2(b) = 1, \quad u_3(b) = 3$$
$$u_1(c) = 1, \quad u_2(c) = 3, \quad u_3(c) = 2$$

This story leads to the following game in strategic form. There are three players, $I = \{1, 2, 3\}$, and each player has as strategies "to vote for a" (A), "to vote for b" (B), or "to vote for c" (C). Hence, $S_i = \{A, B, C\}$ for $i \in I$ are the strategy sets of these players. Given the voting rule described in the story, strategy combinations yield payoffs as summarized in the matrices below.

		Player 3		
		A	B	C
	A	3, 2, 1	3, 2, 1	3, 2, 1
Player 2	B	3, 2, 1	2, 1, 3	3, 2, 1
	C	3, 2, 1	3, 2, 1	1, 3, 2

Player 1: A

		Player 3		
		A	B	C
	A	3, 2, 1	2, 1, 3	2, 1, 3
Player 2	B	2, 1, 3	2, 1, 3	2, 1, 3
	C	2, 1, 3	2, 1, 3	1, 3, 2

Player 1: B

		Player 3		
		A	B	C
	A	3, 2, 1	1, 3, 2	1, 3, 2
Player 2	B	1, 3, 2	2, 1, 3	1, 3, 2
	C	1, 3, 2	1, 3, 2	1, 3, 2

Player 1: C

In the first stage, $t = 1$, the following strategies are dominated and, therefore, eliminated:

for player 1: $\left.\begin{matrix} A \succ B \\ A \succ C \end{matrix}\right\} \Rightarrow D_1(p_1; S_1^0, S_2^0, S_3^0) = \{A\} \equiv S_1^1,$

for player 2: $C \succ B \Rightarrow D_2(p_2; S_1^0, S_2^0, S_3^0) = \{A, C\} \equiv S_2^1,$

for player 3: $B \succ A \Rightarrow D_3(p_3; S_1^0, S_2^0, S_3^0) = \{B, C\} \equiv S_3^1.$

Thus, the second stage game has S_i^1, $i \in \{1, 2, 3\}$, as strategy sets and the following payoff matrix:

Player 3

Player 2		B	C
	A	3, 2, 1	3, 2, 1
	C	3, 2, 1	1, 3, 2

Player 1: A

Dominance reasoning now leads to the following eliminations:

Player 1:

no dominated strategy $\Rightarrow D_1(p_1;\ S_1^1, S_2^1, S_3^1) = \{A\} \equiv S_1^2$,

Player 2: $C \succ A \Rightarrow D_2(p_2;\ S_1^1, S_2^1, S_3^1) = \{C\} \equiv S_2^2$,

Player 3: $C \succ B \Rightarrow D_3(p_3;\ S_1^1, S_2^1, S_3^1) = \{C\} \equiv S_3^2$.

Therefore, for $T = 2$, $S_1^2 = \{A\}$, $S_2^2 = \{C\}$, $S_3^2 = \{C\}$, which satisfies the definition of an iterated dominance equilibrium.

Note that c, the worst outcome for player 1, the president, is chosen. It would have been better for player 1 if someone else had held the tie-breaking vote. Obviously, having a tie-breaking vote is not always a blessing. ■

There are games with several sophisticated equilibria. The following provides an example of multiple iterated dominance equilibria. Notice that these equilibria need not have identical payoffs.

Example 3.7. Consider matching pennies with an outside option for player 1, O.

Player 2

Player 1		H	T
	H	1, −1	−1, 1
	T	−1, 1	1, −1
	O	3, 1	2, 1

Both strategies of matching pennies, H and T, for player 1 are dominated by the outside option O. Therefore, $D_1(p_1;\ S_1^0, S_2^0) = \{O\}$ and $D_2(p_1;\ S_1^0, S_2^0) = S_2^0$ are the sets of undominated strategies after one round of elimination. Thus, the game is dominance solvable and the set of iterated dominance equilibria is $\{(O, H), (O, T)\}$. ■

The definition of an iterated dominance equilibrium required that in each step one delete *all* dominated strategies simultaneously. In example 3.7, this implied the elimination of H and T for player 1 in round 1. This assumption reflects beliefs of the players that opponents will not use dominated strategies. Given this principle, there is no reason to discriminate among different dominated strategies.

Notice however that changing this assumption would alter the elimination procedure considerably. For example, if just one elimination were allowed per stage, the order of elimination would matter for the selection of an equilibrium. If player 1 deleted H first, then player 2 would find strategy T dominated. Deleting this strategy, player 1 would be left with the choice between T and O given that player 2 will play H. Clearly, (O, H) will emerge as the equilibrium after this sequence of elimination. On the other hand, if player 1 begins by deleting T, (O, T) will be the final strategy combination. Thus, for games with multiple iterated dominance equilibria, a sequential elimination procedure may select a particular equilibrium.

For many games, such as matching pennies, no iterated dominance equilibrium exists. Unfortunately, there is no general existence result for an iterated dominance equilibrium in normal form games. For the class of normal form games that are derived from extensive form games with perfect information and no ties in the payoffs, however, dominance solvability is guaranteed. Rather than proving this claim,[2] the following example illustrates this fact.

Example 3.8. Consider the following game in extensive form with perfect information.

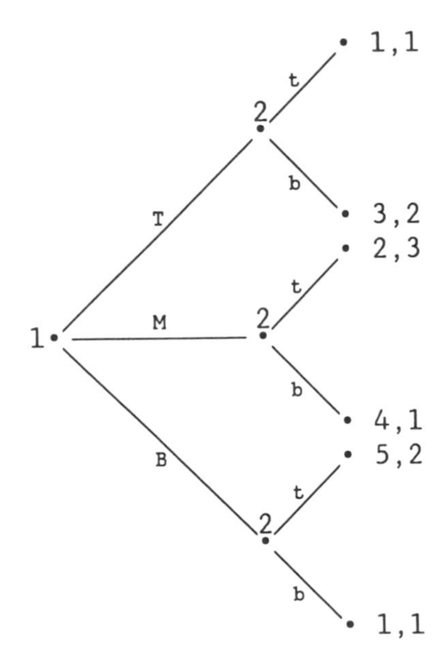

[2]For a proof compare Moulin, (1986, pp. 85–87).

This leads to the following normal form:

$$I = \{1, 2\},$$
$$S_1 = \{T, M, B\} \quad \text{and} \quad S_2 = \{a, b, c, d, e, f, g, h\}$$

where the strategies of player 2 are given in the following table:

		strategies of player 2							
		a	b	c	d	e	f	g	h
if	T	t	t	t	t	b	b	b	b
player 1	M	b	b	t	t	t	t	b	b
plays	B	t	b	t	b	t	b	t	b

and payoff functions represented by the following matrix:

		Player 2							
		a	b	c	d	e	f	g	h
	T	1, 1	1, 1	1, 1	1, 1	3, 2	3, 2	3, 2	3, 2
Player 1	M	4, 1	4, 1	2, 3	2, 3	2, 3	2, 3	4, 1	4, 1
	B	5, 2	1, 1	5, 2	1, 1	5, 2	1, 1	5, 2	1, 1

Consider the first stage of eliminating dominated strategies. One checks easily that player 1 has no dominated strategy, hence $S_1^1 = \{T, M, B\}$. Player 2, in contrast, has a dominant strategy, namely e, hence $S_2^1 = \{e\}$.

After round 1, column e forms the remaining payoff matrix. Now T and M are dominated by B for player 1. Therefore, in round 2, one has $S_1^2 \times S_2^2 = \{B\} \times \{e\}$ and (B, e) is the unique iterated dominance equilibrium.

Note that (B, t) is played in this equilibrium, the same outcome that results from backward induction. ■

The last result of this section shows that a dominant strategy equilibrium can be viewed an a special case of an iterated dominance equilibrium.

LEMMA 3.2. *A dominant strategy equilibrium is an iterated dominance equilibrium.*

Proof. By the definition, a *dominant strategy equilibrium* is a strategy combination $\tilde{s}$ that satisfies, for all players $i \in I$,

$$p_i(\tilde{s}_i, s_{-i}) \geq p_i(s_i, s_{-i})$$

for all $s_i \in S_i$ and all $s_{-i} \in S_{-i}$.

Thus, one of the following two possibilities obtains:

(i) There is some player $i \in I$ with a strategy s_i dominated by $\tilde{s}_i$, that is, such that $\tilde{s}_i \succ s_i$ holds.

(ii) There is no player $i \in I$ with a strategy s_i dominated by $\tilde{s}_i$ and $p_i(\tilde{s}_i, s_{-i}) = p_i(s_i, s_{-i})$ for all $s_{-i} \in S_{-i}$ and all $i \in I$ holds.

In the first round of elimination of dominated strategies, all strategies of type (i) will be eliminated. Hence, all remaining strategy combinations $s \in S_1^1 \times S_2^1 \times \cdots \times S_I^1$ satisfy condition (ii), namely

$$p_i(\tilde{s}_i, s_{-i}) = p_i(s_i, s_{-i})$$

for all $s_{-i} \in S_{-i}^1$ and for all $i \in I$. Thus, $S_1^1 \times \cdots \times S_I^1$ is the set of iterated dominance equilibria. Clearly, $\tilde{s} \in S_1^1 \times \cdots \times S_I^1$ proving the lemma. ■

Note that this result implies that the set of dominant strategy equilibria is a subset of the set of iterated dominance equilibria.

This section showed how a repeated application of the argument that no player plays a dominated strategy, combined with the assumption that this behavior as well as all payoff functions are common knowledge, is in some cases sufficient to determine a solution of a game. The final example in this chapter calls into question the logic of iterated dominance itself.

EXAMPLE 3.9 (*modified from Samuelson*, 1990). Consider the following game.

		Player 2		
		L	C	R
	T	1, 1	0, 1	3, 1
Player 1	M	1, 0	2, 2	1, 3
	B	1, 3	3, 1	2, 2

It is easy to check that M and C are dominated and will be eliminated in stage 1. The remaining game has B and R as dominated strategies that will be deleted in round 2. This leads to (T, L) as the unique iterated dominance equilibrium.

Note however that M was dominated by B because player 2 might choose strategy C or R. Precisely these two strategies are, however, eliminated, thus eliminating the rationale for M being dominated in the first place. ■

Example 3.9 indicates a problem of repeated application of the dominance argument: The reason for not playing a dominated strategy may disappear during the iterated elimination of dominated strategies. This problem, together with the fact that there are many games in strategic form that have no iterated dominance equilibrium, precludes dominance arguments as a general solution concept. The next chapter will discuss a solution concept that is more generally applicable.

3.3 Summary

This chapter introduced two new equilibrium concepts for games in strategic form based on the idea that players will play best responses to other players' behavior. In some games, the best response strategy does not depend on the other players' behavior. If all players have such a dominant strategy, this strategy combination presents a *dominant strategy equilibrium*. Unfortunately, few games have such an equilibrium. In addition, as the prisoners' dilemma shows, dominant strategy equilibria may have suboptimal outcomes.

A weaker notion of equilibrium is built on the idea that players will not play dominated strategies. Applying this principle iteratively may lead occasionally to a single strategy combination (or a set of strategy combinations that are payoff-equivalent for players). If such a strategy combination results from the iterated deletion of dominated strategies, then it forms an *iterated dominance equilibrium*. Once again, many games do not have such an equilibrium. Furthermore, it has been shown that the logic of the argument that players do not play dominated strategies may be questioned by the iterated elimination process itself.

3.4 Remarks on the Literature

The concepts of a dominant strategy equilibrium and of an iterated dominance equilibrium can be found in Moulin (1986). The presentation of these concepts in the first two sections of this chapter follows the basic line of argument used in Moulin (1986). Samuelson (1990) provides a critical assessment of the iterated dominance approach. The notion of an iterated dominance equilibrium, introduced here for games with finite strategy sets, is extended to games with infinite strategy sets in Fudenberg and Tirole (1991, p. 45). Closely related to the notion of the iterated dominance equilibrium is the concept of rationalizability introduced by Bernheim (1984) and Pearce (1984). Fudenberg and Tirole (1991, pp. 45–53) provide an excellent discussion and comparison of these two concepts.

Exercises

3.1 *Consider a sealed-bid second-price auction for a building. Each player makes a sealed bid. The highest bidder receives the building and pays the second highest bid. If two players make the same bid, the one with the highest valuation will get the building.*

(a) Write down the strategic form of this game.

(b) Prove that it is a dominant strategy for a player to bid her true valuation of the building.

3.2 *Consider* 50 *pirates who have to share a booty of* 100 *ducats. The rule for sharing the ducats is as follows: In order of seniority, each pirate makes a proposal for sharing the* 100 *ducats. If the majority (at least* 25*) accept the proposal, it is implemented. If an absolute majority rejects the proposal, then the pirate making the proposal receives nothing, and the next oldest pirate has to propose a distribution among the remaining pirates.*

(a) Determine the partition that will be implemented.

(b) What kind of equilibrium concept is used?

Hint: Consider the game for three and four pirates first.

3.3 *Reconsider the Colonel Blotto game (exercise 2.2).*

(a) Show that the game has neither a dominant strategy equilibrium nor an iterated dominance equilibrium.

(b) There are dominated strategies in this game. Show that a particular equilibrium is selected if one assumes that no player will ever use a dominated strategy.

3.4 *Consider two firms producing and selling a homogeneous product in a market with the inverse demand function,* $p(x) = \max\{0, 100 - x\}$, *where x denotes the total quantity supplied to the market. Suppose that both firms have constant average costs of production equal to* 20.

(a) Derive the profit function of firm 1 and determine the profit-maximizing quantity x_1^m when this firm has a monopoly.

(b) Show that any supply decision $x_1 > x_1^m$ is dominated by the quantity x_1^m.

3.5 *Reconsider the duopoly case of exercise 3.4.*

(a) Determine the sequence of strategy sets (S_1^t, S_2^t) that arises if the two firms iteratively delete dominated strategies.

(b) Does this process end in finite time? If possible, determine the strategies that remain undominated in the limit?

4

Equilibrium Concepts II: Nash Equilibrium

The solution concepts discussed in Chapter 3 are based on the notion of dominance. Strategies are dominant if they are best responses to any strategy opponents take. If all players have a dominant strategy, then such a strategy combination appears as a natural prediction of how the game will be played. In general, however, what is a best response for one player will depend on what the other players do, that is, best responses will vary with the opponents' behavior. The following example illustrates this fact.

EXAMPLE 4.1 (*battle of the sexes*). Consider a couple who want to spend an evening out together. The husband prefers to see a boxing match, the wife would like to go to the theater. Nevertheless, they would rather go to one of these events together than spend the evening separately. The following payoff matrix captures this situation.

		Player 2	
		T	B
Player 1	T	2, 1	0.5, 0.5
	B	0, 0	1, 2

Player 1 is the wife and player 2 the husband, T means "goes to the theater" and B means "goes to the boxing match." Obviously, they are slightly better off if the wife goes to the theater and the husband to the boxing match than if they do the reverse.

There is no dominated strategy for either player in this game. The best response for the wife is to go to the theater if her husband goes too, but she would want to go to the boxing match if her husband goes there. *Mutatis mutandis*, the same holds true for the husband. ■

Clearly there is no iterated dominance equilibrium or, *a fortiori*, no dominant strategy equilibrium in the battle of the sexes game. On the other hand, if players followed a safety first type of strategy, that is, play their maximin strategies, the husband would go to the boxing match and the wife to the theater. This would be one of the worst possible outcomes for the game. Furthermore, it would mean that neither player would be playing a best response to the action of the other.

The solution that suggests itself here is the equilibrium concept introduced by the second argument in Chapter 2: A strategy combination in which each player plays a best response to the opponents' behavior.

DEFINITION 4.1. A strategy combination $s^* \in S$ is a *Nash equilibrium*[1] (or an *equilibrium*) if

$$p_i(s^*) \geq p_i(s_i, s^*_{-i}) \quad \text{for all} \quad s_i \in S_i \quad \text{and all} \quad i \in N$$

holds.

Note that it is now no longer possible to say how a player should behave in a game without simultaneously saying how all the other players behave. In other words, a player must predict correctly the behavior of opponents.

It is easy to check that in the battle of the sexes game there are two Nash equilibria: (T,T) and (B,B). Clearly, if one goes to the theater the best choice for the other is to follow suit, and if one goes to the boxing match the other does best to do the same. Of course, each player has a preference for one particular equilibrium. The husband likes the equilibrium (B,B), whereas the wife prefers the equilibrium (T,T). There is, however, no clue that would help predict which of these two Nash equilibria will actually be played. Nor is it clear how a player would predict how the other player will behave. These are fundamental problems with the Nash equilibrium concept to be discussed in more detail later.

In spite of these problems, the Nash equilibrium concept has some attractive characteristics:

(i) It generalizes the dominant strategy equilibrium concept and the iterated dominance equilibrium concept.
(ii) It yields a payoff for each player that is always at least as good as the maximum value each player could guarantee himself.

[1]The name Nash equilibrium derives from John Nash who was the first to suggest this equilibrium concept in a now famous paper (Nash, 1951).

The following two lemmas will establish these claims.

LEMMA 4.1. *Any Nash equilibrium s^* satisfies for all players $i \in I$*

$$p_i(s^*) \geq \max_{s_i \in S_i} \min_{s_{-i} \in S_{-i}} p_i(s_i, s_{-i}).$$

Proof. Note that, for any $s_i \in S_i$,

$$\min_{s_{-i} \in S_{-i}} p_i(s_i, s_{-i}) \leq p_i(s_i, s_{-i}) \quad \text{for all} \quad s_{-i} \in S_{-i}$$

holds. In particular,

$$\min_{s_{-i} \in S_{-i}} p_i(s_i, s_{-i}) \leq p_i(s_i, s^*_{-i}) \quad \text{for} \quad s_i \in S_i$$

must be true. Hence,

$$\max_{s_i \in S_i} \min_{s_{-i} \in S_{-i}} p_i(s_i, s_{-i}) \leq \max_{s_i \in S_i} p_i(s_i, s^*_{-i}) \equiv p_i(s^*)$$

follows. ■

The battle of the sexes illustrates that maximin strategy combinations are usually not Nash equilibria. For the class of zero-sum games, however, Nash equilibria coincide with maximin strategy combinations (theorem 2.1).

The following lemma shows that an iterated dominance equilibrium is a special case of a Nash equilibrium. Since it has been established already (lemma 3.2) that every dominant strategy equilibrium is an iterated dominance equilibrium, one can conclude that a dominant strategy equilibrium is a Nash equilibrium. The latter claim is obvious since it is clear that a strategy of a player that is a best response to any other strategy combination of the other players must also be a best response to the actual behavior of the other players as is required for a Nash equilibrium.

LEMMA 4.2. *An iterated dominance equilibrium is a Nash equilibrium.*

Proof. An iterated dominance equilibrium is obtained by sequentially eliminating dominated strategies. Denote by $Z \equiv Z_1 \times Z_2 \times \ldots Z_I$ the set of those strategy combinations remaining after successive elimination of dominated strategies. For an iterated dominance equilibrium, all strategy combinations $s, s' \in Z$ must satisfy $p_i(s_i, s_{-i}) = p_i(s'_i, s_{-i})$ for all $s_{-i} \in Z_{-i}$ and all $i \in I$. Thus, s and s' are Nash equilibria of the game with the reduced strategy sets Z. Since, for all players $i \in I$, Z_i was derived from S_i, the original strategy set, by iteratively deleting dominated strategies, it is sufficient to show that any Nash equilibrium $s^* \in Z$ is a Nash equilibrium of the larger set S, if $S \setminus Z$ contains only strategies dominated by some element of Z.

Suppose that z^* is a Nash equilibrium of the game $(I, (Z_i)_{i \in I}, p)$, then, for any player $i \in I$,

$$p_i(z_i^*, z_{-i}^*) \geq p_i(z_i, z_{-i}^*) \quad \text{for all} \quad z_i \in Z_i \quad \text{follows.}$$

Let $(I, (S_i)_{i \in I}, p)$ be a game for which, for all players $i \in I$, $Z_i \equiv D_i(p_i; S)$ holds. This means that all strategies in $S_i \setminus Z_i$ are dominated by some strategy in Z_i. Then, for any strategy $s_i \in S_i \setminus Z_i$, there must be a strategy $z_i \in Z_i$ such that

$$p_i(z_i, z_{-i}^*) \geq p_i(s_i, z_{-i}^*)$$

holds.

Combining these two inequalities yields, for any player $i \in I$,

$$p_i(z_i^*, z_{-i}^*) \geq p_i(s_i, z_{-i}^*) \quad \text{for all} \quad s_i \in S_i.$$

Hence, z^* is a Nash equilibrium of the game $(I, (S_i)_{i \in I}, p)$.

Applying this argument to every step of the iteration proves the claim of this lemma. ■

This lemma establishes that the Nash equilibrium concept is a generalization of iterated dominance equilibrium and dominant strategy equilibrium. For any game in strategic form Γ, denote by DE(Γ) the set of all dominant strategy equilibria that may be the empty set, and by IDE(Γ) and NE(Γ) the sets of iterated dominance equilibria and Nash equilibria respectively that may also be empty sets if none of these equilibria exist. According to lemma 3.2 and lemma 4.2, the following relationship between these sets obtains:

$$\mathrm{DE}(\Gamma) \subseteq \mathrm{IDE}(\Gamma) \subseteq \mathrm{NE}(\Gamma).$$

Thus, if a game Γ has a dominant strategy equilibrium, then it clearly has both an iterated dominance equilibrium and a Nash equilibrium. But a game with a Nash equilibrium need not have a dominant strategy equilibrium. These relationships do not imply that a Nash equilibrium strategy is undominated. Indeed, a Nash equilibrium strategy combination may consist of a dominated strategy for every player. The following example illustrates this possibility.

EXAMPLE 4.2. Consider the following game which is given by its payoff matrix.

		Player 2		
		s_2^1	s_2^2	s_2^3
Player 1	s_1^1	1, 1	0, 1	0, 1
	s_1^2	0, 0	1, 0	0, 1
	s_1^3	1, 0	0, 1	1, 0

It is straightforward to check that (s_1^1, s_2^1) is the only Nash equilibrium of this game. However, s_1^1 is dominated by s_1^3, and s_2^1 is dominated by s_2^2 and s_2^3. This game does not have an iterated dominance equilibrium either. If one deletes (s_1^1, s_2^1) because they are dominated, then the remaining game has no equilibrium in pure strategies. ■

Example 4.2 shows Nash equilibria may have the undesirable property of requiring dominated strategies to be played. The following subsections will probe deeper into characteristics of Nash equilibria.

4.1 Existence of Nash Equilibrium

The previous section demonstrated that Nash equilibrium is a solution concept more general than either iterated dominance equilibrium or dominant strategy equilibrium. Hence, there are more games with a Nash equilibrium than there are games with a dominant strategy equilibrium or an iterated dominance equilibrium. Clearly, the usefulness of a solution concept depends not only on how convincing are its defining criteria but also on the extent of applicability. Although dominant strategy equilibrium provides a very forceful prediction of players' behavior, it fails to exist in most games. Hence, for many games, this concept cannot be applied. To determine under which conditions an equilibrium concept is applicable is the problem of existence of an equilibrium. Naturally, a solution concept that works under all circumstances is desired. Indeed, one reason for the popularity of the Nash equilibrium concept is that it exists in so many games. Nevertheless, the matching pennies game in Chapter 2 did not have a Nash equilibrium in pure strategies. The following example presents another case where no equilibrium exists.

Example 4.3. The following game is known in many countries. Two players choose simultaneously one of the following objects: "stone" (s), "scissors" (sc), or "paper" (p). If both players choose the same object, the game is a draw and the payoff is zero to both of them. Otherwise, one wins a dollar from the other according to the following rules:

- "stone" beats "scissors"
- "scissors" beat "paper"
- "paper" beats "stone"

The following strategic form represents this story. Each of the two players, $i \in \{1, 2\}$, has three strategies, $S_i = \{s, sc, p\}$, with the following payoffs.

		Player 2		
		s	sc	p
Player 1	s	0, 0	1, −1	−1, 1
	sc	−1, 1	0, 0	1, −1
	p	1, −1	−1, 1	0, 0

It is easy to see that no strategy combination exists such that no player has an incentive to deviate (i.e., such that each player is playing a best response to the other player's choice). For example, (s, s) is not an equilibrium because both players would want to deviate: player 1 to strategy p and player 2 to strategy p. Similarly, (s, sc) is not an equilibrium strategy combination, since player 2 would want to deviate to s. Checking all nine possible strategy combinations, one finds that this game has no equilibrium in pure strategies. ■

There is no equilibrium in pure strategies in the game of example 4.3. It is possible, however, to show that it has an equilibrium in mixed strategies which is to play each pure strategy with probability one-third. This raises the question of what properties of the strategy sets and the payoff functions will guarantee the existence of a Nash equilibrium. This section will focus on this problem.

To analyze the question of existence, it is useful to find an alternative definition of a Nash equilibrium in terms of best response mappings. As mentioned previously a Nash equilibrium is characterized by the fact that each player plays a best response to the other players' behavior. Therefore, the mapping that associates with each strategy combination the best responses of an agent is of particular interest. Unfortunately, this mapping is usually not a function. A function associates with elements of one set a unique element of another set. Best responses, however, are often not unique, that is, for a given behavior of the opponents there may be more than one best response strategy.

To describe the class of games for which the Nash equilibrium concept is applicable, it is necessary to characterize sets (strategy sets) and functions (payoff functions). Thus, some formal language is unavoidable here. Whenever necessary, a few remarks regarding this terminology will be given under the heading remarks on mathematical concepts. These remarks do not represent an introduction to these concepts. Rather they should be viewed as hints that may either help to recall existing knowledge or assist readers in finding further references from the specialized literature.[2]

[2]Green and Heller (1981) survey the mathematical concepts necessary for the analysis of existence problems. They present important theorems and give further references to the relevant mathematical literature.

Remark 4.1 (on mathematical concepts). For any set A, denote by $\mathscr{P}(A)$ the set of all subsets of A (or power set of A). A correspondence is a mapping that associates with each element of one set A a set of elements from another set B; this is written formally as $\phi: A \to \mathscr{P}(B)$. A function, on the other hand, associates with each element of one set A just one element of another set B; formally, $f: A \to B$. For any correspondence $\phi: A \to \mathscr{P}(A)$, a point $x \in A$ that satisfies $x \in \phi(x)$ is called *fixed point*. For a function $f: A \to A$, a fixed point is a point $x \in A$ with $x = f(x)$. ■

With this terminology, it is possible to give a precise meaning to the concept of a best response mapping.

DEFINITION 4.2. The *best response mapping* (or *best reply mapping* or *reaction correspondence*) of player i is the mapping r_i that associates with each strategy combination $s \in S$ the set of best response strategies of player i; formally, the correspondence $r_i: S \to \mathscr{P}(S_i)$ defined by

$$r_i(s) \equiv \underset{s_i \in S_i}{\operatorname{argmax}}\, p_i(s_i, s_{-i}).$$

The best response mapping (or best reply mapping or reaction correspondence) is the mapping $r(s) \equiv r_1(s) \times r_2(s) \times \cdots \times r_I(s)$.

Two points of this definition are worth emphasizing:

(i) The best response mapping of player i associates a best response strategy of player i with each strategy combination $s \in S$ instead of each strategy combination of the other players $s_{-i} \in S_{-i}$. A strategy combination $s \equiv (s_1, s_2, \ldots, s_I)$ lists a strategy for each player including player i. This redundancy is of no importance for the analysis but simplifies notation, because one can speak of a best response to a strategy combination s for any player, whereas the opponents' s_{-i} vary with the player considered.

(ii) Best responses $r(s) \equiv r_1(s) \times r_2(s) \times \cdots \times r_I(s)$ are a list of sets, not points, namely one set of best response strategies for each player. Hence, one writes $s \in r(s)$ (instead of $s = r(s)$ which would be appropriate for a function $r(s)$). Of course, $s \in r(s)$ means $s_1 \in r_1(s)$, $s_2 \in r_2(s)$, $s_3 \in r_3(s)$, etcetera.

Definition 4.2 allows the definition of a Nash equilibrium to be cast in a different form more suitable for answering the existence question.

LEMMA 4.3. *A strategy combination $s^* \in S$ is a Nash equilibrium if and only if s^* satisfies $s^* \in r(s^*)$.*

Proof. The proof has two parts: It is shown that a Nash equilibrium strategy combination s^* satisfies $s^* \in r(s^*)$ (" $\Rightarrow$ "); it will be shown that

$s^* \in r(s^*)$ implies that s^* is a Nash equilibrium strategy combination ("$\Leftarrow$").

(i) "$\Rightarrow$" By definition 3.4, $s^* \in S$ is a Nash equilibrium if, for all $i \in I$,

$$p_i(s_i^*, s_{-i}^*) \geq p_i(s_i, s_{-i}^*) \quad \text{for all} \quad s_i \in S_i$$

holds. Hence,

$$s_i^* \in r_i(s^*) \equiv \underset{s_i \in S_i}{\operatorname{argmax}}\, p_i(s_i, s_{-i}^*)$$

must hold for any player $i \in I$. Or, $s^* \in r(s^*)$.

(ii) "$\Leftarrow$" If $s^* \in r(s^*)$ holds, then

$$s_i^* \in r_i(s^*) \equiv \underset{s_i \in S_i}{\operatorname{argmax}}\, p_i(s_i, s_{-i}^*)$$

is true for all $i \in I$. Therefore, $p_i(s_i^*, s_{-i}^*) \geq p_i(s_i, s_{-i}^*)$ for all $s_i \in S_i$ holds for all players $i \in I$. This is, however, the definition of a Nash equilibrium. ■

Lemma 4.3 shows that a game has a Nash equilibrium if and only if the reaction correspondence $r(s)$ has a fixed point. Conditions that guarantee the existence of a fixed point are well-known from mathematical theory. This fixed point theory can be applied to obtain the following theorem.

THEOREM 4.1 (*existence of a Nash equilibrium*). *A game in strategic form* $(I, (S_i)_{i \in I}, p)$ *has at least one Nash equilibrium, if for each player* $i \in I$:

(i) *The strategy set* S_i *is a (nonempty) compact and convex subset of a Euclidean space.*
(ii) *The payoff function* p_i *is continuous and quasi-concave in* s_i.

The proof of this theorem, with some reasons why these assumptions are needed for existence of a Nash equilibrium, will be provided in an appendix to this section. The remainder of this section will concentrate on the application of this theorem. In particular, the meaning of the assumptions on the strategy sets and the payoff functions will be discussed in detail.

According to theorem 4.1, to be certain a game has a Nash equilibrium, it is necessary to check

- whether the strategy set of each agent is a compact and convex subset of a Euclidian space
- whether the payoff function is a continuous function that is quasi-concave in a player's own strategy

Therefore, some consideration of the meaning of compact, convex, Euclidian space, continuous, and quasi-concave is in order.

Remark 4.2 (on mathematical concepts). A Euclidian space is a vector space of real numbers of an arbitrary finite dimension n, formally $\mathbb{R}^n = \mathbb{R} \times \mathbb{R} \times \mathbb{R} \times \cdots \times \mathbb{R}$ (n-times). As the examples will show, any finite set can be viewed as a subset of such a Euclidian space because each element can be associated with a number.

A subset X of Euclidian space, formally $X \subseteq \mathbb{R}^n$, is compact if it satisfies the following two conditions:

(i) There are two vectors $\underline{x}$ and $\bar{x}$ in the Euclidian space $\mathbb{R}^n$ such that $\underline{x} \leq x \leq \bar{x}$ for all $x \in X$ holds

(ii) for any converging sequence of points in X the limit point lies in X as well, that is, if $x^\nu \in X$ for all $\nu = 1, 2, 3, \ldots,$ and $x^0 \equiv \lim_{\nu \to \infty} x^\nu$ exists, then $x^0 \in X$ must hold

Note that for two vectors $x \equiv (x_1, x_2, \ldots, x_n) \in \mathbb{R}^n$ and $y \equiv (y_1, y_2, \ldots, y_n) \in \mathbb{R}^n$ $x \geq y$ means $x_i \geq y_i$ for all components $i = 1, 2, \ldots, n$. Similarly, for a sequence of vectors x^ν, $\nu = 1, 2, 3, \ldots,$ $x^0 \equiv \lim_{\nu \to \infty} x^\nu$ exists if for all components, $i = 1, 2, \ldots, n$, $x_i^0 \equiv \lim_{\nu \to \infty} x_i^\nu$ exists.

Examples of compact subsets of Euclidian spaces include: any finite set of numbers, any interval of the real numbers that contains its end points, and the set of points within and on the border of a circle in the plane, that is, in $\mathbb{R}^2$. By contrast, the set of all non-negative vectors of real numbers, $\mathbb{R}^n_+$, an interval excluding some end point, and the set of all natural numbers provide examples of sets that are not compact.

A subset $X \subseteq \mathbb{R}^n$ is *convex* if for any two points $x \in X$ and $y \in X$, the line joining these two points belongs to the set X as well, formally if for any λ, $0 \leq \lambda \leq 1$, the point $z \equiv \lambda \cdot x + (1 - \lambda) \cdot y$ satisfies $z \in X$. Typical examples of convex sets include: any interval that includes its end points, and the set of points within a circle in $\mathbb{R}^2$. Note that a set of finitely many points in $\mathbb{R}^n$ is never convex if it contains more than one point.

Let $X \subseteq \mathbb{R}^n$ and $Y \subseteq \mathbb{R}^m$ hold, that is, two subsets of two Euclidian spaces not necessarily of the same dimension ($m \neq n$ may hold). A function f: $X \to Y$ is continuous if for any converging sequence x^ν, $\nu = 1, 2, 3, \ldots,$ in X, the induced sequence $f(x^\nu)$, $\nu = 1, 2, 3, \ldots,$ in Y converges as well and, in particular, converges to $f(x^0)$; formally, if $\lim_{\nu \to \infty} x^\nu = x^0$ implies that $\lim_{\nu \to \infty} f(x^\nu) = f(x^0)$ holds. Examples of continuous functions include: all linear functions, all power functions with an integer power, and all functions f: $X \to Y$ where X is a finite set. On the other hand, functions that are piecewise constant at different levels are good examples of discontinuous functions.

Finally, for $X \subseteq \mathbb{R}^n$, a function f: $X \to \mathbb{R}$ is *quasi-concave* in x if all its upper level sets are convex; formally if $\{x \in X | f(x) \geq \alpha\}$ is a convex (or empty) set for all $\alpha \in \mathbb{R}$. Examples of quasi-concave functions include all linear functions, all concave functions, and all increasing functions of one variable $f\colon \mathbb{R} \to \mathbb{R}$. ■

The following examples will illustrate how to check whether the conditions of theorem 4.1 are satisfied in matching pennies.

EXAMPLE 4.4 (*example* 2.1 *resumed*). Recall the matching pennies game.

		Player 2	
		H	T
Player 1	H	$1, -1$	$-1, 1$
	T	$-1, 1$	$1, -1$

First, the strategy sets $S_1 = \{H, T\}$ and $S_2 = \{H, T\}$ are finite in this example. Therefore, each strategy can be identified with a number, for example, one could write $S_i = \{0, 1\}$ with the understanding that strategy 0 is "playing heads" and strategy 1 is "playing tails." Thus, the strategy set S_i becomes a subset of a Euclidian space, $S_i \subset \mathbb{R}$, for both players.

Obviously, like any finite subset of Euclidian space, S_i is compact. To verify this claim, one has to check the two criteria already mentioned:

(i) Take $\underline{x} = -1 \in \mathbb{R}$ and $\bar{x} = 2 \in \mathbb{R}$. Clearly, $-1 \leq s_i \leq 2$ for any strategy $s_i \in S_i = \{0, 1\}$.
(ii) Any sequence s_i^ν, $\nu = 1, 2, 3, \ldots$, that converges and has $s_i^\nu \in S_i$ for all ν must be constant beyond some number N, that is, if it converges it will be to either $s_i^0 = 0$ or $s_i^0 = 1$. Thus, $s_i^0 \in S_i$ holds.

On the other hand, S_i is not convex, because for $s_i = 0$ and $s_i' = 1$, and any λ, $0 < \lambda < 1$, one has $\lambda \cdot s_i + (1 - \lambda) \cdot s_i' = 1 - \lambda$. But a strategy $\tilde{s}_i = 1 - \lambda$ is not an element of S_i for $\lambda \neq 0$.

Since the set of strategy combinations S has just four elements and is therefore finite, both payoff functions $p_i(s)$ must be continuous. To check this claim, one applies the definition as follows: every sequence of strategy combinations s^ν with $s^\nu \in S$ for all $\nu = 1, 2, 3, \ldots$, that converges must become a constant sequence beyond a certain stage N, because S is finite. But then the payoff sequence $p_i(s^\nu)$ must also become constant for all $\nu \geq N$. For example, for player 1, and a sequence of strategy combinations s^ν that converges to $s^0 = (1, 0)$, the following sequence of payoffs $p_1(s^\nu)$ arises that, obviously, converges to a payoff of 0 for player 1.

$$
\begin{array}{rl}
\nu = & 1 \quad 2 \quad 3 \quad 4 \quad \ldots \quad N \quad N+1 \, N+2 \, N+3 \ldots \\
(s^\nu) = & ((0,0), (0,1), (0,1), (1,1), \ldots, (1,0), (1,0), (1,0), (1,0), \ldots) \\
(p_1(s^\nu)) = & (\ 1\ , -1\ , -1\ ,\ 1\ , \ldots, -1, \ -1\ , \ -1\ , \ -1\ , \ldots)
\end{array}
$$

Finally, since S_i is not convex, the payoff function cannot be quasi-concave. To verify this, consider for a given strategy of player 2, the level set of player 1 for $\alpha = -1$ $\{s_1 \in S_1 | p_1(s_1, s_2) \geq -1\}$ which equals S_1 and, as shown earlier, is not convex.

In summary, the strategy sets S_i, $i \in I$, are compact but not convex, and the payoff functions p_i are continuous in s but not quasi-concave in s_i. Thus, the game of matching pennies does not satisfy the conditions of theorem 4.1, which would guarantee existence of an equilibrium. Indeed, matching pennies has no equilibrium in pure strategies.

Consider now the case in which mixed strategies are allowed. In this case, the strategy set M_i for each player is a set of real number pairs $(m_i(0), m_i(1))$ where the first number $m_i(0)$ denotes the probability of playing pure strategy 0 ("heads") and $m_i(1)$ denotes the probability of choosing pure strategy 1 ("tails"). Since $m_i(0)$ and $m_i(1)$ are probabilities, these numbers must satisfy $m_i(0) \geq 0$, $m_i(1) \geq 0$, and $m_i(0) + m_i(1) = 1$, that is,

$$M_i = \{(m_i(0), m_i(1)) | m_i(0) \geq 0, m_i(1) \geq 0, m_i(0) + m_i(1) = 1\}.$$

Clearly $M_i \subset \mathbb{R}^2$ is satisfied, that is, the strategy sets are subsets of a Euclidian space, namely $\mathbb{R}^2$. Furthermore, the sets M_i, $i \in I$, are compact. To check this claim consider that:

(i) For $\underline{x} = (-1, -1)$ and $\bar{x} = (2, 2)$,
$$(-1, -1) \leq (m_i(0), m_i(1)) \leq (2, 2)$$
holds.

(ii) For any sequence $(m_i^\nu(0), m_i^\nu(1))$ with $(m_i^\nu(0), m_i^\nu(1)) \in M_i$ for all $\nu = 1, 2, 3, \ldots$, that converges to some $(m_i^0(0), m_i^0(1))$, $m_i^\nu(0) \geq 0$ and $m_i^\nu(1) \geq 0$ for all ν imply $m_i^0(0) \geq 0$ and $m_i^0(1) \geq 0$ and $m_i^\nu(0) + m_i^\nu(1) = 1$ for all ν implies $m_i^0(0) + m_i^0(1) = 1$. Hence, $(m_i^0(0), m_i^0(1)) \in M_i$ must be true.

In contrast to S_i, the set of mixed strategies M_i is convex as well. To verify this claim, take any two mixed strategies $(m_i(0), m_i(1)) \in M_i$ and $(m'_i(0), m'_i(1)) \in M_i$ and any λ, $0 \leq \lambda \leq 1$:

- $m_i(0) \geq 0$ and $m'_i(0) \geq 0$ imply $\lambda \cdot m_i(0) + (1 - \lambda) \cdot m'_i(0) \geq 0$
- $m_i(1) \geq 0$ and $m'_i(1) \geq 0$ imply $\lambda \cdot m_i(1) + (1 - \lambda) \cdot m'_i(1) \geq 0$
- $m_i(0) + m_i(1) = 1$ and $m'_i(0) + m'_i(1) = 1$ imply $[\lambda \cdot m_i(0) + (1 - \lambda) \cdot m'_i(0)] + [\lambda \cdot m_i(1) + (1 - \lambda) \cdot m'_i(1)] = 1$

Thus, $((\lambda \cdot m_i(0) + (1 - \lambda) \cdot m'_i(0)), (\lambda \cdot m_i(1) + (1 - \lambda) \cdot m'_i(1))) \in M_i$ holds, which proves that M_i is a convex set.

If mixed strategies are considered, the relevant payoff function is the expected payoff

$$\begin{aligned} P_i((m_1(0), m_1(1)), (m_2(0), m_2(1))) \\ = p_i(0,0) \cdot m_1(0) \cdot m_2(0) + p_i(0,1) \cdot m_1(0) \cdot m_2(1) \\ + p_i(1,0) \cdot m_1(1) \cdot m_2(0) + p_i(1,1) \cdot m_1(1) \cdot m_2(1). \end{aligned}$$

This function is continuous in (m_1, m_2), since it is the sum and product of the components of (m_1, m_2), that is, of $m_1(0)$, $m_1(1)$, $m_2(0)$, and $m_2(1)$.

Furthermore, $P_1(m_1, m_2)$ is quasi-concave in m_1, since $P_1(m_1, m_2)$ can be written as follows:

$$P_1(m_1, m_2) = m_1(0) \cdot A + m_1(1) \cdot B$$

where

$$A \equiv p_1(0,0) \cdot m_2(0) + p_1(0,1) \cdot m_2(1)$$
$$B \equiv p_1(1,0) \cdot m_2(0) + p_1(1,1) \cdot m_2(1).$$

Note that, for a given m_2, A and B are constants. Hence, $P_1(m_1, m_2)$ is linear in m_1 and, therefore, quasi-concave in m_1. The same argument shows that $P_2(m_1, m_2)$ is linear and quasi-concave.

In summary, the mixed strategy sets M_i are compact and convex subsets of the Euclidian space $\mathbb{R}^2$. Furthermore, the expected payoff functions $P_i(m_1, m_2)$ are continuous in $m \equiv (m_1, m_2)$, and quasi-concave in m_i. Hence, the game $(I, (M_i)_{i \in I}, (P_i)_{i \in I})$ satisfies all the conditions of theorem 4.1 and must have an equilibrium. In example 2.1, it has been shown that matching pennies has, indeed, an equilibrium in mixed strategies, namely, $(m_1^*, m_2^*) = ((.5, .5), (.5, .5))$ which requires each player to play the pure strategy 1 with probability .5 and the pure strategy 0 with probability .5. ■

Example 4.4 illustrates that

- games with finite pure strategy sets lack convexity of strategy sets and quasi-concavity of payoff functions
- the games with mixed strategies arising from them satisfy these conditions

This explains why von Neumann's theorem (theorem 2.2) that every game with finitely many strategies has an equilibrium in mixed strategies holds. Indeed, all one has to do to demonstrate this result is to verify that the arguments used in example 4.4 to show that the set of mixed strategies was compact and convex and the payoff function continuous and quasi-concave carry over to the case of arbitrary numbers of pure strategies.[3] Thus, theorem 2.2 can be viewed as a special case of theorem 4.1.

It is possible to carry this argument further. From example 4.4, it becomes clear that considering probability distributions on pure strategy sets, that is, mixed strategies, has two consequences:

- Mixed strategy sets become convex even if the pure strategy sets are not convex.
- Expected payoff functions become linear in a player's own mixed strategy and, therefore, quasi-concave.

[3]For a finite pure strategy set, the set of mixed strategies forms a simplex (compare Chapter 1, section 1.2.2). Simplexes are compact and convex subsets of a Euclidian space.

Thus, one may conjecture that in cases in which pure strategy sets are compact but not convex, and payoff functions lack quasi-concavity, existence of an equilibrium may be obtained by considering mixed strategies. As the following theorem demonstrates, this conjecture is in fact correct.

THEOREM 4.2 (*Glicksberg* 1952). *Any game* $(I,(S_i)_{i\in I},(p_i)_{i\in I})$ *that satisfies for all players* $i \in I$

(i) S_i *is a compact subset of a Euclidian space, and*
(ii) p_i *is a continuous function,*

has at least one equilibrium in mixed strategies.

This theorem is stated without proof.[4] A comparison with the statement of theorem 4.1 shows that the conditions of convexity and quasi-concavity can be relaxed if mixed strategies are considered. This shows that if one allows for mixed strategies, Nash equilibria exist in a wide range of games.

It is harder to relax the assumption of compactness regarding the strategy sets S_i and of continuity of the payoff functions p_i. If either compactness of S_i or continuity of p_i fails to hold, then it is difficult to guarantee that maximizing behavior is well defined. The following modified version of example 3.4 illustrates this problem.

EXAMPLE 4.5 (*example* 3.4 *modified*). Consider the duopoly problem of example 3.4 with the following demand functions:

$$d_1(p_1,p_2) = \frac{10\cdot p_2}{p_1} \quad \text{and} \quad d_2(p_1,p_2) = \frac{20\cdot p_1}{p_2}.$$

In addition, assume that each firm i, $i \in I$, has the same cost function $c_i(x_i) = 5\cdot x_i$, where x_i denotes the quantity produced by firm i.

Note that it is now no longer possible to argue that prices above some maximum level will yield zero profits because demand becomes zero. Hence, the following game arises:

- player set: $I = \{1,2\}$
- strategy sets: $S_1 = \{p_1 \in \mathbb{R} | p_1 \geq 0\}$ and $S_2 = \{p_2 \in \mathbb{R} | p_2 \geq 0\}$
- payoff functions:

$$\begin{aligned}\pi_1(p_1,p_2) &= p_1\cdot d_1(p_1,p_2) - c_1(d_1(p_1,p_2))\\ &= \left[10 - \frac{50}{p_1}\right]\cdot p_2,\\ \pi_2(p_1,p_2) &= \left[20 - \frac{100}{p_2}\right]\cdot p_1.\end{aligned}$$

[4] For a proof compare Glicksberg (1952).

The following diagram shows the strategy set and payoff function of firm 1.

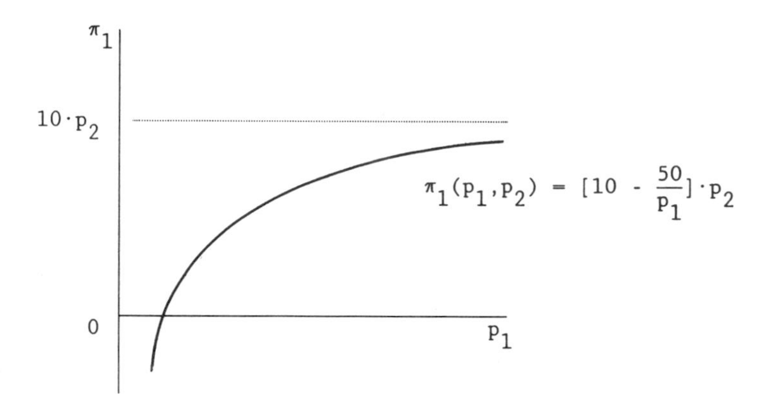

Notice that an optimal price for firm 1, p_1, does not exist. The reason lies in the unit-elastic demand function. It implies a constant revenue of $10 \cdot p_2$ for firm 1 no matter what price p_1 it charges. By increasing its price, the firm observes a fall in sales and, therefore, production costs, but no decline in revenues. This provides an incentive for the firm to increase its price further and further. Thus, no optimal decision exists for firm 1, and its best response is not well defined. This problem arises because the strategy set is not bounded above and, therefore, not compact. If there were an upper bound, say a maximum price $\bar{p}_1$ that the firm could charge, the optimal choice of the firm would be to charge this maximum price $\bar{p}_1$. In addition, notice that the payoff function is discontinuous at the price $p_1 = 0$. This discontinuity does not cause problems in this example since the firm would always want to charge a higher price rather than a lower one. ■

The following remark indicating a special case of theorem 4.1 that is useful for many economic applications concludes this section.

Remark 4.3. There is a useful corollary to theorem 4.1. In economics, many models assume that consumers or producers are identical. This assumption implies some symmetry for the game. A game is symmetric if, for any two players $i, j \in I$,

$$S_i = S_j \quad \text{and}$$

$$p_i(s_1, \ldots, s_{i-1}, s_i, s_{i+1}, \ldots, s_{j-1}, s_j, s_{j+1}, \ldots, s_I)$$
$$= p_j(s_1, \ldots, s_{i-1}, s_j, s_{i+1}, \ldots, s_{j-1}, s_i, s_{j+1}, \ldots, s_I)$$

holds. In symmetric games, all players have the same pure strategy sets and it is possible to exchange the strategies of any two players without changing any payoff except for the payoff of these two players.

Under the conditions of theorem 4.1, a symmetric game has a symmetric Nash equilibrium. Thus, there is a Nash equilibrium strategy combination s^* such that $s_i^* = s_j^*$ for all players $i, j \in I$ holds. This implies that a symmetric Cournot duopoly has an equilibrium at which both firms produce and sell the same quantity provided that the firms' profit functions satisfy the conditions of theorem 4.1. ■

Appendix to Section 4.1

In this appendix the Kakutani fixed point theorem is stated, its assumptions are explained and motivated by examples, and the proof of theorem 4.1 is given. The proof consists of a demonstration that the assumptions of theorem 4.1 are in fact sufficient to apply Kakutani's fixed point theorem to the best response correspondence.

The following theorem is the cornerstone of most existence theorems in economics and game theory. Fixed point theorems for functions have a long tradition in mathematics. For game-theoretic applications, however, where best response mappings are usually derived from optimizing behavior of agents and, therefore, are correspondences rather than functions, a more general theorem was required. The fixed point theorem of Kakutani provides such a generalization. Before stating the theorem a generalized concept of continuity, suitable for correspondences, has to be introduced.

Remark 4.4 (*on mathematical concepts*). Recall that a correspondence associates with each point of a set X a (non-empty) subset of another set Y, rather than a single point of Y as in the case of a function. For functions, continuity relates sequences of points in the set X to sequences of points in the set Y induced by the function. In the case of correspondences, however, many sequences in Y may correspond to a particular sequence in X. This motivates the following concept of continuity.[5]

A correspondence ϕ: $X \to \mathscr{P}(Y)$ is called *upper hemi-continuous* (*u.h.c.*) if, for any sequence $x^\nu \in X$, $\nu = 1, 2, 3, \ldots,$ that converges to $x^o \in X$ and for any converging sequence $y^\nu \in \phi(x^\nu)$, $\nu = 1, 2, 3, \ldots,$ the sequence y^ν converges to some $y^o \in \phi(x^o)$.

[5]The definition of u.h.c. given here is appropriate only if the range of ϕ is compact and if ϕ itself is compact valued. These conditions are, however, satisfied in this book. For a more rigorous definition compare Green and Heller (1981).

The definition of upper hemi-continuity for a correspondence is different from the concept of continuity for a function because there may be more than one sequence $y^\nu \in \phi(x^\nu)$ associated with a particular sequence x^ν. If the correspondence ϕ has a single point in $\phi(x)$ for each $x \in X$, that is, ϕ is essentially a function, then the concept of upper hemi-continuity will coincide with the concept of continuity of a function.

The following diagrams will provide two simple examples of correspondences, one u.h.c. and another not u.h.c.

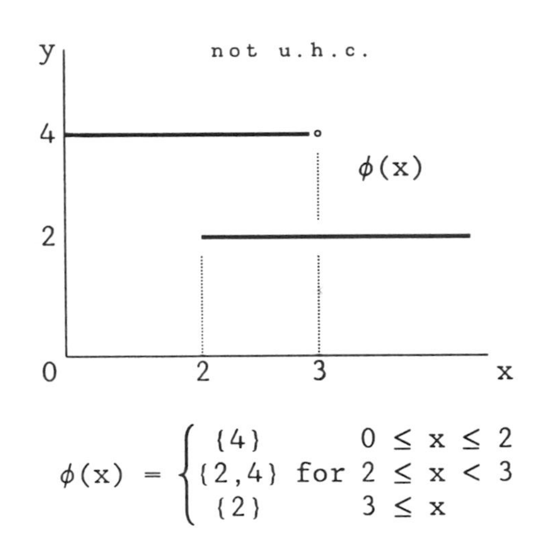

$$\phi(x) = \begin{cases} \{4\} & 0 \leq x \leq 2 \\ \{2,4\} \text{ for} & 2 \leq x < 3 \\ \{2\} & 3 \leq x \end{cases}$$

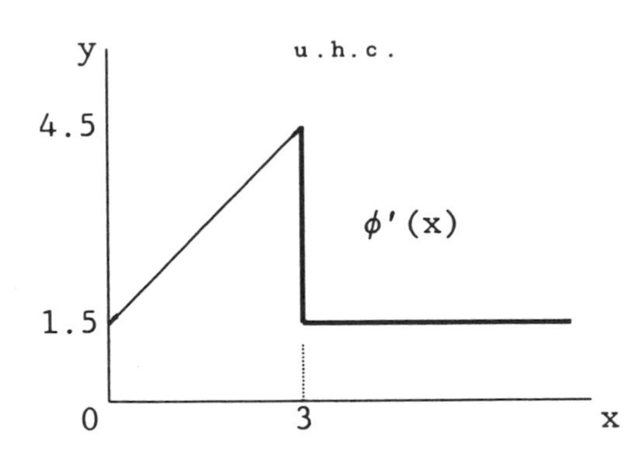

$$\phi'(x) = \begin{cases} \{1.5 + x\} & 0 \leq x < 3 \\ [1.5, 4.5] \quad \text{for} & x = 3 \\ \{1.5\} & 3 \leq x \end{cases}$$

■

Theorem (*Kakutani*, 1941). *Let X be a (non-empty) compact, convex subset of $\mathbb{R}^n$, and let $\gamma\colon X \to \mathscr{P}(X)$ be a u.h.c. correspondence. If $\gamma(x)$ is a convex set for every $x \in X$, then there exists an $x^* \in X$ such that $x^* \in \gamma(x^*)$ holds.*

No proof will be given here for this theorem.[6] To provide some intuition for why the assumptions of the theorem are required, the following example is considered. For the special case in which X is a subset of $\mathbb{R}$, this example shows in a series of diagrams the meaning of the premises made in the theorem.

Example. In this example correspondences $\gamma\colon X \to \mathscr{P}(Y)$ are considered where X and Y are subsets of $\mathbb{R}$. Hence, X can be depicted on the horizontal axis of a diagram and Y on the vertical axis. Note that a fixed point is a point of the correspondence γ which lies on the 45° line.

Kakutani's theorem guarantees that there is a fixed point if the following conditions are satisfied:

- $Y = X$
- X is a convex set
- X is a compact set
- γ is u.h.c.
- $\gamma(x)$ is a convex set for all $x \in X$

The following diagrams show examples illustrating why the theorem may fail to hold if any one of these conditions is violated.

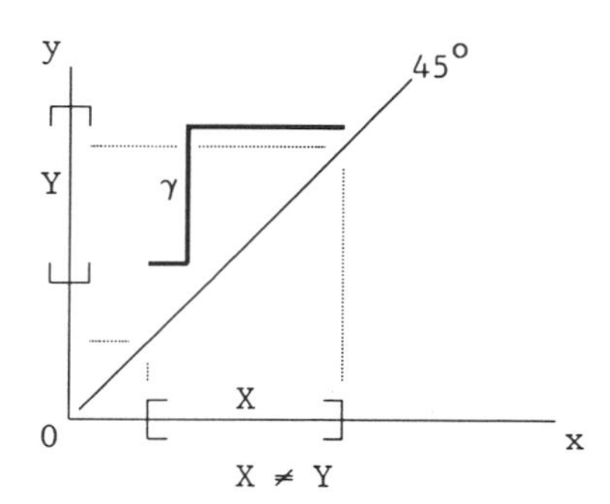

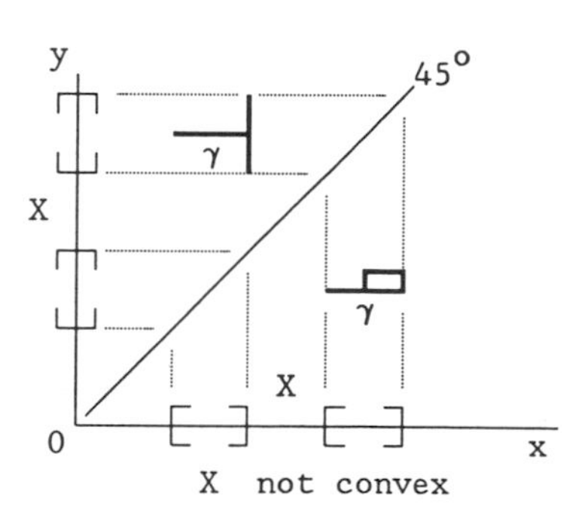

Note that a set is compact if it is both closed and bounded.

[6] A proof of Kakutani's fixed point theorem can be found in Berge (1963).

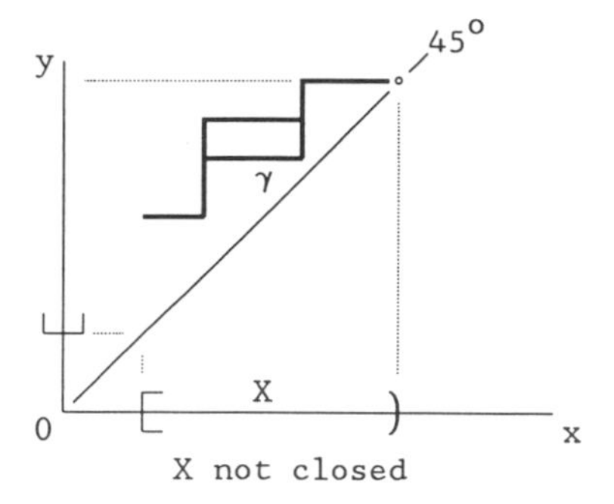

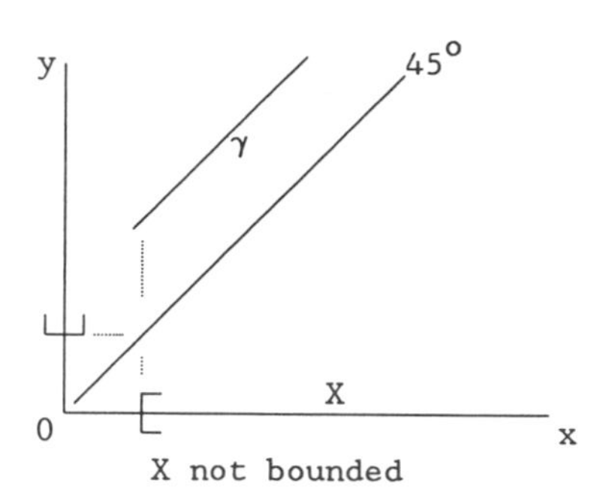

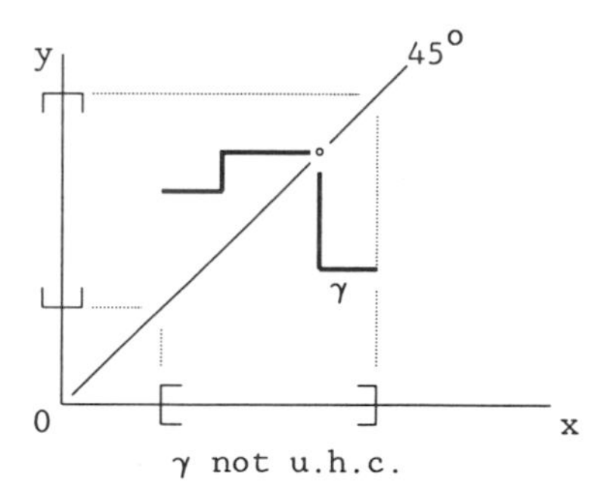

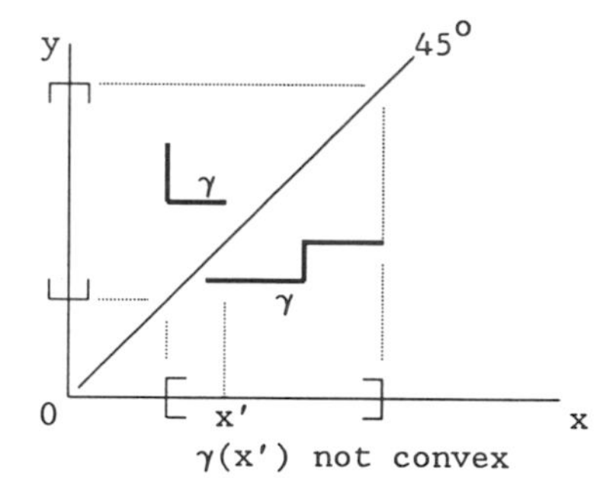

These examples are designed to provide some intuition for why the conditions of the Kakutani fixed point theorem are needed to guarantee the existence of a fixed point. ■

This completes the discussion of the Kakutani fixed point theorem. For the following proof of theorem 4.1 this result will be taken as given.

Proof of theorem 4.1. To prove this theorem it is sufficient to check that, given the premises of theorem 4.1, the reaction correspondence $r(s)$ satisfies all the conditions of the Kakutani fixed point theorem. If this can be established, then the Kakutani theorem guarantees a fixed point $s^* \in r(s^*)$ of the reaction correspondence. By lemma 4.3, it is clear that such a fixed point s^* is a Nash equilibrium.

To apply Kakutani's fixed point theorem the following conditions must hold:

- S is a compact and convex subset of $\mathbb{R}^n$
- r is a mapping from S to subsets of S

- r is an upper hemi-continuous correspondence
- $r(s)$ is a convex set for all $s \in S$

By assumption (i) of the theorem, each S_i is a compact and convex subset of a Euclidian space. Therefore, the set of strategy combinations S must be a compact and convex subset of a Euclidian space as well. This follows because S is the Cartesian product of the sets S_i.

The correspondence r associates with each strategy combination s the set of those strategy combinations that have best response strategies as components for each player. Thus, r associates with each $s \in S$ a set of strategy combinations, that is, a subset of S. Note that best responses for each player are guaranteed to exist because the strategy sets S_i are compact and the payoff functions are continuous. Example 4.5 illustrates this point.

Next, it is checked that $r(s)$ is a convex set for all $s \in S$. If a player i has just one best response strategy for some $s \in S$, then the set $r_i(s)$ is trivially convex. Otherwise, there are two best responses for player i, say $\bar{s}_i$, $\bar{\bar{s}}_i \in r_i(s)$. For an arbitrary λ, $0 \leq \lambda \leq 1$, consider the strategy $s_i^\lambda \equiv \lambda \cdot \bar{s}_i + (1-\lambda) \cdot \bar{\bar{s}}_i$ and assume that $s_i^\lambda \notin r_i(s)$ holds, that is, that s_i^λ is not a best response to the strategy combination s. Thus, $p_i(s_i^\lambda, s_{-i}) < p_i(\bar{s}_i, s_{-i}) = p_i(\bar{\bar{s}}_i, s_{-i})$ must hold. But this contradicts assumption (ii) of the theorem that p_i is quasi-concave. To see this, let $\alpha \equiv p_i(\bar{s}_i, s_{-i}) = p_i(\bar{\bar{s}}_i, s_{-i})$ be the payoff level of $p_i(\bar{s}_i, s_{-i})$ and $p_i(\bar{\bar{s}}_i, s_{-i})$. By quasi-concavity, any convex combination of $\bar{s}_i$ and $\bar{\bar{s}}_i$ must belong to the upper level set of α, that is,

$$s_i^\lambda \in \{s_i \in S_i | p_i(s_i, s_{-i}) \geq \alpha\}$$

must hold. Hence, $p_i(s_i^\lambda, s_{-i}) \geq \alpha \equiv p_i(\bar{s}_i, s_{-i}) = p_i(\bar{\bar{s}}_i, s_{-i})$ holds, contradicting the assumption that s_i^λ is not a best response to strategy combination s. In conclusion, quasi-concavity of p_i in s_i implies that $r_i(s)$ is a convex set for each player $i \in I$, and $r(s)$ is a convex set since it is the Cartesian product of the individual sets $r_i(s)$.

To see that $r_i(s)$ is u.h.c., consider a sequence $s^\nu \in S$, $\nu = 1, 2, 3, \ldots$, that converges to some strategy combination $s^o \in S$. In addition, take an arbitrary sequence $\tilde{s}_i^\nu \in r_i(s^\nu)$, $\nu = 1, 2, 3, \ldots$, that converges to $\tilde{s}_i^o$. It remains to show that $\tilde{s}_i^o \in r_i(s^o)$ holds. Suppose this is not the case, so that there must be a strategy $\bar{s}_i \in S_i$ for player i yielding a higher payoff than strategy $\tilde{s}_i^o$, that is, for which $p_i(\bar{s}_i, s_{-i}^o) > p_i(\tilde{s}_i^o, s_{-i}^o)$ holds. By assumption (ii) of theorem 4.1, payoff functions p_i are continuous. Therefore, the sequence $p_i(\tilde{s}_i^\nu, s_{-i}^\nu)$ converges to $p_i(\tilde{s}_i^o, s_{-i}^o)$, and the sequence $p_i(\bar{s}_i, s_{-i}^\nu)$ has $p_i(\bar{s}_i, s_{-i}^o)$ as its limit. If $p_i(\bar{s}_i, s_{-i}^o) > p_i(\tilde{s}_i^o, s_{-i}^o)$ holds, however, then, by continuity of p_i, there must be a number N such that, for all $\nu > N$, $p_i(\bar{s}_i, s_{-i}^\nu) > p_i(\tilde{s}_i^\nu, s_{-i}^\nu)$ holds as well. This would imply that $\tilde{s}_i^\nu$ is not a best response to s^ν, that is, $\tilde{s}_i^\nu \notin r_i(s^\nu)$ holds. But this contradicts the assumption $\tilde{s}_i^\nu \in r_i(s^\nu)$ for all ν.

Therefore, all the assumptions of the Kakutani fixed point theorem are satisfied. This completes the proof of theorem 4.1. ■

4.2 Uniqueness of Nash Equilibrium

In the previous section, conditions on strategy sets and payoff functions that guaranteed existence of at least one Nash equilibrium were given. But as example 4.1 showed, there may be more than one strategy combination with the property that each agent chooses a best response to the other players' behavior and, hence, more than one Nash equilibrium. If there are multiple Nash equilibria, then it remains unclear which of these possible equilibria will eventuate in the game. Non-uniqueness of Nash equilibrium raises a number of difficult questions. Some of these problems will be dealt with in the following sections.

In addition, many applications in economics use comparative static analysis to draw conclusions on what would happen to equilibrium behavior if some exogenous parameter were changed. For comparative static analysis, uniqueness of an equilibrium is often necessary. The following example illustrates this point.

EXAMPLE 4.6 (*example* 3.4 *modified*). Consider a duopoly with zero production cost and the following demand functions:

$$d_1(p_1, p_2) = \begin{cases} 1 + p_2^2 - .5 \cdot p_1 & \text{for } 0 \leq p_2 \leq 1 \\ 2 \cdot p_2 - .5 \cdot p_1 & \text{for } 1 \leq p_2 \leq 4 \\ 4 \cdot \sqrt{p_2} - .5 \cdot p_1 & \text{for } 4 \leq p_2 \end{cases}$$

and

$$d_2(p_1, p_2) = p_1 - .5 \cdot a \cdot p_2$$

for some a, $0 \leq a \leq 4$.

This model can be translated into a game in strategic form in the same way as in example 3.4. In particular, the payoff functions (profit functions) are obtained as described in example 3.4. It is not difficult to derive the following best response functions:

$$r_1(p_2) = \begin{cases} 1 + p_2^2 & \text{for } 0 \leq p_2 \leq 1 \\ 2 \cdot p_2 & \text{for } 1 \leq p_2 \leq 4 \\ 4 \cdot \sqrt{p_2} & \text{for } 4 \leq p_2 \end{cases} \quad \text{and} \quad r_2(p_1) = \frac{1}{a} \cdot p_1.$$

The diagram below shows these best response functions.

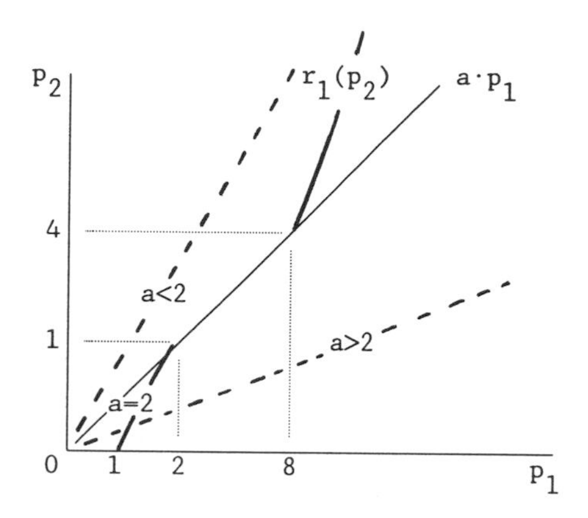

In this diagram, three possible locations of the best response function $r_2(p_1)$ are indicated by $a > 2$, $a = 2$, and $a < 2$. Nash equilibria are intersections of the best response functions. Hence, for $a \neq 2$, there is a unique Nash equilibrium. For $a = 2$, however, there is a continuum of equilibria, that is, all strategy combinations (p_1, p_2) that satisfy $2 \leq p_1 \leq 8$ and $p_2 = .5 \cdot p_1$ form a Nash equilibrium for $a = 2$. ■

Example 4.6 shows a particular type of non-uniqueness of Nash equilibrium, namely a continuum of equilibria in the case of $a = 2$. This case is particularly worrisome for comparative static analysis. If one were to analyze how market prices vary in response to the demand parameter a, one would find that a tiny change in the parameter away from $a = 2$ would result in a large change of the equilibrium prices.

For this reason, it is worth studying under what conditions uniqueness can be guaranteed. Unfortunately, there are virtually no results regarding what type of strategy sets S_i and payoff functions p_i will lead to a unique Nash equilibrium. The only results on uniqueness give conditions on best response functions. Therefore, the following sufficient conditions for uniqueness implicitly assume already that best responses are unique, that one deals with best response functions and not best response correspondences.[7]

If best response mappings are functions, then a Nash equilibrium is simply the solution to the following system of equations: $s^* = r(s^*)$. The system of best response functions $r(s) \equiv (r_1(s), r_2(s), \ldots, r_I(s))$ may be

[7]A sufficient condition for unique best responses of a player i and, therefore, for best response mappings to be functions is to assume that payoff functions $p_i(s_i, s_{-i})$ are strictly quasi-concave in s_i.

written equivalently in the following implicit form: $g(s) \equiv r(s) - s$. In this formulation, a Nash equilibrium is a strategy combination s^* such that $g(s^*) = 0$ holds. This is a convenient formulation, because it allows the application of some well-known mathematical results.

Assuming that the best response functions $r_i(s)$, $i \in I$, are all twice continuously differentiable, one can give a sufficient condition for uniqueness of Nash equilibrium, provided that one exists, in terms of the slope of these best response functions. Differentiating each of the best response functions in implicit form $g_i(s_1, s_2, \ldots, s_I) \equiv r_i(s_1, s_2, \ldots, s_I) - s_i$, $i = 1, 2, \ldots, I$, with respect to all its arguments and arranging them in matrix form, one obtains $J(g)$, the Jacobian matrix of the function $g(s)$:

$$J(g) \equiv \begin{bmatrix} \frac{\partial g_1}{\partial s_1} & \cdots & \frac{\partial g_1}{\partial s_I} \\ \vdots & & \vdots \\ \frac{\partial g_I}{\partial s_1} & \cdots & \frac{\partial g_I}{\partial s_I} \end{bmatrix}$$

Obviously, $J(g)$ is a square matrix.

Remark 4.5 (*on mathematical concepts*). A $(n \times n)$-matrix A is called negative quasi-definite if, for any vector $x \in \mathbb{R}^n$, $x \neq 0$, $x \cdot (A + A^T) \cdot x^T < 0$ holds. This implies that all the elements on the main diagonal of $A + A^T$ must be negative and that all the off-diagonal elements of $A + A^T$ have to be small in absolute value. ■

The following result on uniqueness of Nash equilibrium holds.[8]

THEOREM 4.3. *If all the best response functions $r_i(s)$, $i \in I$, are twice continuously differentiable and if the Jacobian matrix $J(g)$ of $g(s) \equiv r(s) - s$ is negative quasi-definite, then there is at most one Nash equilibrium.*

A proof of this theorem can be found in Rosen (1965) and will be omitted here. The following variation of example 3.4 will show how one can use theorem 4.3 to check for uniqueness of an equilibrium.

EXAMPLE 4.7 (*example* 3.4 *modified*). Consider the price setting duopoly once again. Assume that there are no production costs and that

[8]In requiring that $r_i(s)$ should be twice continuously differentiable, one has to take care that these derivatives are well defined on the boundary of the set S. For more details see Rosen (1965).

the demand functions have the following form ($a_i > 0$, $0 \leq b_i < 1$, for $i = 1, 2$):

$$d_1(p_1, p_2) = a_1 + b_1 \cdot p_2 - 0.5 \cdot p_1 \text{ and } d_2(p_1, p_2) = a_2 + b_2 \cdot p_1 - 0.5 \cdot p_2.$$

This is a two-player game with the set of possible prices as the strategy set for each player, $S_i \equiv \mathbb{R}_+$ for $i = 1, 2$. The payoff functions (or profit functions) are easily deduced from this model as $\pi_i(p_1, p_2) = p_i \cdot d_i(p_1, p_2)$, $i = 1, 2$. This yields the following best response functions:

$$r_1(p_1, p_2) = a_1 + b_1 \cdot p_2 \quad \text{and} \quad r_2(p_1, p_2) = a_2 + b_2 \cdot p_1.$$

The implicit form of the best response functions takes the following form:

$$g_1(p_1, p_2) = a_1 + b_1 \cdot p_2 - p_1,$$
$$g_2(p_1, p_2) = a_2 + b_2 \cdot p_1 - p_2.$$

Partial differentiation results in the following Jacobian matrix:

$$J(g) \equiv \begin{bmatrix} -1 & b_1 \\ b_2 & -1 \end{bmatrix}.$$

This matrix is negative quasi-definite if for any vector $(x_1, x_2) \in \mathbb{R}^2$,

$$(x_1, x_2) \cdot \left(\begin{bmatrix} -1 & b_1 \\ b_2 & -1 \end{bmatrix} + \begin{bmatrix} -1 & b_2 \\ b_1 & -1 \end{bmatrix} \right) \cdot (x_1, x_2) < 0$$

holds or, equivalently: $-2 \cdot [x_1^2 + x_2^2 - (b_1 + b_2) \cdot x_1 \cdot x_2] < 0$. Clearly, this condition is satisfied if the expression in the square bracket is positive. This is however guaranteed, since $0 \leq b_i < 1$, for $i = 1, 2$, has been assumed. To see this, note that

$$0 \leq (x_1 + x_2)^2 = \left[x_1^2 + x_2^2 - 2 \cdot x_1 \cdot x_2\right] < \left[x_1^2 + x_2^2 - (b_1 + b_2) \cdot x_1 \cdot x_2\right]$$

holds, whenever $x_1 \cdot x_2$ is positive, otherwise the expression in square brackets is trivially positive. ■

In fact, the result of example 4.7 can be generalized to the proposition that a sufficient condition for uniqueness of an equilibrium is a slope of the best response functions that is consistently less than one in absolute value.

As mentioned previously, theorem 4.3 on uniqueness does not guarantee existence of an equilibrium. There is, however, one approach that simultaneously establishes existence and uniqueness of a Nash equilibrium.

Remark 4.6 (on mathematical concepts). For any two vectors of real numbers x and y, that is, $x, y \in \mathbb{R}^n$, denote by

$$d(x, y) \equiv \max_{i \in \{1,2,\ldots n\}} |x_i - y_i|$$

the distance between the vectors x and y. Note that the distance between the two vectors is measured by the longest distance between any component of the vectors.

A function f that associates vectors of real numbers from a set A with vectors of real numbers from the same set A, formally $f: A \to A$, $A \subset \mathbb{R}^n$, is called *contraction* if there exists some real number λ, $0 < \lambda < 1$, such that for all $x, x' \in A$

$$d(f(x), f(x')) \leq \lambda \cdot d(x, x')$$

holds. Thus, a contraction is a function that shrinks if it is iteratively applied, that is, the sequence x^ν, $\nu = 1, 2, 3, \ldots$, for an arbitrary x^1 defined by $x^{\nu+1} = f(x^\nu)$, has the property that the distance between $x^{\nu+1}$ and x^ν shrinks as ν increases. Therefore, the sequence converges to a fixed point. One can show that a contraction is always a continuous function. ■

With the help of these concepts the following theorem can be stated.

THEOREM 4.4. *If the best reply function r is a contraction, then there exists a unique equilibrium.*

The fact that a contraction mapping has a unique fixed point is a well-known result and its proof can be found, for example, in Lang (1968, pp. 338–339). Note that, in contrast to theorem 4.3, this theorem simultaneously guarantees existence and uniqueness of a Nash equilibrium. A great advantage of this theorem is that it requires no conditions on the set of strategies S. However, not many functions have the contraction property.

This section concludes by demonstrating that the best response functions in example 4.7 are contractions.

EXAMPLE 4.7 (*resumed*). Consider two price vectors (p_1, p_2) and (p'_1, p'_2) and consider the difference between the value of the best response function at these two price vectors:

$$\begin{aligned} r(p_1, p_2) - r(p'_1, p'_2) &= ((r_1(p_1, p_2), r_2(p_1, p_2)) - (r_1(p'_1, p'_2), r_2(p'_1, p'_2)) \\ &= (r_1(p_1, p_2) - r_1(p'_1, p'_2), r_2(p_1, p_2) - r_2(p'_1, p'_2)) \\ &= (b_1 \cdot (p_2 - p'_2), b_2 \cdot (p_1 - p'_1)) \end{aligned}$$

Let b^* be the larger of the two coefficients b_1 and b_2. Since both are strictly smaller than 1, b^* must also be strictly smaller than 1. Thus, there must be a number λ^* which is larger than b^* and strictly smaller than 1. Consider the distance between $r(p_1, p_2)$ and $r(p_1', p_2')$:

$$\begin{aligned}
&d(r(p_1, p_2), r(p_1', p_2')) \\
&\quad \equiv \max\{|r_1(p_1, p_2) - r_1(p_1', p_2')|, |r_2(p_1, p_2) - r_2(p_1', p_2')|\} \\
&\quad = \max\{|b_1 \cdot (p_2 - p_2')|, |b_2 \cdot (p_1 - p_1')|\} \\
&\quad = \max\{b_1 \cdot |(p_2 - p_2')|, b_2 \cdot |(p_1 - p_1')|\} \\
&\quad \leq b^* \cdot \max\{|(p_2 - p_2')|, |(p_1 - p_1')|\} \\
&\quad < \lambda^* \cdot \max\{|(p_2 - p_2')|, |(p_1 - p_1')|\} \\
&\quad \equiv \lambda^* \cdot d((p_1, p_2), (p', p_2')).
\end{aligned}$$

Thus, in this example, the best response function $r(\cdot)$ is a contraction. ■

4.3 Two Interpretations of the Nash Equilibrium Concept

In the previous section, several sufficient conditions for uniqueness of a Nash equilibrium were given. There are, however, many games of interest to economists for which the best response functions are neither contractions nor do they have a negative quasi-definite Jacobian matrix. In these cases, a selection among the existing equilibria has to take place. This section does not attempt to survey the growing literature on selection mechanisms,[9] but tries to show how two different interpretations of the Nash equilibrium concept suggest two different selection procedures. Two examples will be used to make this point.

Example 4.8. Consider two players who bargain about three (indivisible) dollars according to the following procedure: each player $i \in I \equiv \{1, 2\}$ announces what share s_i he wants $s_i \in S_i \equiv \{1, 2, 3\}$. If their demands are feasible, if $s_1 + s_2 \leq 3$ holds, then this outcome is implemented; otherwise each gets zero.

[9]Kreps (1990b) provides an introduction to these problems and the ongoing research in this regard.

Demands of three dollars will never be feasible and, therefore, will be dropped from consideration. This leaves the following payoff matrix.

		Player 2	
		1	2
Player 1	1	1, 1	1, 2
	2	2, 1	0, 0

Obviously, there are two Nash equilibria in this game, namely (1, 2) and (2, 1). Player 1 prefers equilibrium (2, 1) and player 2 prefers equilibrium (1, 2). ■

This example makes it clear that one cannot take a Nash equilibrium as the solution to the game, because it is not clear how players should behave in such a game. The requirement that each player should play a best response to the other player's behavior does not help in predicting an outcome of the game, because there are two strategy combinations that satisfy this requirement.

Furthermore, each player prefers a different equilibrium. If player 1 decides to claim 2, hoping that the other player "accepts" the equilibrium (2, 1), and player 2 does the same thing, goes for his best equilibrium and asks for 2, the undesirable non-equilibrium outcome of a zero payoff for each player would result. On the other hand, to avoid such a bad outcome, each player may decide to accept the less favorable equilibrium; in fact, this would be each player's maximin strategy in this game. But once again a non-equilibrium outcome would result that provides incentives for the players to deviate.

Thus, noting that there are some Nash equilibria does not tell what can be expected to happen in the game. In this sense, a Nash equilibrium is not a solution for the game. Not surprisingly, therefore, two alternative interpretations of Nash equilibrium have appeared in the literature. In one, players get together and communicate in some way about how to play a game before they actually play it. During this negotiation process, they may agree on a strategy combination to be played. In contrast to the basic assumption of cooperative game theory, however, players cannot make binding agreements. Thus, they are unable to commit themselves to any agreement that they may reach. It seems natural in such a situation that an agreement must be self-enforcing, that is, following the agreed-upon behavior must be optimal for each agent given that everyone else sticks to the agreement. Otherwise, agents would have an incentive to break the agreement.

If one views a Nash equilibrium as a self-enforcing agreement, then selection of the equilibrium has to occur during pregame negotiations. Rather than answering the question of what one should expect to happen in a given game, this approach shifts the question to the pregame negotiation process. The only claim made by this approach requires that any solution must be a Nash equilibrium because otherwise it will not be self-enforcing. No indication is given which equilibrium will be selected.

This interpretation views Nash equilibria as potential outcomes of pregame negotiation. From this perspective, it appears natural to model this negotiation process by extending the game to include details of the pregame negotiation such as sequences of proposals and counterproposals of players. In this way one hopes to obtain a prediction of the outcome of the game. Of course, the selection of the equilibrium so obtained will depend on the assumed bargaining procedure.[10]

A second interpretation of a Nash equilibrium is related to another kind of application. According to this view, the outcome of a game is considered a rational expectations equilibrium. Players are assumed to behave optimally in regard to their beliefs about their opponents' behavior, and in equilibrium these beliefs have to be correct. This interpretation is particularly appealing if the game involves a large number of agents, making pregame negotiation senseless. In this case, the description of how agents acquire certain beliefs about their opponents' behavior may provide the necessary additional information to select among equilibria. Nash equilibria can be viewed as "standards of behavior" that govern the interaction of many agents. In this view, multiplicity of equilibria simply means a multiplicity of possible "customs" among a given set of players. It may be more appropriate to trace certain existing "customs" or equilibria back to different beliefs of the agents together with some learning procedure. The following example may illustrate this interpretation.

EXAMPLE 4.9. Consider a large group of people. Any two of them can combine their labor inputs, e_1 and e_2 respectively, to produce some public good according to the production function

$$f(e_1, e_2) \equiv 2 \cdot \min\{e_1, e_2\}.$$

Each agent has a disutility cost of providing labor, $c(e_i) = e_i$. For simplicity, just two levels of effort, 10 and 20, are assumed available, that is, $S_i = \{10, 20\}$, $i = 1, 2$, are the strategy sets of the two players. The net payoff for player i is simply $p_i(e_1, e_2) \equiv f(e_1, e_2) - c(e_i)$, $i = 1, 2$. One can summarize the game between any two agents about the provision of an

[10]Chapter 9 will analyze bargaining problems.

adequate amount of labor in the following matrix:

		Player 2	
		10	20
Player 1	10	10, 10	10, 0
	20	0, 10	20, 20

It is easy to check that there are two Nash equilibria in pure strategies, namely (10, 10) and (20, 20). Note that these equilibria can be ranked in the sense that the payoff from (20, 20) is strictly better for both players than the payoff from (10, 10). This is the prototype of a *coordination game*. It is, however, a coordination game in which a failure to coordinate on the better equilibrium (20, 20) is costly, whereas a failure to coordinate on the low-level equilibrium (10, 10) is virtually costless. ■

Consider now the case in which in a sequence of periods two agents are randomly paired to play this game. Suppose, for some unexplained reason, that the established "standard of behavior" of the agents in this population requires them to play the equilibrium (20, 20). If a new agent enters this game believing there is a high probability ($\geq .5$) that agents tend to provide 20 units of labor, then he will find it optimal to choose a level of 20 and his beliefs will be confirmed. This will increase his confidence that a labor input of 20 is the current standard of behavior.

If his beliefs, however, put a low probability on agents choosing 20 units of labor input, he may decide to work only 10 units. Of course, he will learn that he was wrong in his assessment and may decide to change behavior in the future, but his opponent will also have been frustrated in his beliefs, and may in turn decide to reassess them and to behave differently in the future. Thus, this unexpected behavior would initialize a learning process in the population that in the long run may reconfirm the existing "custom" or establish a new "standard of behavior."

These two interpretations of a Nash equilibrium suggest different ways to select among multiple equilibria. In many cases, multiplicity of equilibria poses a problem for the applicability of the Nash equilibrium concept. The following section provides another method for reducing the number of Nash equilibria by requiring that additional criteria hold for an equilibrium.

4.4 Perfect Equilibrium

So far it has been assumed that players carry out their planned strategies without error. This assumption allowed for Nash equilibria where a player's strategy was optimal only if the other players follow exactly their equilibrium strategies. The slightest deviation of any other player might render the chosen strategy suboptimal. In these cases, the Nash equilibrium is not robust to small errors or deviations of some player. Based on the assumption that there is always at least a small chance of a player making a mistake, one may want to exclude these Nash equilibria. The following example illustrates this case.

EXAMPLE 4.10. Consider the following two-player game.

		Player 2	
		L	R
Player 1	T	1, 1	0, 0
	B	0, 0	0, 0

Player 1 has two strategies "top" (T) and "bottom" (B), $S_1 = \{T, B\}$; player 2 can choose between "left" (L) and "right" (R), $S_2 = \{L, R\}$.

This game has two Nash equilibria, namely (T, L) and (B, R). Note, however, that strategy T dominates strategy B for player 1 and strategy L dominates strategy R for player 2. The Nash equilibrium (B, R) involves playing dominated strategies for both players. ■

In example 4.10, it seems natural to discard the equilibrium (B, R) as unreasonable, because neither player 1 nor player 2 would want to play his strategy if there was any chance that the other player would deviate from his equilibrium strategy. If player 1 considers that there is a probability ϵ that player 2 may play his first strategy L, then his expected payoff from playing T would be $\epsilon \cdot 1 + (1 - \epsilon) \cdot 0 > 0$ compared to an expected payoff of $\epsilon \cdot 0 + (1 - \epsilon) \cdot 0 = 0$ from choosing B. Hence, it is optimal for player 1 to play T no matter how small the probability ϵ is. A similar argument shows that L is the best response for player 2 if there is the smallest probability of player 1 choosing T. This suggests (T, L) as the only Nash equilibrium robust against small errors. The rest of this section provides a formalization of this notion of robustness in general strategic form games.

Consider a game in strategic form, $\Gamma = (I, (M_i)_{i \in I}, (P_i)_{i \in I})$ where M_i is the set of mixed strategies, that is, the set of probability distributions on the set of pure strategies S_i, and P_i is the expected payoff function of player i. Playing a particular pure strategy s_i corresponds to the special case in which this strategy s_i is played with probability one, and any other strategy with probability zero.

The basic idea for modeling errors consists in the assumption that no pure strategy will ever be played with probability zero. A particular strategy may be played with a vanishingly small probability but no player can be certain that this strategy will never be played even if it is against the interest of the player using it. Since this deviation occurs randomly and unintentionally, it is justifiable to call it a *mistake* or *error*. A Nash equilibrium is called *trembling-hand perfect* if it is robust against small random deviations. A game in which every strategy must be played with a positive probability is called a *perturbed game*. To formally define the notion of a perturbed game, considerable notation is necessary.

For each player $i \in I$ with n_i pure strategies, let $\eta_i = (\eta_i^1, \eta_i^2, \ldots, \eta_i^n)$ be a vector of strictly positive numbers that sum to a number strictly less than 1, that is, such that $\eta_i \gg 0$ and $\Sigma_{j=1}^{n_i} \eta_i^j < 1$ holds. Denote by $\eta = (\eta_1, \ldots, \eta_I)$ the list of such vectors.

DEFINITION 4.3. The set of η-perturbed mixed strategies is the set

$$M_i^{\eta} = \{m_i \in M_i | m_i^k \geq \eta_i^k, \text{ for } k = 1 \ldots n_i\},$$

where player i, $i \in I$, chooses his kth pure strategy at least with probability η_i^k. The η-perturbed game $\Gamma(\eta) = (I, (M_i^{\eta})_{i \in I}, (P_i)_{i \in I})$ is the game in which all mixed strategy sets M_i are replaced by the η-perturbed mixed strategy sets M_i^{η}.

In an η-perturbed game, players actually have to play each strategy with a positive probability. Hence, every pure strategy will have a positive probability. If a player finds it optimal to play a particular pure strategy s_i^k with probability zero, then in a perturbed game the constraint $m_i^k \geq \eta_i^k$ will be binding. Thus, errors are modeled as constraints on the set of mixed strategies that a player may choose. The following example shows an η-perturbed mixed strategy set.

EXAMPLE 4.10 (*resumed*). The pure strategy sets in this example are $S_1 = \{T, B\}$ and $S_2 = \{L, R\}$. Letting $m_1^T, m_1^B, m_2^L, m_2^R$ represent the probabilities that the pure strategies T, B, L, R, are played by the respective players, then the sets of mixed strategies have the following form:

$$M_1 = \{(m_1^T, m_1^B) \in \mathbb{R}^2 | m_1^T \geq 0, m_1^B \geq 0, m_1^T + m_1^B = 1\},$$
$$M_2 = \{(m_2^L, m_2^R) \in \mathbb{R}^2 | m_2^L \geq 0, m_2^R \geq 0, m_2^L + m_2^R = 1\}.$$

Now, consider the η-perturbation $\eta_1^T = .1$, $\eta_1^B = .05$, $\eta_2^L = .2$, $\eta_2^R = .1$. Note that these perturbations satisfy the condition that η_i^k must be positive for all k, and, for each player, sum over k to a number less than 1. Hence, $\eta = (\eta_1, \eta_2) = (\eta_1^T, \eta_1^B, \eta_2^L, \eta_2^R) = (.1, .05, .2, .1)$ holds. The η-perturbed mixed strategy sets M_i^{η} for this perturbation η have the following form:

$$M_1^{\eta} = \{(m_1^T, m_1^B) \in \mathbb{R}^2 | m_1^T \geq .1, m_1^B \geq .05, m_1^T + m_1^B = 1\},$$
$$M_2^{\eta} = \{(m_2^L, m_2^R) \in \mathbb{R}^2 | m_2^L \geq .2, m_2^R \geq .1, m_2^L + m_2^R = 1\}.$$

The following diagram shows these two η-perturbed mixed strategy sets.

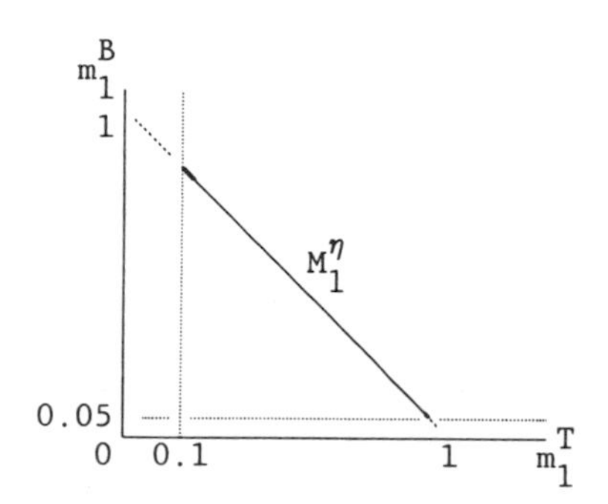

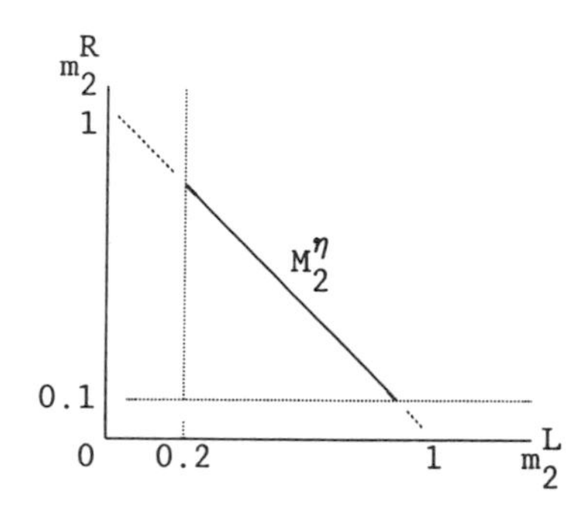

The constraints on the set of mixed strategies imposed by the η-perturbation are indicated by the dotted lines. The sets M_i^η, $i = 1, 2$, are indicated by the solid lines. The η-perturbed game is the game with these η-constrained mixed strategy sets. ■

An η-perturbed game is a game in strategic form with a Nash equilibrium $m^*(\eta)$ that depends on the perturbation η. To capture the notion that errors should be small, sequences of perturbations that get smaller and smaller are considered. A Nash equilibrium is robust against such small trembles if it can be obtained as a limit of the Nash equilibria of perturbed games $m^*(\eta)$ as the η-perturbations converge to zero.

Definition 4.4. An equilibrium $m^* \in M$ is (*trembling-hand*) *perfect*, if it is a limit point of some sequence $m^*(\eta)$, where $m^*(\eta)$ is an equilibrium of $\Gamma(\eta)$, for all η, and η converges to zero.

Example 4.10 can again be used to illustrate this definition.

Example 4.10 (*resumed*). Consider the same η as before. The expected payoff for both players from playing a mixed strategy combination $(m_1, m_2) = (m_1^T, m_1^B, m_2^L, m_2^R)$ is particularly simple in this example:

$$P_1(m_1^T, m_1^B, m_2^L, m_2^R) = m_1^T \cdot m_2^L \quad \text{and} \quad P_2(m_1^T, m_1^B, m_2^L, m_2^R) = m_1^T \cdot m_2^L.$$

Note that all pure strategy combinations have a zero payoff for both players, except for (T, L). Since the η-perturbation rules out the possibility of either m_1^T or m_2^L being zero, that is, $m_1^T \geq .1$ and $m_2^L \geq .2$ must hold, the best response for each player is to play the pure strategy T, respectively L, with the highest possible probability. Hence, the best response functions of both players are simply

$$r_1(m_1^T, m_1^B, m_2^L, m_2^R) = (1 - \eta_1^B, \eta_1^B),$$
$$r_2(m_1^T, m_1^B, m_2^L, m_2^R) = (1 - \eta_2^R, \eta_2^R).$$

Note that these best response functions are valid as long as m_1^T and m_2^L are strictly positive. This is, however, guaranteed by the η perturbation. If either m_1^T or m_2^L were zero, then any choice of (m_1^T, m_1^B) and (m_2^L, m_2^R) would be optimal, in particular $(m_1^T, m_1^B) = (0, 1)$ and $(m_2^L, m_2^R) = (0, 1)$. It is precisely this latter case that is ruled out by the perturbation.

Given any positive perturbation η, it is clear from the best response functions that $(m_1^*(\eta), m_2^*(\eta)) = (1 - \eta_1^B, \eta_1^B, 1 - \eta_2^R, \eta_2^R)$ is the unique Nash equilibrium of $\Gamma(\eta)$. Hence, as η converges to zero in all its components, the sequence of Nash equilibria $(m_1^*(\eta), m_2^*(\eta)) = (1 - \eta_1^B, \eta_1^B, 1 - \eta_2^R, \eta_2^R)$ converges to $(m_1^*, m_2^*) = (1, 0, 1, 0)$. Thus, in the limit player 1 plays T with probability one and player 2 plays L with probability one. Since this is obviously true for all sequences of perturbed games in this example, the Nash equilibrium (B, R) cannot be approximated by slightly perturbed games. This shows that the equilibrium (T, L) is indeed perfect and (B, R) is not perfect. ■

As the following theorem shows, there is no existence problem for trembling-hand perfect equilibria in games with finitely many pure strategies.

THEOREM 4.5. *Every normal form game with finite pure strategy sets has at least one perfect equilibrium.*

Proof. Note that the perturbed mixed strategy sets M_i^η are compact and convex sets of real numbers. Furthermore, the payoff functions P_i are continuous in mixed strategy combinations m and linear in m_i. Thus, all the conditions of theorem 4.1 are satisfied, and at least one Nash equilibrium $m^*(\eta)$ must exist in the game $\Gamma(\eta)$ for all η.

Therefore, for any sequence of η^ν, $\nu = 1, 2, 3, \ldots, m^*(\eta^\nu)$ defines a sequence of mixed strategies. Note that, for any ν, $m^*(\eta^\nu)$ lies in the set M of mixed strategy combinations that is compact. Hence, any sequence $m^*(\eta)$ will converge (or contain a converging subsequence). Let m^* be the mixed strategy combination to which $m^*(\eta)$ converges as η goes to zero.

To see that m^* is an equilibrium of Γ, note that just two possibilities can arise:

(i) If, for all η close to zero, $P_i(m^*(\eta)) \geq P_i(m_i, m_{-i}^*(\eta))$ holds for all $m_i \in M_i$ and all players $i \in I$, then $P_i(m^*) \geq P_i(m_i, m_{-i}^*)$ follows for all $m_i \in M_i$ and all $i \in I$. Hence, m^* is a Nash equilibrium of Γ.

(ii) If, for all η close to zero, there is some $i \in I$ and some $m_i \in M_i$ such that $P_i(m_i, m_{-i}^*) > P_i(m^*(\eta))$ holds, then $m_i^*(\eta) = (\eta_i^1, \ldots, 1 - \Sigma_{j \neq k} \eta_i^j, \ldots, \eta_i^n)$ must hold for some pure strategy k, that is, the player would want to play some pure strategy k with a higher probability than the constraints η_i allow. Hence, $m_i^*(\eta)$ will converge to $m_i^* = (0, \ldots, 1, \ldots, 0)$, the mixed strategy that plays the pure strategy k with probability one. The strategy m_i^* must be a best response to m_{-i}^* because any other mixed strategy that involves playing the pure strategy k with a probability less than 1 would have become feasible for some positive η and, therefore, would have been a best response to some $m_{-i}^*(\eta)$. This contradicts the premise that a better strategy could be found for all η. But if m_i^* is a best response to m_{-i}^* for all players $i \in I$, then m^* is a Nash equilibrium. ■

Theorem 4.5 shows that every game in strategic form has at least one perfect equilibrium. On the other hand, example 4.10 shows that a game may have equilibria that are not perfect. Hence, for games with finite pure strategy sets, the set of perfect equilibria is a non-empty subset of the set of Nash equilibria. Requiring perfectness of an equilibrium may reduce the multiplicity of equilibria.

Perfect equilibria have another attractive feature. As the following theorem shows, a perfect equilibrium strategy is never dominated.

THEOREM 4.6. *Every perfect equilibrium strategy is undominated.*

Proof. First, note that a dominated strategy can never be a best response to a mixed strategy combination m_{-i} of the other players that puts some positive probability on all pure strategy combination s_{-i} of the opponents.

To see this, assume that, for some player $i \in I$, the pure strategy s_i^ℓ is dominated by the pure strategy s_i^k, that is, $p_i(s_i^k, s_{-i}) \geq p_i(s_i^\ell, s_{-i})$ for all $s_{-i} \in S_{-i}$ with strict inequality for some $s'_{-i} \in S_{-i}$. Let $\tilde{m}_i^{s\ell}$ be the mixed strategy that assigns probability one to the pure strategy s_i^ℓ and probability zero to all other pure strategies of this player. Similarly, $\tilde{m}_i^{sk}$ is the mixed strategy that plays the pure strategy s_i^k with probability one. Denote by $m_{-i}(s_{-i})$ the probability that the pure strategy combination s_{-i} will be played according to the mixed strategy combination m_{-i}.

$$m_{-i}(s_{-i}) \equiv m_1(s_1) \cdot m_2(s_2) \cdot \ldots \cdot m_{i-1}(s_{i-1}) \cdot m_{i+1}(s_{i+1}) \cdot \ldots \cdot m_I(s_I).$$

If $\overline{m}_{-i}$ is a mixed strategy combination of the other players that satisfies $\overline{m}_{-i}(s_{-i}) > 0$ for all $s_{-i} \in S_{-i}$, then

$$\begin{aligned} P_i\left(\tilde{m}_i^{sk}, \overline{m}_{-i}\right) &= \sum_{s_{-i} \in S_{-i}} p_i\left(s_i^k, s_{-i}\right) \cdot \overline{m}_{-i}(s_{-i}) \\ &> \sum_{s_{-i} \in S_{-i}} p_i\left(s_i^\ell, s_{-i}\right) \cdot \overline{m}_{-i}(s_{-i}) = P_i\left(\tilde{m}_i^{s\ell}, \overline{m}_{-i}\right) \end{aligned}$$

holds. Hence, playing the dominated pure strategy, s_i^ℓ, with probability one can never be a best response to a choice of a mixed strategy combination by opponents that puts a positive probability on all pure strategy combinations. Putting it differently, playing a dominated strategy can be a best response only if a player can be certain that the opponents will never play a pure strategy combination against which the dominated strategy performs strictly worse than the dominating strategy.

Second, since in an η-perturbed game no pure strategy can be played with probability zero by the opponents, it follows from this argument that playing a mixed strategy that assigns probability one to a dominated strategy can never be a best response. Furthermore, it cannot even be a best response to play a mixed strategy that puts the highest possible probability on a dominated strategy and, at the same time, the lowest possible probability on the dominating strategy.

Since in a Nash equilibrium each player must play a best response strategy, it is clear that a mixed strategy that gives probability one to a dominated strategy, or any other mixed strategy close to it, cannot be part of a Nash equilibrium of any η-perturbed game $\Gamma(\eta)$. Hence, there cannot be a sequence of Nash equilibrium strategy combinations that converges to a dominated strategy of the unperturbed game Γ. Since the existence of such a sequence is the defining criterion for a perfect equilibrium, this proves that no perfect equilibrium uses dominated strategies. ■

Though no player will use a dominated strategy in a perfect equilibrium, it does not follow, in general, that every Nash equilibrium that does not use a dominated strategy is a perfect equilibrium. The following example demonstrates that, in games with more than two players, an undominated Nash equilibrium may fail to be perfect.

EXAMPLE 4.11 (*van Damme*, 1991, *p.* 29). Consider the three-player game given by the following two matrices. Each player has two strategies s_i^1 and s_i^2, $i \in \{1, 2, 3\}$, player 1 chooses rows, player 2 chooses columns, and player 3 chooses the matrix. The first payoff in each cell of a matrix belongs to player 1, the second to player 2, and the third to player 3.

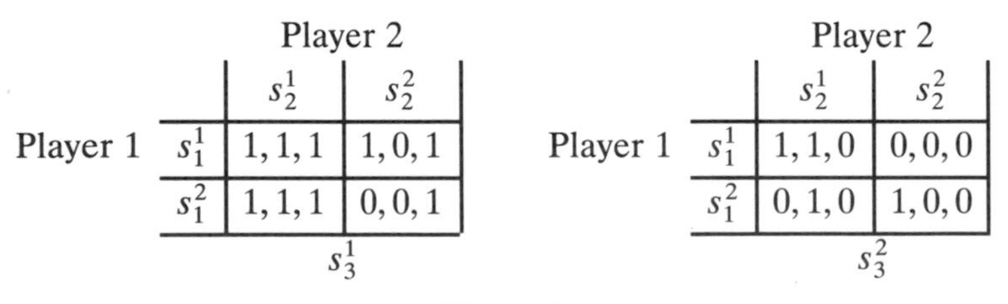

s_3^1:

Player 1 \ Player 2	s_2^1	s_2^2
s_1^1	1, 1, 1	1, 0, 1
s_1^2	1, 1, 1	0, 0, 1

s_3^2:

Player 1 \ Player 2	s_2^1	s_2^2
s_1^1	1, 1, 0	0, 0, 0
s_1^2	0, 1, 0	1, 0, 0

Player 3

It is easy to see that both strategy combinations (s_1^1, s_2^1, s_3^1) and (s_1^2, s_2^1, s_3^1) are undominated Nash equilibria, that is do not contain a dominated strategy. Actually, the strategies of player 2 and player 3 in these combinations, s_2^1 and s_3^1, are dominant strategies for these players. But player 1's best response depends on the strategy choice of the other players. Nevertheless, (s_1^1, s_2^1, s_3^1) is a *perfect equilibrium*, whereas (s_1^2, s_2^1, s_3^1) is *not perfect*. This is easy to see, since player 1 would play s_1^1 with the highest possible probability if there were some chance that player 2 might choose s_2^2. ■

In example 4.11, a three-player game was used to show that undominated Nash equilibria need not be perfect equilibria. As a matter of fact, for two-player games this is not true. The following lemma, proved in van Damme (1991, theorem 3.2.2), makes this point.

LEMMA 4.4. *A Nash equilibrium of a two-player game is undominated if and only if it is a perfect equilibrium.*

This section concludes with an example showing that the robustness criterion embodied in the perfect equilibrium concept may rule out socially optimal Nash equilibria.

EXAMPLE 4.12 (*example* 4.10 *modified*). Consider the following two-player game.

		Player 2	
		L	R
Player 1	T	1, 1	10, 0
	B	0, 10	10, 10

It is straightforward to check that (T, L) and (B, R) are the only pure strategy Nash equilibria of this game. However, only (T, L) is a perfect equilibrium. To see this, note that L dominates R and T dominates B. Since there are dominated strategies used in the equilibrium (B, R), it follows from theorem 4.6 that (B, R) is not a perfect equilibrium. ■

Example 4.12 clarifies that the possibility of errors provides an argument for the exclusion of any equilibrium not robust to minor deviations by opponents. This may, however, preclude equilibria desirable for other reasons.

4.5 Summary

This chapter introduced equilibrium concepts based on the idea that players will choose best responses to other players' behavior. In this context the most popular equilibrium concept is the Nash equilibrium. It requires an equilibrium strategy combination to be such that every player chooses a best response to the other players' equilibrium strategies.

It could be shown that, under fairly general conditions, a game will have at least one Nash equilibrium. A pervasive problem of the Nash equilibrium concept is the possibility of multiple equilibria. Conditions for uniqueness of Nash equilibrium, however, are quite demanding and few games have a single equilibrium strategy combination. Two interpretations of a Nash equilibrium were presented indicating different approaches for the selection among multiple equilibria.

An additional criterion, robustness to small errors of the players, was also discussed. The notion of a perfect equilibrium tries to capture this idea. In some games one can reduce the set of Nash equilibria by appealing to this principle. However, there are many games with many perfect equilibria.

The interpretation of Nash equilibrium as well as the search for reasonable selection criteria among a multiplicity of equilibria is a field of intensive research. The list of equilibrium concepts discussed in this chapter is, therefore, far from complete.[11]

4.6 Remarks on the Literature

The Nash equilibrium concept was proposed in Nash (1951) and has been applied extensively. The well-known results on existence and uniqueness of a Nash equilibrium can be found in Friedman (1986). Dasgupta and Maskin (1986) analyze a special class of games that includes several well-known economic examples in which existence of a Nash equilibrium can be established under even weaker conditions.

The interest in conceptual problems with the Nash equilibrium approach is more recent. Kreps (1990b) provides an excellent assessment of strengths and weaknesses of the Nash equilibrium concept and discusses many of the most recent developments in research on this issue. The concept of a perfect equilibrium was introduced by Selten (1975). Van Damme (1991, Chapter 2) provides a comprehensive discussion of perfect equilibrium in context with other related concepts.

Exercises

4.1 *Consider two firms competing in the same market offering the quantities* x_1 *and* x_2 *to consumers. Consumers' demand can be described by the*

[11]Further concepts are proper equilibrium, robust equilibrium, persistent equilibrium, etcetera. The interested reader is referred to van Damme (1991, Chapter 2) where most of these concepts are defined and discussed in detail.

following demand function:

$$X(p) = \max\{0, 10 - p\}.$$

Both firms are assumed to have the same technology represented by the following cost function $(i = 1, 2)$:

$$c(x_i) = \begin{cases} 2 \cdot x_i + k & \text{for} \quad x_i > 0 \\ 0 & \text{for} \quad x_i = 0 \end{cases},$$

where $k > 0$ *denotes some fixed cost of production.*

(a) Analyze the profit maximization problem of firm 1 given the supply of firm 2. Derive the best response mapping of this firm and draw a diagram of it. Is the best response mapping a function?

(b) Derive the Nash equilibria of this duopoly market and draw a diagram of it. Determine the profit levels of the two firms in equilibrium.

(c) Draw a diagram showing the equilibrium quantities of firm 1 for various levels of the fixed cost k. Draw a diagram showing the equilibrium profits of firm 1 for different values of k.

4.2 *Two firms supply the same commodity in quantities* x_1 *and* x_2 *respectively. The inverse demand function for this product has been estimated as* $p(x) = \max\{0, 1 - x\}$. *Each firm has a constant returns to scale technology reflected by the cost function* $c_i(x_i) = .5 \cdot x_i$ *and faces a capacity constraint of* .5.

(a) Derive the symmetric Nash equilibria of this duopoly. Is this equilibrium unique?

(b) How does the analysis under (a) change if the cost functions take the following form:

$$c_i(x_i) = .5 \cdot x_i - .75 \cdot (x_i)^2.$$

4.3 *Consider two firms supplying slightly different products produced with identical constant marginal costs of* 10. *The demand functions of these two firms have been estimated as*

$$d_1(p_1, p_2) = \max\{0, 100 + 10 \cdot (p_2)^2 - 5 \cdot p_1 \cdot p_2\},$$

$$d_2(p_1, p_2) = \max\{0, 30 + 20 \cdot (p_1)^2 - .5 \cdot p_1 \cdot p_2\},$$

where p_1 *and* p_2 *denote the prices of the two firms.*

(a) Write down the normal form of this game, if both firms are supposed to maximize profits.

(b) Draw a diagram with the reaction functions of the two firms.
(c) Check whether, in this game, the sufficient conditions for existence of a Nash equilibrium are satisfied.

4.4 *Consider an economy with one public good and one private good. There are n identical consumers with preferences represented by the utility function* $v(q, x_i) = q \cdot x_i$, *where* q *denotes the quantity of the public good and* x_i *the quantity of the private good consumed by individual* i. *Each consumer has the same quantity of the private good,* $x^o > 0$. *The public good can be produced from the private good according to the production function* $q = \alpha \cdot x$.
(a) Suppose that each consumer has to decide how much of the private good to contribute to production of the public good. Formalize this situation as a game in strategic form and derive the symmetric Nash equilibrium allocation.
(b) Derive the symmetric Pareto optimal allocation and compare it to the Nash equilibrium allocation derived under (a).

4.5 *(Bertrand equilibrium) Consider a duopoly in which both firms sell perfect substitute products, and have constant average costs of 2 per unit. Both firms simultaneously set prices and the firm with the lowest price gets the whole market. If both charge the same price they share the market equally. Market demand is given by the following function:* $D(p) = \max\{0, 15 - p\}$. *Assume that firms can charge full dollar prices only, that is,* p *has to be a natural number.*
(a) Derive the price and quantity in this market for the case in which one firm is a monopolist and for the case in which the firms behave as price takers.
(b) Assume that no firm will ever charge a price above the monopoly price or below the competitive (price-taking) price. Derive the payoff matrix for this game.
(c) Show that there are two Nash equilibria in pure strategies and show that one of them is a perfect equilibrium.
(d) The conventional Bertrand analysis shows that both firms charge a price equal to their marginal costs. Compare the equilibria obtained in this game with the traditional result and explain the differences.

4.6 *Consider 3 potential buyers at a house auction. The auction proceeds according to the principles of a first-price open auction, that is, an auctioneer raises the price after each bid until no further bids occur. The house goes to the last bidder at the price he accepted. Assume that the following order obtains for the buyers' maximum willingness to pay:* $a_1 > a_2 > a_3$.
(a) Write down the strategy sets and payoff functions of the players.

(b) Show that there is a Nash equilibrium at which player 1 gets the house at a price of a_2.

(c) Is the Nash equilibrium under (b) unique? Is the Nash-equilibrium payoff under (b) unique?

Hint: Notice that this is a game with perfect information in which each player has just one decision to make, namely when to stop bidding.

4.7 *Consider two firms that produce the same output* x. *Firm* 1 *has constant unit costs of production* c_1 *and firm* 2 *has constant unit costs of production* c_2. *The inverse demand function for the market is given as* $p(x) = max\{0, A - \frac{1}{3} \cdot x\}$.

(a) Show that the Cournot equilibrium for a duopoly is a Nash equilibrium. Draw a diagram with the reaction function.

(b) Derive the outputs of the two firms, the market price, and the profits of the two firms in equilibrium.

(c) Show that the reaction functions are contraction mappings.

5

Games with Incomplete Information

The analysis of games has been conducted so far under the assumption that the complete description of the game, its extensive or strategic form, is common knowledge. Games for which such an assumption is appropriate are called games of complete information.

In games with complete information, all players are supposed to know in particular the exact payoffs that opponents can obtain. Information about the opponents' payoffs is not important for determining maximin strategies or dominant strategies; it is crucial, however, for predicting which strategies are best responses of other players. Thus, complete payoff information is needed to determine a Nash equilibrium.

The assumption that every player knows the other players' payoff functions is demanding. Even if, in some applications, monetary payoffs can be assigned to players, for example, profits from certain actions, it is necessary to know the expected utility each player obtains from these monetary rewards to determine their best responses. Expected utilities, however, capture such unobservable individual characteristics as attitudes towards risks. It would be a severe limitation for the game-theoretic approach to the analysis of social interactions if games for which this information is unavailable could not be analyzed at all.

In this chapter, a first approach to the analysis of games with incomplete information will be presented. This approach considers games in strategic form only but will lead to an interpretation of incompleteness of information that carries over to games in extensive form. To simplify analysis it will be assumed without loss of generality[1] that agents are incompletely informed about their opponents' payoffs but fully informed about the strategy sets of all players.

The following two examples illustrate the type of problem that arises from incomplete information about payoffs.

EXAMPLE 5.1. Consider a potential entrant to a monopolist's market. Without entry, the monopolist earns three units of profit and the entrant no profit. Should the potential entrant enter the market of the monopolist, two reactions of the incumbent have to be considered:

- either the monopolist will accommodate the entrant, in which case its profit will be reduced to one unit and the entrant will also make one unit of profit
- or it may fight entry, which will cause the entrant a loss of one unit.

The crucial question for the entrant concerns the likelihood of the monopolist fighting entry. This will depend on the outcome for the monopolist of fighting with the entrant. If the entrant knew, for example, that the monopolist would suffer heavy losses from fighting, it would be reasonable to enter the market trusting that the monopolist would not fight. Alternatively, if the monopolist were to suffer little or no loss from fighting, entry would probably be met by a fight, and consequently, the entrant would find it optimal to stay out of the monopolist's market.

The following game in strategic form captures this situation.

		Monopolist	
		a	f
Entrant	e	$1, 1$	$-1, k$
	ne	$0, 3$	$0, 3$

Here, e means the decision to enter and ne the decision not to enter; a denotes the decision of the monopolist to accommodate the entrant and f the decision to fight. The letter k represents the crucial payoff parameter about which the entrant is not informed. Without knowing whether k is larger or smaller than one it is impossible for the entrant to predict the monopolist's reaction to entry. ■

A second example illustrates that incomplete information need not be one-sided as it is in example 5.1.

[1]Harsanyi (1967) has shown how one can transform a game with incomplete information about strategy sets into a game with incomplete information about payoffs by expanding the strategy spaces of the players appropriately.

EXAMPLE 5.2 (*example* 3.4 *modified*). Once again, consider the price-setting duopoly discussed in example 3.4. Both firms, $I = \{1, 2\}$, have no production costs, but face the following demand functions for their goods:

$$d_1(p_1, p_2) = \max\left\{0, a + .5 \cdot \frac{p_2}{p_1} - p_1\right\},$$

$$d_2(p_1, p_2) = \max\{0, b + .5 \cdot p_1 - p_2\}.$$

Prices are denoted by p_1 and p_2 respectively, and a and b are positive demand parameters unknown to the respective competitor.

The two firms compete by setting prices and the strategy sets are therefore $S_1 = S_2 = \mathbb{R}_+$, the set of non-negative real numbers. Since firm 1 does not know the demand parameter b and firm 2 does not know the demand parameter a, neither can predict the profit of its competitor:

$$\pi_1(p_1, p_2) = \left(a + .5 \cdot \frac{p_2}{p_1} - p_1\right) \cdot p_1,$$

$$\pi_2(p_1, p_2) = (b + .5 \cdot p_1 - p_2) \cdot p_2.$$

In contrast to example 5.1, however, both firms are incompletely informed about the other firm's payoff. ■

Examples 5.1 and 5.2 illustrate the problem arising from incomplete information about opponents' payoffs. Since players can no longer predict what would be a best response for the other players, they cannot determine what constitutes optimal behavior for themselves. In a seminal series of articles, Harsanyi (1967) suggested a method for transforming games of incomplete information into games of imperfect information for which best responses and equilibrium behavior are well defined.

The basic idea is simple. A player with incomplete information about some other player's payoff will be treated as if she were uncertain of the type of player she will face. For example, in example 5.1, the entrant may approach the situation by assuming there are various possible types of monopolists who may oppose her, each type being characterized by a different parameter value of k. Thus, the entrant can be viewed as uncertain regarding the type of the incumbent firm. If one also assumes that there is an artificial player, called *nature*, that chooses according to some probability distribution the particular type of monopolist that will play the game, then the entrant and the monopolist face the familiar environment of a game with imperfect information. The entrant cannot observe the move of nature. Thus, incompleteness of information about payoffs is transformed into uncertainty about the move of nature. The probability distribution over types that nature uses is the unique mixed strategy of this player. Thus, no payoff function has to be introduced for nature.

EXAMPLE 5.1 (*resumed*). Suppose there are two possible types of monopolists, identified by $k_1 = -1$ and $k_2 = 2$. Type k_1 represents a monopolist with heavy losses from fighting, whereas type k_2 is a monopolist facing no serious cost from fighting an entrant. In addition, nature chooses with probability p type k_1 and with probability $(1-p)$ type k_2. Thus, nature decides which game is actually played.

		Monopolist	
		a	f
Entrant	e	1, 1	-1, k_1
	ne	0, 3	0, 3

with probability p

		Monopolist	
		a	f
Entrant	e	1, 1	-1, k_2
	ne	0, 3	0, 3

with probability $(1-p)$

■

It is an important assumption of this approach that players are informed about the possible types of all other players and about the associated probability distribution. In fact, this distribution and the set of all possible types must be common knowledge. On the other hand, the assumption that all players hold identical initial beliefs about possible types can be relaxed at the cost of a more complex notation.[2]

Many applications require only a limited degree of diversity in terms of types to gain insights into the effect that incomplete information has on the outcome of the game. For instance, in example 5.1, one could have introduced many more types than just k_1 and k_2, but it is not difficult to see that, from a strategic point of view, all that matters is whether a monopolist is of a type with a payoff k larger than 1 or less than 1.

In games with incomplete information, the type combination t that nature chooses determines a player's payoff. Payoff functions do not depend on the strategy combination s alone. Hence, for any player $i \in I$, $p_i(s, t)$ is the payoff of player i if strategy combination s has been chosen by the players and type combination t has been chosen by nature. Note that player i's payoff may depend not only on her own type t_i but also on the type of other players t_{-i}. Most applications in economics, however, deal with the special case in which player i's payoff is independent from t_{-i}, the types chosen by nature for the other players, as in

[2]It is possible to assume that players differ in their initial beliefs about the probability of types. With such an assumption, however, one can explain any equilibrium behavior by differences in prior beliefs. Harsanyi (1967, Part III) provides an extensive discussion of this issue.

example 5.1 in which the monopolist's payoff was a function of this player's type, k_1 or k_2, only.

Denote by T_i, $i \in I$, the set of possible types for player i, and by μ the probability distribution on the set $T = T_1 \times T_2 \times \cdots \times T_I$ of type combinations. In principle, type sets may be finite or infinite. There are, however, some conceptual and technical problems if infinite type sets are considered so this case will be treated separately in section 5.3. For the exposition of concepts in section 5.1 and 5.2, it will be assumed that type sets T_i are finite. If all T_i are finite, then T is a finite set and it is possible to write $\mu(t)$ for the probability that type combination $t = (t_1, t_2, \ldots, t_I)$ will be chosen by nature.

Besides player set I, strategy sets S_i and payoff functions $p_i(s,t)$, the description of a game with incomplete information must include type sets for each player T_i and a probability distribution on type combinations μ:

$$\Gamma = (I, (S_i)_{i \in I}, (p_i(s,t))_{i \in I}, (T_i)_{i \in I}, \mu).$$

Example 5.2 illustrates these concepts.

EXAMPLE 5.2 (*resumed*). Suppose that firm 1 can be one of two types, having either a high demand parameter a_H or a low demand parameter a_L. Similarly, b_H and b_L will be the types (demand parameters) of firm 2. The distribution μ that determines the probability that a particular type-pair will be selected by nature is given as follows:

$$\mu(a_H, b_H) = .5, \qquad \mu(a_H, b_L) = \mu(a_L, b_H) = .125, \qquad \mu(a_L, b_L) = .25.$$

Hence, $T_1 = \{a_H, a_L\}$ and $T_2 = \{b_H, b_L\}$ are the type sets in this example. With strategy sets $S_1 = S_2 = \mathbb{R}_+$ and profit functions

$$\pi_1(s,t) = \left(a_i + .5 \cdot \frac{p_2}{p_1} - p_1\right) \cdot p_1,$$

$$\pi_2(s,t) = (b_j + .5 \cdot p_1 - p_2) \cdot p_2,$$

for any strategy combination $s = (p_1, p_2) \in S_1 \times S_2$ and any type combination $t = (a_i, b_j) \in T_1 \times T_2$, one can write this game as

$$\Gamma = (\{1,2\}, (S_1, S_2), (\pi_1(s,t), \pi_2(s,t)), (T_1, T_2), \mu). \qquad \blacksquare$$

A useful way to think about a game of incomplete information is as a two-stage procedure:

(i) At the beginning of the game, before players make a decision, nature chooses a particular player-type combination and each player learns her own type, but not the types of the other players.
(ii) Then players make a choice of strategy knowing their own type and the initial type distribution.

Depending on the character of the initial type distribution μ, a player may learn something about the other players' types from the information she receives about her own type. Thus, players may be able to update their beliefs in regard to the other players' types in the light of their own type assignment.

This learning or updating follows a procedure known as *Bayesian learning* (or *Bayesian updating*). In fact, the concept of having nature make a random choice first, then having the decision maker receive information about nature's choice and update the initial beliefs is derived from statistical decision theory. For this reason, it is useful to consider briefly Bayesian decision theory before returning to the discussion of games of incomplete information.

5.1 Bayesian Decision Theory

Bayesian decision theory is concerned with the question of how a decision maker should choose a particular action from a set of possible choices A if the outcome of the choice also depends on some unknown state of the world ω. Consider therefore a finite set of possible states of the world Ω, and let $u(a, \omega)$, u: $A \times \Omega \to \mathbb{R}$, be the expected utility of the decision maker if action $a \in A$ is chosen and state $\omega \in \Omega$ occurs.

Assume also that, before choosing a, the decision maker receives some information w correlated with the state ω. Let W be a finite set of signals and denote by ϑ the joint probability distribution on $\Omega \times W$. Before receiving the signal, the probability of a state ω will be given by $\nu(\omega) \equiv \Sigma_{w \in W} \vartheta(\omega, w)$, the marginal distribution of ϑ on Ω. Given the marginal distribution ν on Ω, Bayesian decision theory recommends choosing an action $a \in A$ that maximizes expected utility, $\Sigma_{\omega \in \Omega} u(a, \omega) \cdot \nu(\omega)$.

Suppose the decision maker learns that a particular value of the signal w' has been observed, how will this affect the optimal decision? Given her knowledge of the joint probability distribution ϑ, the information that some signal $w' \in W$ has occurred allows the decision maker to update her beliefs about the likelihood of the states in Ω. In particular, one can derive the probability $\nu'(\omega|w')$ of any state ω conditional on the observed signal w' as

$$\nu'(\omega|w') \equiv \vartheta(\omega, w') \Big/ \Big(\sum_{\omega \in \Omega} \vartheta(\omega, w') \Big). \qquad (*)$$

The difference between the conditional probability ν' and the marginal distribution ν reflects the informational content of the signal. Note that

$\nu' = \nu$ holds if signals and states are independent, that is, if $\vartheta(\omega, w) = f(\omega) \cdot g(w)$ holds for some probability distributions f on Ω and g on W. To check this claim, just substitute $f(\omega) \cdot g(w)$ into the definition of ν' given in equation $(*)$.

Bayesian decision theory assumes that the updated (conditional) distribution ν' (rather than the unconditional distribution ν) will be used to evaluate the expected utility of the choice of action a:

$$\sum_{\omega \in \Omega} u(a, \omega) \cdot v'(\omega | w). \qquad (**)$$

It is clear that the optimal choice of action obtained from the maximization of the expected utility $(**)$ will depend on the signal w that the decision maker observes. In general, for different signals $w, w' \in W$, the decision maker will choose different actions $a(w)$ and $a(w')$. Hence, one can call the function $a(\cdot), a: W \to A$, which indicates the optimal action choice for each signal $w \in W$ that may be observed, a *Bayesian decision function*. One can call this Bayesian decision function a *signal-contingent* plan of action.

EXAMPLE 5.3. Suppose an oil extraction company has to decide whether to buy the concession to drill for oil in a certain field. If the oil content of the field turns out to be high, a profit rate of 10% on the required capital investment can be achieved, whereas for a low oil content a return of 2% can be expected. The opportunity cost of the firm is 7% return on an alternative investment project.

To obtain additional information, exploratory geological studies of the potential field were undertaken, which showed a high slate content. Past experience indicates that the probability of finding a high slate content together with a high oil content is 67%, and the probability of each other slate-oil content combination equals 11%.

This is a classic example for the application of Bayesian decision theory. The set of states Ω has just two elements (ω_H, ω_L), namely a high or a low oil content. Suppose also that one can distinguish only two levels of slate content. Hence, the set of possible signals $W = \{w_H, w_L\}$ contains two elements, a high or a low level of slate. The joint probability distribution ϑ has the following form:

$$\vartheta(\omega_H, w_H) = .67 \quad \text{and} \quad \vartheta(\omega_H, w_L) = \vartheta(\omega_L, w_H) = \vartheta(\omega_L, w_L) = .11.$$

If the firm tries to maximize the expected return rate by choosing either to buy the oil concession, $a = 1$, or not to do so, $a = 0$, then one obtains the following expected return situation (in percentages).[3]

		signal	
		w_H	w_L
action	1	8.9	6.0
	0	7.0	7.0

The firm will bid for the oil field in this example, but would not have done so if the exploration had indicated a low slate level. ■

To summarize this section, Bayesian decision theory suggests that a rule be chosen that describes an optimal action conditional on the signal that the decision maker receives. Thus, the result of a Bayesian decision problem is a *decision function* and not a particular action. The decision function indicates the optimal action conditional on the information the agent receives.

The relationship between Bayesian decision theory and the incomplete information approach is now easy to see. The set of type combinations for the other players T_{-i} takes the role of the set of states of the world Ω, and the set of a player's own types T_i corresponds to the set of signals W. The probability distribution μ according to which nature chooses a type combination $t=(t_i, t_{-i})$ forms the joint probability distribution on signals and states $T_i \times T_{-i}$. Every player $i \in I$ can use the information on her own type t_i as a signal to update the initial probability distribution μ, before choosing a strategy s_i. Thus, each player chooses a type-contingent strategy $s_i(t_i)$, a Bayesian decision function, instead of a single strategy. Examples 5.1 and 5.2 illustrate this point.

Example 5.1 (*resumed*). Let T_E and T_M be the sets of types for the entrant and the monopolist respectively. Since there is complete information about all characteristics of the entrant, just one type of entrant, say t, needs to be considered, that is, T_E is a one-element set. The set of types for the monopolist T_M, however, has two elements k_1 and k_2. In this case, the joint probability distribution μ on $T_E \times T_M = \{(t, k_1), (t, k_2)\}$ takes the following form:

$$\mu(t, k_1) = p \quad \text{and} \quad \mu(t, k_2) = 1 - p.$$

Clearly, no player can learn anything from the revelation of her type in this example. For the entrant, learning the type t leads to the following

[3]The expected return rates in the table can be calculated as follows: $\nu'(\omega_H | w_i) \cdot 10 + \nu'(\omega_L | w_i) \cdot 2$ for $a = 1$, and $\nu'(\omega_H | w_i) \cdot 7 + \nu'(\omega_L | w_i) \cdot 7$ for $a = 0$, where $i = H, L$ holds.

conditional distribution on T_M:

$$\mu'_E(k_1|t) \equiv \mu(t, k_1)/(\mu(t, k_1) + \mu(t, k_2)) = p$$
$$\mu'_E(k_2|t) \equiv \mu(t, k_2)/(\mu(t, k_1) + \mu(t, k_2)) = 1 - p.$$

Hence, the entrant learns nothing about the monopolist's type by observing her own type. Since the type of the entrant t is common knowledge, there is nothing to be learned for the monopolist of any type either,

$$\mu'_M(t|k_1) \equiv \mu(t, k_1)/\mu(t, k_1) = 1$$
$$\mu^*_M(t|k_2) \equiv \mu(t, k_2)/\mu(t, k_2) = 1. \quad \blacksquare$$

In example 5.1, learning her type did not convey any information to the player. The following example shows that this is not necessarily so.

EXAMPLE 5.2 (*resumed*). Here $T_1 = \{a_H, a_L\}$ and $T_2 = \{b_H, b_L\}$ are the type sets of the two firms. Using the updating rule $(*)$, it is easy to see that, for firm 1, learning the type a_k, $k = H, L$, provides some information regarding the type of the opponent:

$$\mu'_1(b_H|a_H) \equiv \mu(a_H, b_H)/(\mu(a_H, b_H) + \mu(a_H, b_L)) = .5/(.5 + .125) = 4/5,$$
$$\mu'_1(b_L|a_H) \equiv \mu(a_L, b_H)/(\mu(a_H, b_H) + \mu(a_H, b_L)) = .125/(.5 + .125) = 1/5,$$
$$\mu'_1(b_H|a_L) \equiv \mu(a_H, b_L)/(\mu(a_L, b_H) + \mu(a_L, b_L)) = .125/(.125 + .25) = 1/3,$$
$$\mu'_1(b_L|a_L) \equiv \mu(a_L, b_L)/(\mu(a_L, b_H) + \mu(a_L, b_L)) = .25/(.125 + .25) = 2/3.$$

Similarly, for firm 2 learning that its type is b_k, $k = H, L$, leads to an adjustment of her beliefs in regard to the opponent's type a_k, $k = H, L$:

$$\mu'_2(a_H|b_H) \equiv \mu(a_H, b_H)/(\mu(a_H, b_H) + \mu(a_L, b_H)) = .5/(.5 + .125) = 4/5,$$
$$\mu'_2(a_L|b_H) \equiv \mu(a_L, b_H)/(\mu(a_H, b_H) + \mu(a_L, b_H)) = .125/(.5 + .125) = 1/5,$$
$$\mu'_2(a_H|b_L) \equiv \mu(a_H, b_L)/(\mu(a_H, b_L) + \mu(a_L, b_L)) = .125/(.125 + .25) = 1/3,$$
$$\mu'_2(a_L|b_L) \equiv \mu(a_L, b_L)/(\mu(a_H, b_L) + \mu(a_L, b_L)) = .25/(.125 + .25) = 2/3.$$

In this example both players gain information from observing their own type. Note that without the information about its own type, firm 1 would have assumed that the probability of a high demand parameter for firm 2, b_H, is simply $\mu(a_H, b_H) + \mu(a_L, b_H) = .625$, and for b_L, $\mu(a_H, b_L) + \mu(a_L, b_L) = .375$. Hence, learning that its demand parameter is a_H makes firm 1 increase her belief that the other firm has a high demand parameter, whereas observation of a_L leads the firm to revise these beliefs downward. A symmetric argument applies for firm 2. $\blacksquare$

5.2 Bayes–Nash Equilibrium

Once a player has received information about her own type, she chooses a particular strategy to maximize her expected payoff. Of course, the expected payoff from any of her strategies depends on the strategies chosen by her opponents, which in turn, will vary with the types of these players. Though a player may not know the type of her opponents, she can derive an updated probability assessment of the type combination of opponents, based on the initial probability distribution over types and the information about her own type.

A player's strategy choice depends on her type, since it determines her payoff function and her information about the opponents' payoff functions. Thus, she chooses a type-contingent strategy profile. This type-contingent strategy profile is the Bayesian decision function of this player. Since every player chooses a decision function that specifies her choice for every type she may become, one must apply the Nash equilibrium concept to these decision functions rather than to a single strategy combination: every player chooses a type-contingent best response to the decision functions of the other players.

Denote by $s_i(\cdot)$ a decision function of player $i \in I$ that, for each type $t_i \in T_i$, specifies the strategy $s_i(t_i) \in S_i$ this player will choose if her type turns out to be t_i. In addition, let $\mu'_i(t_{-i}|t_i)$ be the updated probability of a particular type combination for the opponents t_{-i}, given that player i has type t_i. For each type profile $(t_1, t_2, \ldots, t_I)$ that nature selects, there are updated beliefs for each player, that is, a list of conditional probability distributions $(\mu'_1(t_{-1}|t_1), \ldots, \mu'_I(t_{-I}|t_I))$. Beliefs of agents are, therefore, no longer identical after they have received the information of their type.

Finally, given $s_{-i}(\cdot) = (s_1(\cdot), \ldots, s_{i-1}(\cdot), s_{i+1}(\cdot), \ldots, s_I(\cdot))$, a list of decision functions for all players other than player i, and $t_{-i} = (t_1, \ldots, t_{i-1}, t_{i+1}, \ldots, t_I)$, a type combination for the other players, $s_{-i}(t_{-i})$ denotes the strategy combination of all players except player i that will be played according to the decision functions $s_{-i}(\cdot)$ if type combination t_{-i} occurs, that is,

$$s_{-i}(t_{-i}) = (s_1(t_1), \ldots, s_{i-1}(t_{i-1}), s_{i+1}(t_{i+1}), \ldots, s_I(t_I)).$$

With this notation, one can define a Bayes–Nash equilibrium as follows:

Definition 5.1 (*Bayes–Nash equilibrium*). A *Bayes–Nash equilibrium* is a list of decision functions $(s_1^*(\cdot), \ldots, s_I^*(\cdot))$ such that for all players $i \in I$ and all types $t_i \in T_i$,

$$\sum_{t_{-i} \in T_{-i}} p_i(s_i^*(t_i), s_{-i}^*(t_{-i}), t_i, t_{-i}) \cdot \mu_i(t_{-i}|t_i)$$
$$\geq \sum_{t_{-i} \in T_{-i}} p_i(s_i, s_{-i}^*(t_{-i}), t_i, t_{-i}) \cdot \mu_i(t_{-i}|t_i)$$

for all strategies $s_i \in S_i$ holds.

In a Bayes–Nash equilibrium, the decision function $s_i^*(\cdot)$ associates with each type that player i may become, a strategy that maximizes the expected payoff of this player provided:

(i) The other players use their decision functions $s_{-i}^*(\cdot)$.
(ii) The conditional probability functions $\mu_i(t_{-i}| \cdot)$ reflect the information of player i.

The following examples illustrate this notation and the Bayes–Nash concept.

EXAMPLE 5.1 (*resumed*). Recall the payoff structure of this example and that $k_1 = 2$ and $k_2 = -1$ has been assumed.

		Monopolist	
		a	f
Entrant	e	1, 1	$-1, k_1$
	ne	0, 3	0, 3

with probability p

		Monopolist	
		a	f
Entrant	e	1, 1	$-1, k_2$
	ne	0, 3	0, 3

with probability $(1-p)$

There is just one type of entrant, t, in this example. A type-contingent strategy for this player is, therefore, simple a strategy choice $s_E(t)$. This strategy must maximize the entrant's expected payoff given a type-contingent strategy of the monopolist $(s_M(k_1), s_M(k_2))$ and the updated probability distribution $(\mu_E(k_1|t), \mu_E(k_2|t))$. The expected payoff of the entrant is

$$p_E(s_E(t), s_M(k_1), t, k_1) \cdot \mu_E(k_1|t) + p_E(s_E(t), s_M(k_2), t, k_2) \cdot \mu_E(k_2|t)$$

where $p_E(s_E(t), s_M(k_i), t, k_i)$, $i = 1, 2$, denotes the payoff to the entrant given that (t, k_i) is the type-pair selected by nature and given that $s_M(k_i)$ is the strategy chosen by the monopolist of type k_i according to its decision function $s_M(\cdot)$.

In maximizing this expected payoff function, the entrant chooses a best response to the monopolist's type-contingent strategy $s_M(\cdot)$. The following table shows the expected payoff of the entrant for any of the four possible decision functions of the monopolist and any strategy of the entrant. Recall that $(k_1, k_2) = (2, -1)$, and $(\mu_E(k_1|t), \mu_E(k_2|t)) = (p, 1-p)$ holds.

		$s_E(t)$: e	ne
$(s_M(k_1), s_M(k_2))$	(a, a)	1	0
	(a, f)	$2 \cdot p - 1$	0
	(f, a)	$1 - 2 \cdot p$	0
	(f, f)	-1	0

It is easy to see that the best response of the entrant depends on whether p is greater or smaller than .5. If both types of the monopolist choose the same strategy, (a, a) or (f, f), then the beliefs of the entrant about the monopolist type do not matter. In this case, the best strategy of the entrant is to enter in response to (a, a) and to stay out against (f, f). On the other hand, if the probability of a monopolist of type k_1 is low enough, $p \leq .5$, then the entrant will find it optimal to enter the market in response to (f, a) and to stay out against (a, f). For $p \geq .5$, these best replies are reversed.

Turning to the monopolist, note that there is no uncertainty about the type of the entrant. For this reason, the expected payoff for each type of monopolist with respect to a type-contingent strategy of the entrant $s_E(\cdot)$ can be read off the payoff matrices directly.

$s_E(t)$ \ $s_M(k_1)$	a	f
e	1	2
ne	3	3

$s_E(t)$ \ $s_M(k_2)$	a	f
e	1	-1
ne	3	3

A monopolist of type k_1 has a dominant strategy to fight, whereas a monopolist of type k_2 has a dominant strategy to accommodate. The fact that both types of monopolist have a dominant strategy greatly simplifies the computation of the Bayes–Nash equilibrium. One checks easily that

$$(s_E^*(\cdot), s_M^*(\cdot)) = ((ne), (f, a)) \quad \text{for} \quad p \geq .5$$
$$(s_E^*(\cdot), s_M^*(\cdot)) = ((e), (f, a)) \quad \text{for} \quad p \leq .5$$

is the unique Bayes–Nash equilibrium in this game. Thus, the monopolist of type k_1 will always fight when the entrant enters and the monopolist of type k_2 will always accommodate if the entrant enters. As a best response, the entrant does not enter if the probability is high that the monopolist is of type k_1 ($p \geq .5$), but enters if the probability of a monopolist of type k_1 is sufficiently low ($p \leq .5$).

The type-contingent strategy combination $(s_E^*(\cdot), s_M^*(\cdot))$ just given is a Bayes–Nash equilibrium since $s_M^*(\cdot) \equiv (s_M^*(k_1), s_M^*(k_2)) = (f, a)$ is a best response to any type-contingent strategy of the entrant $s_E(\cdot)$, and

$$s_E^*(\cdot) = \begin{cases} e & \text{if} \quad p \leq .5 \\ ne & \text{if} \quad p \geq .5 \end{cases}$$

is a best response for the entrant against $s_M^*(\cdot) = (f, a)$. ■

Even without careful analysis, most economists would intuitively predict that the entrant would enter (or not enter) depending on the probability that she faces a monopolist willing to fight entry. This equilibrium outcome is, however, so predictable only because example 5.1 has two special features, namely one-sidedness of incomplete information and a dominant strategy for each type of the monopolist. On the other hand, example 5.1 shows that the concept of a Bayes–Nash equilibrium complies with intuitive notions about behavior under incomplete information if the structure of the situation is simple enough. Example 5.2 shows how this concept works in a more general context.

EXAMPLE 5.2 (*resumed*). Reconsider the duopolists with unknown demand parameters and, therefore, unknown payoff functions. As explained earlier, this economic situation can be modeled as the following game:

$$\Gamma = (\{1,2\}, (S_1, S_2), (p_1(s,t), p_2(s,t)), (T_1, T_2), \mu),$$

with the set of possible prices as strategy sets, $S_1 = S_2 = \mathbb{R}_+$, and the following payoff functions:

$$\pi_1(p_1, p_2, a_i) = \left(a_i + .5 \cdot \frac{p_2}{p_1} - p_1\right) \cdot p_1,$$

$$\pi_2(p_1, p_2, b_j) = (b_j + .5 \cdot p_1 - p_2) \cdot p_2.$$

The sets of possible types were assumed to be $T_1 = \{a_H, a_L\}$ and $T_2 = \{b_H, b_L\}$, and the initial probability distribution on $T = T_1 \times T_2$, the set of type profiles, was given as

$$\mu(a_H, b_H) = .5, \qquad \mu(a_H, b_L) = \mu(a_L, b_H) = .125, \qquad \mu(a_L, b_L) = .25.$$

In a Bayes–Nash equilibrium, each duopolist is supposed to choose a type-contingent strategy, that is, a decision function $p_1(\cdot) \equiv (p_1(a_H), p_1(a_L))$ and $p_2(\cdot) \equiv (p_2(b_H), p_2(b_L))$ respectively, which is a best response to the opponent's decision function.

Take $p_2(\cdot) \equiv (p_2(b_H), p_2(b_L))$ as given and suppose that firm 1 has just learned that it has a high demand parameter a_H. This leads firm 1 to update its beliefs about the other firm's demand parameter to $\mu_1(b_H|a_H) = 4/5$ and $\mu_1(b_L|a_H) = 1/5$ respectively. Hence, the expected payoff of firm 1 can be written as follows:

$$\begin{aligned}
&\pi_1(p_1, p_2(b_H), a_H) \cdot \mu_1(b_H|a_H) + \pi_1(p_1, p_2(b_L), a_H) \cdot \mu_1(b_L|a_H) \\
&= \left[\left(a_H + .5 \cdot \frac{p_2(b_H)}{p_1} - p_1\right) \cdot p_1\right] \cdot .8 + \left[\left(a_H + .5 \cdot \frac{p_2(b_L)}{p_1} - p_1\right) \cdot p_1\right] \cdot .2 \\
&= a_H \cdot p_1 + [.4 \cdot p_2(b_H) + .1 \cdot p_2(b_L)] - p_1^2.
\end{aligned}$$

Note that this payoff function is continuously differentiable and strictly concave in the strategy of firm 1, p_1. Therefore, any p_1 satisfying the first-order condition for a maximum will be a best response to the type-contingent strategy $p_2(\cdot) \equiv (p_2(b_H), p_2(b_L))$.

Solving the first-order condition for a maximum of the expected payoff function $a_H \cdot p_1 + [.4 \cdot p_2(b_H) + .1 \cdot p_2(b_L)] - p_1^2$, one obtains the following best response $p_1(a_H)$ for a firm of type a_H to the decision function $(p_2(b_H), p_2(b_L))$ of the other firm: $p_1(a_H) = .5 \cdot a_H$.

Similarly, one obtains for the other type of firm 1, $p_1(a_L) = .5 \cdot a_L$. Thus, firm 1 will choose a type-contingent strategy that does not depend on the type-contingent strategy of the other firm.[4]

Now consider that firm 2 learns its type is b_L. It will revise its beliefs about firm 1's type $\mu_2(a_H|b_L) = 1/3$ and $\mu_2(a_L|b_L) = 2/3$. In addition, for a type-contingent strategy of firm 1 $p_1(\cdot) \equiv (p_1(a_H), p_1(a_L))$ the expected payoff of firm 2 will be as follows:

$$\begin{aligned}
&\pi_2(p_1(a_H), p_2, b_L) \cdot \mu_2(a_H|b_L) + \pi_2(p_1(a_L), p_2, b_L) \cdot \mu_2(a_L|b_L) \\
&\quad = [(b_L + .5 \cdot p_1(a_H) - p_2) \cdot p_2] \cdot \tfrac{1}{3} + [(b_L + .5 \cdot p_1(a_L) - p_2) \cdot p_2] \cdot \tfrac{2}{3} \\
&\quad = b_L \cdot p_2 + \left[\tfrac{1}{3} \cdot p_1(a_H) + \tfrac{2}{3} \cdot p_1(a_L)\right] \cdot p_2 - p_2^2.
\end{aligned}$$

Once again, the first order condition gives the best response function for a firm of type b_L:

$$p_2(b_L) = .5 \cdot \left(b_L + \left[\tfrac{1}{6} \cdot p_1(a_H) + \tfrac{1}{3} \cdot p_1(a_L)\right]\right).$$

A similar calculation yields the best response of a firm of type b_H to a type-contingent strategy choice of firm 1:

$$p_2(b_H) = .5 \cdot (b_H + [.4 \cdot p_1(a_H) + .1 \cdot p_1(a_L)]).$$

To find a Bayes–Nash equilibrium, one has to solve the equation system given by the best response functions. With two players and two types for each player, this involves four equations and, even for linear best response functions, working out the solution is a tedious, though straightforward task. In this example, the computation is simplified because firm 1 has a dominant strategy. Substituting the solutions for $p_1(a_H)$ and $p_1(a_L)$ into the best response functions of firm 2 for each type yields the following Bayes–Nash equilibrium pair of type-contingent strategies $(p_1^*(\cdot), p_2^*(\cdot))$:

$$\begin{aligned}
&p_1^*(a_H) = .5 \cdot a_H, \quad p_1^*(a_L) = .5 \cdot a_L, \\
&p_2^*(b_H) = .1 \cdot a_H + .025 \cdot a_L + .5 \cdot b_H, \\
&p_2^*(b_L) = \tfrac{1}{24} \cdot a_H + \tfrac{1}{12} \cdot a_L + .5 \cdot b_L. \quad \blacksquare
\end{aligned}$$

[4]Indeed, as shown in example 3.4, firm 1 has a dominant strategy, that is, a best response independent of the opponent's choice of strategy.

These two examples show how a Bayes–Nash equilibrium can be computed. They also make clear, however, that computations rapidly become more difficult. With just two players, with each player having three different possible types, one has to solve a system of six equations. Thus, the greater the uncertainty about the types that agents have, the more complicated the problem even for a two-player game.

5.3 Bayes–Nash Equilibrium: Two Alternative Views

There are two ways of looking at a Bayes–Nash equilibrium that prove to be useful in economic applications. In one view, a game of incomplete information is simply a game in strategic form with an expanded set of players. This approach makes it particularly easy to show that games of incomplete information have a Bayes–Nash equilibrium under similar conditions to standard games in strategic form.

In the second view, a game with incomplete information is an extensive form game with nature as a player. At the beginning of the game nature makes a move that selects a particular list of types for each player. From this perspective, the type-contingent strategies of a Bayes–Nash equilibrium can be regarded as behavior strategy combinations.

5.3.1 Games of Incomplete Information as Games with Expanded Player Sets

A game with incomplete information is given by a player set, strategy sets, payoff functions and type sets for each player, and a probability distribution for player types: $\Gamma = (I, (S_i)_{i \in I}, (p_i(s,t))_{i \in I}, (T_i)_{i \in I}, \mu)$. In examples 5.1 and 5.2, however, it was possible to derive a Bayes–Nash equilibrium by computing a solution to a set of best response functions the same way one would derive a solution to a standard game in normal form. The only difference was that best responses were given in terms of type-contingent strategies. A best response of a particular type of player had to be a payoff-maximizing strategy for given strategies of each type of the other players.

These examples suggest treating each type of player as a separate agent. In this view, one has a player set $\mathcal{T} = \bigcup_{i \in I} T_i$ comprising all possible types of all players. In addition to this simple expansion of the player set from I to $\mathcal{T}$, it is necessary to associate with each type-player a strategy set and a payoff function in an appropriate way to create a "new" game equivalent to the game of incomplete information given earlier.

Any player $\tau \in \mathcal{T}$ in the "new" game belongs to a particular type set T_i of the original game of incomplete information, that is, $\tau \in T_i$ for some $i \in I$ must hold. Hence, one can define the strategy set S_τ and the payoff function $\mathfrak{p}_\tau(s_\tau, s_{-\tau})$ as follows:

$$S_\tau = S_i \quad \text{for} \quad \tau \in T_i,$$
$$\mathfrak{p}_\tau(s_\tau, s_{-\tau}) = \sum_{t_{-i} \in T_{-i}} p_i(s_\tau, s_{-i}(s_{-\tau}, t_{-i}), \tau, t_{-i}) \cdot \mu_i(t_{-i}|\tau) \quad \text{for} \quad \tau \in T_i,$$

where $s_{-i}(s_{-\tau}, t_{-i})$ denotes the vector of strategies for the other players of types t_{-i} selected from $s_{-\tau}$, that is, $s_j(s_{-\tau}, t_{-i}) = s_{t_j}$ for any $j \neq i$. This procedure yields a game in strategic form

$$\Gamma' = (\mathcal{T}, (S_\tau)_{\tau \in \mathcal{T}}, (\mathfrak{p}_\tau)_{\tau \in \mathcal{T}}),$$

which is, by construction, identical to the game with incomplete information Γ that formed the starting point of the transformation.

Note in particular that, given finite type sets T_i, it is a game with a finite player set. In fact, if n_i is the number of types of player $i \in I$, then the number of players in $\mathcal{T}$ is $\sum_{i \in I} n_i$. Therefore, one can apply theorem 4.1, to obtain the following result.

THEOREM 5.1. *Every game of incomplete information with finite type and strategy sets* $(T_i)_{i \in I}, (S_i)_{i \in I}$ *has a Bayes–Nash equilibrium in mixed strategies.*

Proof. Recall that for finite pure strategy sets S_i, the set of mixed strategies is the simplex of a dimension equal to the number of pure strategies in S_i. Since simplexes are compact and convex sets, and since the expected payoff functions are continuous and linear, hence quasi-concave, in the mixed strategy of each player, all the conditions of theorem 4.1 are satisfied. ■

Note that one of the premises of this theorem is a finite set of types T_i for each player. Since each type-player τ has a finite strategy set S_i and a well-defined payoff function $\mathfrak{p}_\tau(\cdot)$, mixed strategies and payoffs from mixed strategy combinations can be defined as in Chapter 1. For games with infinite type sets however the notion of a mixed strategy needs to be reformulated. Section 5.4 will consider these issues.

5.3.2 Games of Incomplete Information as Extensive Form Games

Games with incomplete information have been discussed so far as games in strategic form only. It is possible, however, to describe a game of incomplete information as a special case of a game in extensive form in which "nature" acts like a player and assigns particular types to agents. In this view, nature is the player that moves first and decides which types of players will actually play the game. The other players move simultaneously at the second stage with imperfect information about the move of nature.

The fact that each player is informed about her own type but remains ignorant of other players' types is reflected in a particular structure of the information sets in the extensive form: each type that a player may have leads exactly to one information set of this player. A player with three types will control three information sets and a player with five types will have five information sets. Since all types of the same player i have the same strategy set S_i, each player has the same set of actions available at each of her information sets. A behavior strategy specifying a random choice of action at each information set of a player can, therefore, be regarded as a type-contingent mixed strategy.

For games with incomplete information that have few types and a small number of pure strategies, as in example 5.1, it is possible to capture the extensive form in a game tree.

EXAMPLE 5.1 (*resumed*). The following game tree describes the game between the entrant e whose only type t is common knowledge and the monopolist m who may be either of type k_1 or of type k_2.

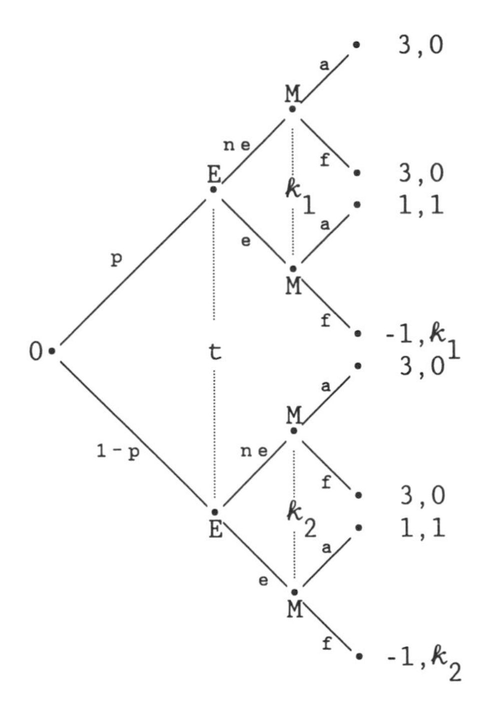

Nature, denoted by "0" in the tree, moves with probability p to the type pair (t, k_1) and with probability $(1-p)$ to the pair (t, k_2). Nature is treated like a player with two exceptions: it has no payoffs associated and chooses a mixed strategy with an exogenously given probability $(p, 1-p)$.

In fact, a payoff assignment to nature is unnecessary since it always plays the same exogenously given strategy.

In any other respect, the description of a game with incomplete information represents an extensive form game as introduced in Chapter 1. In line with the type structure, there are two information sets for the monopolist, k_1 and k_2, but only one for the entrant t. This reflects the information structure of the game:

- A monopolist knows its own type as well as the only possible type of the entrant. Hence, it has two information sets with two nodes, one information set for each type. The two nodes in the information sets indicate that it is a simultaneous move game in which the monopolist does not know the choice of strategy of the entrant. Since there is just one type of entrant, this is the only uncertainty the monopolist faces.
- The entrant has one information set corresponding to her only type. Since the entrant is ignorant of the monopolist's type, the only information set of the entrant contains two nodes that this player cannot distinguish.

A pure strategy in an extensive form game has to specify an action at each information set of a player. The set of possible actions at each information set of a player is identical to the strategy set of this player in the normal form. Hence, a pure strategy of the entrant is simply some $s_E(t) \in S_E$, and a pure strategy for the monopolist is a pair $(s_M(k_1), s_M(k_2))$ with $s_M(k_1), s_M(k_2) \in S_M$. Note that this pure strategy in the extensive form is the same as a type-contingent strategy. ■

This example shows that a game of incomplete information in normal form can be viewed as a special case of an extensive form game. The next chapter will show that incomplete information in extensive form games can be approached in exactly the same way as in games in strategic form. Indeed, the notion of players learning by updating their beliefs in the light of new information as the game proceeds will play a much larger role in the context of extensive form games.

5.4 Infinite Type Sets

So far the exposition of the analysis of games of incomplete information has been restricted to games with finite type sets T_i. For such games, mixed strategies and payoffs from mixed strategy combinations can be defined as in Chapter 1. If one assumes that a player chooses a mixed strategy after observing her type, then a type-contingent mixed strategy is viewed as a behavior strategy, namely a randomized choice of a pure strategy at each information set. On the other hand, a type-contingent strategy $s_i(\cdot)$ itself can be considered as a pure strategy. In this case a mixed strategy is a randomization on the set of all type-contingent strategy combinations, that is on the set S_i^n where n denotes the number of types of player i. Since each player comes to move only once, a game with incomplete information is necessarily of perfect recall. Hence, applying Kuhn's theorem (theorem 1.2), one concludes that both notions of a mixed strategy are in fact equivalent.

With infinite type sets T_i, however, a type-contingent strategy $s_i(\cdot)$ that associates a pure strategy with each type T_i can no longer be viewed as a vector but has to be treated as a general function $s_i: T_i \rightarrow S_i$. A mixed strategy would be a random choice of such a function $s_i(\cdot)$. In general, it is, however, impossible to choose a function randomly. How to define a mixed strategy in such a case is a subtle problem not discussed here. Aumann (1964) and Milgrom and Weber (1985) suggest two definitions of a mixed strategy for the case of infinite type sets. Milgrom and Weber (1985) show in addition that existence of an equilibrium can be guaranteed under rather weak assumptions on the type distribution μ.

Despite the problems with the concept of a mixed strategy, games with infinite type sets are often useful in economic applications. The assumption of a continuous type space in particular is appropriate in many applications for the problem under consideration. In addition, it may even simplify the analysis because equilibria in pure strategies may exist in games with infinite type sets in cases in which the same game with finite type sets would have no equilibrium in pure strategies. For this reason a brief extension of the Bayes–Nash equilibrium concept to games with infinite type sets is provided here. Only pure strategies are considered, however, to avoid the problems mixed strategies pose.

The most common case of continuous type sets in economic applications is a range of parameters. Assume, therefore, that type sets, T_i, are intervals of the real line. The set of type combination $T = T_1 \times \cdots \times T_I$ is the Cartesian product of these intervals. For two players, the set of type combinations $T = T_1 \times T_2$ is a rectangle in the plane.

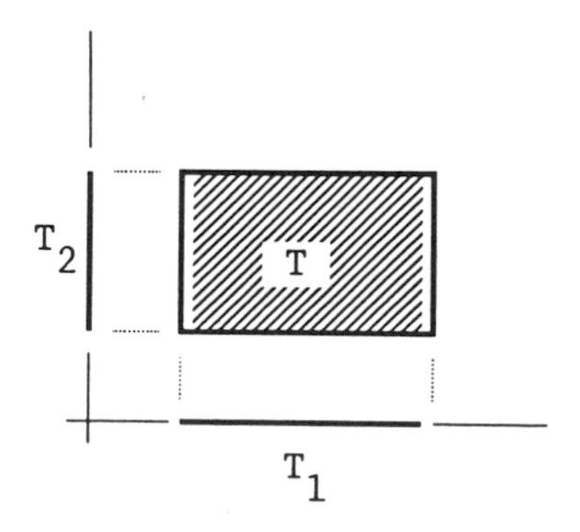

Suppose in addition that the probability distribution of type combinations can be represented by a distribution function $\mu(\cdot)$ with density function $\mu'(\cdot)$. Notice that $\mu(t)$ denotes the probability that a type combination $\tau \in T$ with $\tau \le t$ is chosen, not the probability of the type combination t. Given player i's information about her own type t_i, the conditional distribution function on T_{-i}, the set of other players' types, will be written as $\mu(t_{-i}|t_i)$.

For given type-contingent strategies $s_{-i}(\cdot)$ of the other players, the expected payoff of player i from choosing strategy s_i is obtained by integrating over all possible types of the other players:

$$\int_{T_{-i}} p_i(s_i, s_{-i}(t_{-i}), t_i, t_{-i}) \mu'_i(t_{-i}|t_i)\, dt_{-i}.$$

Substituting this integral for the respective sums, definition 5.1 can be applied. A Bayes–Nash equilibrium is a type-contingent strategy combination $(s_1^*(\cdot), \ldots, s_I^*(\cdot))$ such that, for all players $i \in I$ and all types $t_i \in T_i$,

$$\int_{T_{-i}} p_i(s_i^*(t_i), s_{-i}^*(t_{-i}), t_i, t_{-i}) \cdot \mu'_i(t_{-i}|t_i)\, dt_{-i}$$
$$\ge \int_{T_{-i}} p_i(s_i, s_{-i}^*(t_{-i}), t_i, t_{-i}) \cdot \mu'_i(t_{-i}|t_i)\, dt_{-i}$$

holds for all strategies $s_i \in S_i$. Note, however, that type-contingent strategy combinations are general functions, not vectors, in this case.

The theory of auctions is a field of economics in which the Bayes–Nash equilibrium concept has found important applications. Though it is generally difficult to derive the equilibrium strategies, there are special cases in which they can be determined easily as the following example shows.

Example 5.4 (*first-price sealed-bid auction*). Consider n potential buyers of a building sold according to the rules of a first-price sealed-bid auction. Each bidder submits a bid and the building goes to the highest bidder for the price offered. If the potential buyers knew the willingness to pay of the other bidders, it would be optimal for the player with the highest valuation to bid slightly more than the highest valuation of her competitors.

If the buyers' willingness to pay is private information, however, the problem becomes a game with incomplete information. Assume that the valuation of all players lies in the interval $[0, \bar{a}]$, and that the probability of having a particular valuation a_i is the same for all players. This assumption implies each bidder's valuation of the building is independently drawn from the interval $[0, \bar{a}]$ according to some probability distribution. Finally, assume that this probability distribution is uniform and thus represented by the distribution function $G(a) = a/\bar{a}$ for $a \in [0, \bar{a}]$.

This is a game with incomplete information because the bidders do not know the valuations of their opponents. Players' types are their valuations of the building. The type set for any player i is therefore $T_i = [0, \bar{a}]$, and the set of all type combinations is $T = [0, \bar{a}]^n$. The probability of a type combination below $(a_1, \ldots, a_n)$ is $\mu((a_1, \ldots, a_n)) = G(a_1) \cdot \cdots \cdot G(a_n)$. A pure strategy for player i is a bid $b_i \geq 0$. Each bidder has, therefore, the same strategy set $S_i = \mathbb{R}_+$. The payoff of bidder i from bids $(b_1, b_2, \ldots, b_n)$ is

$$p_i(b_1, \ldots, b_n, a_i, a_{-i}) = \begin{cases} a_i - b_i & \text{for } b_i = \max\{b_1, b_2, \ldots, b_n\} \\ 0 & \text{otherwise.} \end{cases}.$$

Specifying the payoff for the case in which more than one player makes the highest bid is unnecessary since a particular valuation a occurs with probability zero. The game is symmetric, therefore it appears reasonable to look for a symmetric Bayes–Nash equilibrium in which each player has the same type-contingent bidding strategy $b_i^*(\cdot) = b^*(\cdot)$, $i = 1, \ldots, n$.

Begin the analysis with the conjecture that the symmetric equilibrium strategy is proportional to the bidder's valuation a. That means that $b^*(a) = \beta \cdot a$ holds for some number β. This assumption needs verification. Given that all other players follow proportional type-contingent strategies, if the optimal strategy for an arbitrary player i turns out to be proportional, this will confirm the conjecture, otherwise one has to reject it.

Player i wins the auction with bid b_i if all other bidders have valuations a_j, which makes them bid less than b_i, that is if $b_i \geq \beta \cdot a_j$ for all bidders $j \neq i$ holds. Since there are $n - 1$, other players whose valuation is independently drawn according to the probability distribution $G(a) = a/\bar{a}, G(a)^{n-1} = (a/\bar{a})^{n-1}$ is the distribution function for $n - 1$ valuations less than a. The probability of winning the auction with bid b_i is therefore $G(b_i/\beta) = (b_i/(\beta \cdot \bar{a}))^{n-1}$. Notice that the probability of winning the auction rises with higher bids b_i while the net gain of a successful bid, $a_i - b_i$, falls. This is the tradeoff that the bidder has to consider.

Differentiating $G(a)^{n-1}$ with respect to a yields the associated density function $(n - 1) \cdot G'(a)^{n-2} = (n - 1) \cdot (a/\bar{a})^{n-2} \cdot (1/\bar{a})$. The expected payoff of player i, if she has type a_i and bids b_i, and if all other players follow the strategy $\beta \cdot a_j$, can be computed as follows:

$$\begin{aligned} &\int_{T_{-i}} p_i(b_i, b_{-i}^*(a_{-i}), a_i, a_{-i}) \cdot \mu_i'(a_{-i}|a_i)\, da_{-i} \\ &= \int_0^{b_i/\beta} [a_i - b_i] \cdot \left[(n-1) \cdot G'(a)^{n-2}\right] da \\ &\quad + \int_{b_i/\beta}^{\bar{a}} [0] \cdot \left[(n-1) \cdot G'(a)^{n-2}\right] da \\ &= [a_i - b_i] \cdot G(b_i/\beta)^{n-1} = [a_i - b_i] \cdot (b_i/(\beta \cdot \bar{a}))^{n-1}. \end{aligned}$$

If player i with valuation a_i chooses b_i to maximize this payoff, then the first-order condition

$$[a_i - b_i^*] \cdot (n-1) \cdot (b_i^*/(\beta \cdot \bar{a}))^{n-2} \cdot (1/(\beta \cdot \bar{a})) = (b_i^*/(\beta \cdot \bar{a}))^{n-1}$$

must hold for the optimal bid b_i^*. This can be easily solved to yield

$$b_i^* = [(n-1)/n] \cdot a_i,$$

confirming that the optimal bidding strategy is proportional to the bidder's valuation with parameter $\beta = [(n-1)/n]$. The equilibrium type-contingent bidding strategy is therefore

$$b^*(a) = [(n-1)/n] \cdot a.$$

The optimal strategy requires a bid that is a fraction $1/n$ below the valuation of a player. Thus, the more players take part in the auction the smaller the surplus that each bidder can appropriate. As competition increases with the number of bidders, the equilibrium bid approaches the valuation of the player. ■

The derivation of the Bayes–Nash equilibrium strategies in example 5.4 was greatly simplified by the assumption that all valuations were independently drawn from a uniform distribution. This assumption guaranteed symmetric players and the linearity of the equilibrium strategies. Both properties were used to derive the equilibrium. For more general assumptions about type distributions, characterizing the equilibrium strategies is much harder. It is not difficult to see, however, that no equilibrium in pure strategies exists if the type sets are finite (exercise 5.1). The continuity of types provides enough randomization to guarantee existence of an equilibrium, even without mixed strategies. This observation indicates a relationship between type-contingent strategies and mixed strategies that will be explored further in the following section.

5.5 Mixed Strategies as Bayes–Nash Equilibrium Decision Functions

The final section of this chapter will show that the Bayes–Nash equilibrium concept for games with incomplete information can provide an interesting interpretation of mixed strategies. Mixed strategies play an important role in game theory, mainly to guarantee existence of a Nash equilibrium. Yet there is little evidence in any social context that agents would toss a coin to make decisions.

This is not surprising if one considers that, in a Nash equilibrium in mixed strategies, there is no incentive for an agent to choose the particular equilibrium randomization over her strategies. In fact, each player is indifferent about her strategy choice in a mixed strategy equilibrium. The only reason for postulating these mixed strategy combinations lies in the fact that these particular randomizations are mutually consistent.

In informal arguments, therefore, mixed strategies were often considered as an approximate description of observed behavior. According to this view, agents actually choose pure strategies but, in repetitions of the game, randomly matched players alternate their choices so that the relative frequency of the pure strategies played equals the randomizations required for a Nash equilibrium in mixed strategies. Indeed, as this section will show, this intuition can be justified if one considers games with incomplete information.

Consider a Bayes–Nash equilibrium $(s_1^*(\cdot), \ldots, s_I^*(\cdot))$. According to the decision function $s_i^*(\cdot)$, a player $i \in I$ will choose a strategy $s_i^*(t_i)$ with the same probability with which nature will choose its type to be t_i. Nature's choice, however, follows a random draw from the probability distribution on types μ. Formally, let μ_i be the marginal distribution on the type set T_i, that is, $\mu_i(t_i) \equiv \sum_{t_{-i} \in T_{-i}} \mu(t_i, t_{-i})$, and, for any strategy $s_i \in S_i$, denote by

$$T_i(s_i | s_i^*(\cdot)) = \{t_i \in T_i | s_i^*(t_i) = s_i\}$$

the set of types of player i who will play strategy s_i according to the decision function $s_i^*(\cdot)$. Note that the set $T_i(s_i | s_i^*(\cdot))$ may be empty if a particular strategy s_i is not played at all for the given decision function $s_i^*(\cdot)$. Nature's random choice of types makes the actual play of any strategy $s_i \in S_i$ a random choice given the decision function $s_i^*(\cdot)$. Thus, the game proceeds as if player i plays the following mixed strategy:

$$m_i(s_i) = \sum_{t_i \in T_i(s_i | s_i^*(\cdot))} \mu_i(t_i).$$

Consequently, one can view mixed strategies as the result of type-contingent strategies combined with a random choice by nature according to some given probability distribution on types.

There remains, however, the question of whether there is some relationship between the "induced" mixed strategies of a Bayes–Nash equilibrium and the Nash equilibrium in mixed strategies of the actual game. Harsanyi (1973) showed that one can answer affirmatively, at least under certain assumptions. If players' payoffs are randomly disturbed[5] and if players have incomplete information about this disturbance, then a

[5]Some assumptions on the distribution of the disturbances are required. In particular, the support of the disturbances is not allowed to be a discrete set. Van Damme (1991) provides a good exposition of the assumptions needed.

Bayes–Nash equilibrium will approximate a mixed strategy equilibrium of the undisturbed game.

A formal proof of this result requires substantial additional notation and will be omitted here. Instead, an example will illustrate the result and indicate the type of conditions needed to establish it.

EXAMPLE 5.5. Reconsider example 4.9 with the following modification: the cost of providing labor for the public good is $c_i(e_1, e_2) = (e_1 + e_2)/2$. The payoff matrix of the modified game is easily derived and, for easy reference, given below on the left. If players are uncertain about their opponents' exact payoffs, this game becomes one of incomplete information. Assume, for example, that the players are not completely certain of the opponents' payoffs in the case of unsuccessful coordination on the high-level equilibrium. In particular, as shown in the right-hand matrix, the payoffs may be $5+\alpha$, for player 1, and $5+\beta$, for player 2. These payoff parameters α and β are the types of player 1 and player 2 unknown to the opponent. It is common knowledge, however, that α and β are independently and uniformly distributed on the interval $[0, \epsilon]$.

		Player 2	
		10	20
Player 1	10	10, 10	5, 5
	20	5, 5	20, 20

		Player 2	
		10	10
Player 1	10	10, 10	$5, 5+\beta$
	20	$5+\alpha, 5$	20, 20

Note that the original game on the left has three Nash equilibria, two in pure strategies (10, 10) and (20, 20) and one in mixed strategies where each player plays 10 with probability .75,

$$((m_1^*(10), m_1^*(20)), (m_2^*(10), m_2^*(20))) = ((.75, .25), (.75, .25)).$$

To derive a Bayes–Nash equilibrium for the game of incomplete information on the right, one has to work out type-contingent strategies for each player that are best responses to each other given the beliefs after each player learns her type. This derivation is greatly simplified by some special features of this example.

First, prior beliefs coincide with updated beliefs because it is assumed that types are independently distributed. Second, types are only different in regard to one payoff parameter, and the game is symmetric otherwise. Given that each player's type is drawn from the same interval $[0, \epsilon]$, the game is completely symmetric and, therefore, both players will find the same type-contingent strategy optimal.

As explained before, any type-contingent strategy of a player in combination with the distribution on types induces a "mixed strategy" in response to which the other agent will react. Hence, given a type-contingent strategy of player 2 and the induced mixed strategy $(q_2, 1 - q_2)$, player 1 will find it optimal to choose strategy 10 if $q_2 \cdot 10 + (1 - q_2) \cdot 5 > q_2 \cdot (5 + \alpha) + (1 - q_2) \cdot 20$ holds and will choose strategy 20 otherwise. The decision function of player 1 is, therefore,

$$s_1(q_2|\alpha) = \begin{cases} 10 \text{ for } \alpha < 20 - 15/q_2 \\ 20 \text{ for } \alpha > 20 - 15/q_2 \end{cases}.$$

Note that the player would be indifferent between strategy choices if $\alpha = 20 - 15/q_2$ held. With a uniform distribution, however, this is an event with zero probability. Thus, this borderline case needs no special consideration.

Similarly, one obtains for player 2 the following type-contingent best response to a mixed strategy $(q_1, 1 - q_1)$ of player 1:

$$s_2(q_1|\beta) = \begin{cases} 10 \text{ for } \beta < 20 - 15/q_1 \\ 20 \text{ for } \beta > 20 - 15/q_1 \end{cases}.$$

According to the uniform distribution on $[0, \epsilon]$, the probability that nature will choose some type smaller than x is simply x/ϵ. Hence, the mixed strategy of player 1, induced by her type-contingent strategy $s_1(q_2|\alpha)$ and the uniform type distribution, can be calculated as

$$q_1 = \text{Prob}\{\text{player 1 plays 10}\} = \text{Prob}\{\alpha < [20 - 15/q_2]\} = \frac{1}{\epsilon} \cdot [20 - 15/q_2].$$

Similarly, for player 2, one derives

$$q_2 = \frac{1}{\epsilon} \cdot [20 - 15/q_1].$$

A simultaneous solution to these equations must obviously satisfy

$$q_1 = q_2 \equiv q \quad \text{and} \quad q = \frac{1}{\epsilon} \cdot [20 - 15/q].$$

This yields a quadratic equation with two roots, of which only the smaller one makes sense. A straightforward computation shows that

$$q_1 = q_2 = \left[20 - \sqrt{400 - 60 \cdot \epsilon}\right]/(2 \cdot \epsilon)$$

is the solution to the quadratic equation. Substituting this solution into the type-contingent best responses $s_1(q_2|\alpha)$ and $s_2(q_1|\beta)$ yields the

Bayes–Nash equilibrium decision functions ($t_1 = \alpha, t_2 = \beta$):

$$s_i(t_i) = \begin{cases} 10 \text{ for } t_i < 20 - 30 \cdot \epsilon / \left[20 - \sqrt{400 - 60 \cdot \epsilon}\right] \\ 20 \text{ for } t_i > 20 - 30 \cdot \varepsilon / \left[20 - \sqrt{400 - 60 \cdot \epsilon}\right] \end{cases}.$$

Notice that the players play pure type-contingent strategies and do not randomize. Nevertheless, given the randomness of the type distribution, an outside observer, as well as each player, will perceive the other player's behavior as random. Furthermore, as the range of the uncertainty about the opponent's payoff shrinks, the average behavior generated by the Bayes–Nash equilibrium strategies will approximate the mixed strategy of the true game. To see this, note that q_1 and q_2 converge to .75 as ϵ goes to zero. For this result one has to apply l'Hôpital's rule since both the nominator and the denominator go to zero. ■

Example 5.5 shows that one can interpret a mixed strategy as the random behavior arising in games in which the payoffs of players are subject to "small" random deviations and in which players have incomplete information about the precise payoffs of their opponents. Note, however, that players do not actually randomize. In fact, they play well-defined pure strategies for nearly every payoff type they may have. It is the randomness of nature's choice of types that makes it appear they are playing mixed strategies.

5.6 Summary

This chapter investigated how to model social situations in which agents are incompletely informed about characteristics of their opponents. The basic method of dealing with this problem assumes that this ignorance about agent's characteristics can be captured by considering different types of agents together with a probability distribution on the set of possible type combinations. Thus, incomplete information about players' characteristics can be handled analytically if one can assume that the likelihood of different type combinations is exogenously given (by nature) and common knowledge of all players.

The analytic device for transforming a game of incomplete information into one with imperfect information takes the actual type combination as randomly drawn according to the initial probability distribution and assumes that each player learns her own type only. In the light of this information, a player can revise what she knows about the likelihood of the other players' types and try to find an optimal response strategy to the strategy choice of the other player types.

Naturally, strategic choices of agents depend on their types. Hence a player has to consider type-contingent strategies of other players rather than single strategy choices. If these type-contingent strategy choices are consistent, that is, if each type-contingent strategy is optimal given the other players' type-contingent strategies and given the information of the agent, then they form a Bayes–Nash equilibrium.

An appropriate reformulation of games of incomplete information as games with one player for each type shows that, for finite type sets at least, existence of a Bayes–Nash equilibrium is guaranteed in mixed strategies. Furthermore, the type-structure of a game with incomplete information can be captured by the extensive form of such a game, if one treats the random allocation of types as a random move of an initial player, called *nature*. This artificial player chooses a particular type combination with an exogenously determined probability and informs each player about her own type only. The resulting information sets correspond exactly to the type sets in the game. Thus, type-contingent strategies can be viewed as pure strategies in the associated extensive form of the game.

In conclusion, the approach to capture incompleteness of information by a variety of types makes analysis of games with incomplete information possible. On the other hand, however, it shifts the informational problem from knowledge about other players' characteristics to knowledge about the joint probability distribution on type combinations. Thus, it leaves open the question of how agents come to agree in their beliefs about the prior probability distribution on types.

5.7 Remarks on the Literature

The fundamental idea to model incomplete information by specifying types of players and the concept of a Bayes–Nash equilibrium was developed by Harsanyi (1967) in a series of three papers. It has gained widespread acceptance since the late 1970s because of its successful application in the context of extensive form games. Binmore (1992), Friedman (1990), Fudenberg and Tirole (1991), and Myerson (1991) introduce the Bayes–Nash equilibrium concept with different levels of rigor and provide further examples.

The interpretation of types of players as additional agents in a normal form game with an extended player set goes back to Harsanyi's (1967) original article. Milgrom and Weber (1985) show existence of a Bayes–Nash equilibrium in a game with an infinite number of player types. The idea of considering mixed strategies as "average" behavior in incomplete information games was advanced in Harsanyi (1973) and further developed in Aumann (1987). Van Damme (1991) provides a more general proof for this result.

Exercises

5.1 *Reconsider example 5.4 in which a building was sold in a first-price sealed-bid auction. Assume, however, that each bidder's valuation is randomly drawn from the set* $\{a_H, a_L\}$, *with* $a_H > a_L$. *Note that there is a positive probability that two or more bidders have the same valuation. If several bidders make the same bid, each of them has the same probability of obtaining the building.*

Show that no symmetric equilibrium in pure strategies exists in this case.

5.2 *Consider example 5.1 with the following modification: the monopolist is uncertain as to the payoff that the entrant receives if a fight occurs. The following payoff matrix shows this situation where* ξ *denotes the payoff of the entrant in a fight.*

		Monopolist	
		a	f
Entrant	e	$1,1$	$\xi,\ k$
	ne	$0,3$	$0,3$

Assume that there are just two different payoff values for each player, ξ_1, ξ_2 *and* $k_1,\ k_2$ *respectively with the following joint probability distribution:*

$$\text{prob}\{(\xi_1,\ k_1)\} = q, \text{prob}\{(\xi_2,\ k_2)\} = (1-q), \text{prob}\{(\xi_1,\ k_2)\} = \text{prob}\{(\xi_2,\ k_1)\} = 0.$$

(a) Show that Bayesian updating implies that each learns the opponent's type from the information about her own type.

(b) Prove that any Nash equilibrium strategy combination for the game with payoffs $(\xi_i,\ k_i)$, $i = 1, 2$, forms a type-contingent strategy combination of the game.

5.3 *Consider a car insurance company. Experience shows that one can distinguish two groups of customers: A fraction* μ *of all applicants for a policy have a poor safety record and incur damage with probability* q_k; *the rest have a damage probability* $q_\ell, q_k > q_\ell$. *Assume that customers for a policy know their safety record, but the insurer does not. An insurance contract* $c = (\delta, \rho)$ *specifies the payments of the insurance company to the insured in a damage case,* δ, *and a premium payment of the insured to the company in the case of no damage,* ρ. *Apart from their safety record, all applicants are identical with risk-preferences described by a von Neumann–Morgenstern utility function*

$u(w)$, that is strictly increasing in wealth w, and concave and with the same initial wealth W.

(a) Given several insurance contracts on offer, write down the payoff functions, type, and strategy sets of the customers.

(b) Suppose the insurer tries to design a pair of contracts c_h and c_ℓ such that high-risk customers select the contract c_h and low-risk customers choose the contract c_ℓ. Write down the Bayes–Nash equilibrium conditions for an equilibrium contract pair (c_h, c_ℓ).

(c) Show that the incentive-compatability constraints imposed on the insurer's contract problem guarantee that the chosen contract pair forms a Bayes–Nash equilibrium.

5.4 *Consider two producers of a homogeneous product. Firm 1 was the sole supplier of the market in the past, whereas firm 2 is about to enter the market. The technology of the incumbent is known to all but the cost function of the new firm is private information. Assume the following information is common knowledge:*

(*i*) *the inverse demand function*: $g(x) = \max\{0, A - b \cdot x\},\ A, b > 0$

(*ii*) *the cost function of the incumbent*: $C_1(x_1) = \gamma \cdot x_1$

(*iii*) *the cost function of the new firm*:

$$C_2(x_2) = \begin{cases} \gamma_H \cdot x_2 & \text{with probability } q \\ \gamma_L \cdot x_2 & \text{with probability } 1-q \end{cases}, \ \gamma_H > \gamma > \gamma_L > 0.$$

(a) Write down the payoff functions and type sets for $q \in [0, 1]$. Derive the Cournot equilibrium strategies for $q = 1$ and $q = 0$.

(b) Derive the Bayes–Nash equilibrium of this game. Show that the complete information Nash equilibria derived under (a) are the limits of the Bayes–Nash equilibrium as q converges to zero or one respectively.

5.5 *In Section 5.3.1 it was shown how one can transform a game with incomplete information and finite type sets into a normal form game with an extended player set.*

Prove that any Nash equilibrium of the game with the extended player set is a Bayes–Nash equilibrium of the incomplete information game from which it is derived.

6

Refinements of the Nash Equilibrium Concept: Backward Induction

In previous chapters, equilibrium concepts were discussed in the context of the strategic form of a game. It was shown that most games with complete information have Nash equilibria and most games of incomplete information have Bayes–Nash equilibria. Yet, in general, games have many equilibria. Hence, the question arises whether some equilibria have additional properties that justify eliminating them as possible outcomes of the game. Such additional properties are called *refinements*. The strategic form of a game abstracts completely from the fact that moves occur in a sequence. This chapter analyzes whether sequentiality of moves will make it possible to discriminate between the potentially large number of Nash equilibria.[1]

As explained in Chapter 1, it is possible to represent any game given in extensive form by its normal form. Kuhn's theorem guarantees that, for games with perfect recall, every mixed strategy combination in the strategic form of the game has a corresponding behavior strategy combination in its extensive form achieving the same payoff distribution. Thus, the two ways of representing a game appear equivalent. In particular, any mixed strategy that is a best response in a game in its normal form has a

[1]Kohlberg and Mertens (1986) argue that refinements should be applied to the strategic form only because all elements of the extensive form that matter in a strategic sense are captured in the strategic form (or even in a reduced form of it).

corresponding behavior strategy in its extensive form that is a best response to the other players' behavior strategies derived from their mixed strategies. This implies that the set of Nash equilibria will be the same, whether the game is considered in its extensive or in its normal form.

EXAMPLE 6.1 (*example* 5.1 *reconsidered*). Reconsider the example of an entrant E confronting a monopolist M. In contrast to Chapter 5, consider a particular sequence of moves. The entrant acts first entering the market (e) or not entering it (ne). The monopolist responds by fighting (f) or accommodating the entrant (a). The following diagram shows both the extensive and the strategic form.

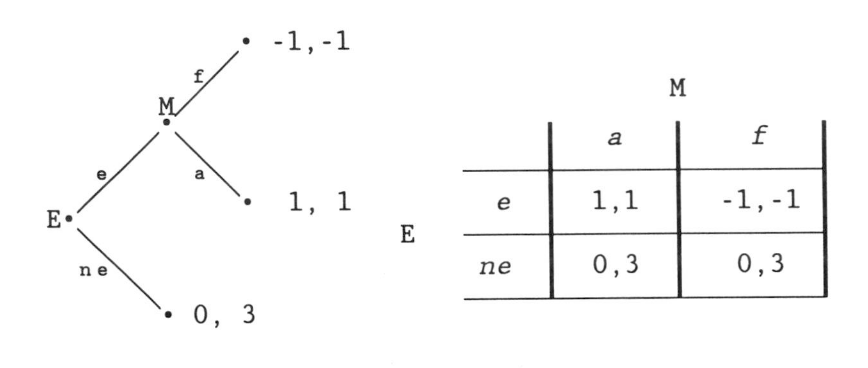

	a	f
e	1,1	-1,-1
ne	0,3	0,3

extensive form *strategic form*

Given payoffs as indicated, it is easy to see there are two (pure strategy) Nash equilibria in this game

- either the entrant enters and the monopolist accommodates (e, a)
- or the entrant stays out of the market and the monopolist fights (ne, f)

Notice that the strategic and the extensive form have the same set of Nash equilibria. ■

If one considers these equilibria solely in terms of best responses, then both equilibria are equally convincing predictions for the outcome of the game. Such a perspective, however, neglects the fact that moves are carried out in a particular sequence. In the equilibrium (ne, f), f is a best reply to ne, because "fighting" is as good a response to "no entry" as is "accommodation," and ne is a best response to f because "not to enter" is optimal if the monopolist fights. But would the monopolist really fight if entry occurs? This is an open question since the monopolist's resolve will not be tested if the entrant stays out of the market. Clearly, for the monopolist, "fighting" is equivalent to "accommodating" only if the entrant will not enter. Considering this situation, the entrant may enter, thus creating a *fait accompli*, and rely on the monopolist accommodating the entry. Hence, considering the sequence of moves helps to distinguish between equilibria.

The equilibrium (ne, f) in example 6.1 can be interpreted as the monopolist threatening to fight entry to keep potential entrants out of the market. Obviously, the outcome of this equilibrium would be preferred by the monopolist. The decision of the entrant not to enter and of the monopolist to fight if entry occurs lacks consistency because the monopolist would have to follow a suboptimal policy if entry actually occurred. Such a suboptimal action in equilibrium is only possible if the behavior of the opponent makes it unnecessary to act this way. Thus, there is good reason to disregard Nash equilibria lacking such dynamic consistency.

In example 6.1, the backward induction method discussed in Chapter 1 in the context of the chess theorem (theorem 1.1) can be applied to derive a Nash equilibrium. The backward induction procedure begins at a decision node at which the action chosen concludes the game. It determines the best action of the player who moves at this node and assigns the resulting payoff combination to this decision node, which becomes a terminal node of the reduced game. Repeating this the initial node will be reached in a finite sequence of reductions. In example 6.1 just one reduction is necessary to derive the strategy combination (e, a). The resulting strategy combination is a Nash equilibrium and satisfies, by construction, dynamic consistency.

The backward induction procedure, however, can only be applied to finite extensive form games with perfect information. For general games in extensive form it may not be applicable because they may either be infinite and have no terminal nodes from which to work backward or have imperfect information and, therefore, players who have to choose an action at some information set without certainty about its payoff. The rest of this chapter will present different approaches that try to extend the backward induction idea to arbitrary games in extensive form. Before proceeding to the first concept however, the following remark briefly summarizes the notation needed for extensive form games.

Remark 6.1 (notation for games in extensive form). Recall some of the notation in Chapter 1. A game in extensive form specifies the player set I, the set of nodes N and the predecessor node function σ, the set of actions A and the predecessor action function α, the player partition $(N_i)_{i \in I}$, the set of information sets U, the set of choices $A(u)$ for each information set

$u \in U$, and the payoff function r that associates a payoff $r_i(n)$ for each player with each terminal node $n \in \mathscr{T}(N)$.

$$\Gamma = \big(I, (N, \sigma), (A, \alpha), (N_i)_{i \in I}, U, (A(u))_{u \in U}, r\big)$$

A behavior strategy of player i, b_i, specifies for each information set u of player i a random choice among the actions available. Note that $b_i^u(a)$ denotes the probability that action a is played by player i at the information set u. A pure strategy is the special case in which one action is chosen at each information set with probability one and all others with probability zero. Any behavior strategy combination $b = (b_1, b_2, \ldots, b_I)$ induces a probability distribution on the set of terminal nodes $\mathscr{T}(N)$. For the special case of a pure strategy combination, one terminal node is reached with probability one and all others with probability zero. Thus, each behavior strategy combination b yields, for each player i, an expected payoff $R_i(b)$.

A behavior strategy combination b, however, induces not only a probability distribution on the set of terminal nodes $\mathscr{T}(N)$ starting from the initial node (the root of the game tree) but from any decision node $x \in N \backslash \mathscr{T}(N)$ onward. For any decision node x, denote by $q_b(n|x)$ the probability that terminal node n is reached if the behavior strategy combination b is played and if the game has already reached the decision node x. This probability is computed easily by multiplying all the probabilities given by the behavior strategy combination b for the sequence of moves leading from node x to node n. If there is no such sequence of moves, set $q_b(n|x) = 0$. The expected payoff of player i, if the behavior strategy combination b is played and the game has already reached node x, is given as

$$R_i(b|x) = \sum_{n \in \mathscr{T}(N)} r_i(n) \cdot q_b(n|x).$$

In the entry game of example 6.1, there are just two decision nodes, one for each player.

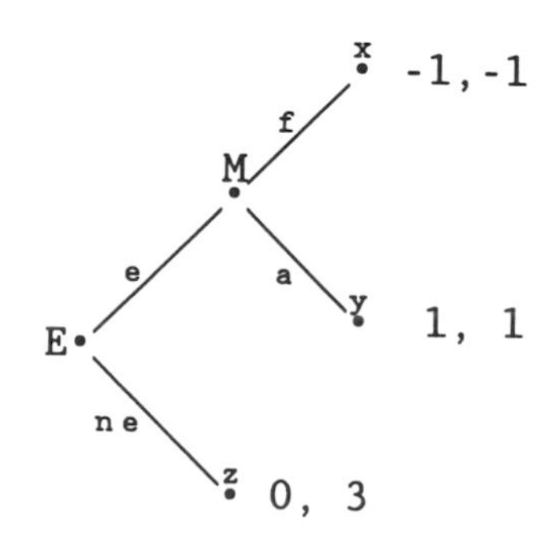

Given a behavior strategy $b = ((b_E(e), b_E(ne)), (b_M(a), b_M(f)))$, it is straightforward to determine the probabilities of the terminal nodes from decision node E and decision node M as

$$q_b(x|E) = b_E(e) \cdot b_M(f), \qquad q_b(y|E) = b_E(e) \cdot b_M(a), \qquad q_b(z|E) = b_E(ne),$$

$$q_b(x|M) = b_M(f), \qquad q_b(y|M) = b_M(a), \quad \text{and} \quad q_b(z|M) = 0.$$

The expected payoffs of the entrant from nodes E and M, given the behavior strategy combination b, are therefore

$$\begin{aligned} R_E(b|E) &= (-1) \cdot q_b(x|E) + (1) \cdot q_b(y|E) + (0) \cdot q_b(z|E) \\ &= b_E(e) \cdot (b_M(a) - b_M(f)), \\ R_E(b|M) &= (-1) \cdot q_b(x|M) + (1) \cdot q_b(y|M) + (0) \cdot q_b(z|M) \\ &= (b_M(a) - b_M(f)). \end{aligned}$$

Similarly, one obtains the expected payoffs of the monopolist. ■

6.1 Subgame Perfect Equilibrium

Attempting to generalize the backward induction notion to general games in extensive form, Selten (1965) suggested the concept of a *subgame perfect equilibrium*. This concept requires an equilibrium strategy combination to be a Nash equilibrium of the whole game and of any part of a game tree that can be considered a well-specified game in its own right. A part of an extensive form game that can be viewed as a separate game is called a *subgame*. A subgame must start at an information set with a single node and must contain all information sets that follow the initial node. The following example illustrates that not every information set containing a single node constitutes an initial node of a subgame.

EXAMPLE 6.2. Consider a game with an incumbent monopolist M and a potential entrant E. The entrant has the choice to enter the market without consulting firm C, which specializes in market observation, or to take advice from this firm. After consultation, the entrant may decide to stay out of the market or to enter it. The monopolist M observes the entry but not whether it follows after consultation or from independent action. In response to entry the monopolist may decide to accommodate or to fight.

The consultant, if employed, will earn a payoff of 1 if no entry is recommended, or a payoff of 2 and 0 respectively, depending on whether entry is successful or not. The monopolist and the entrant have the same payoffs as in example 6.1 except that the entrant's payoff after successfully entering the market after consultation has been raised to 2. The following game tree describes this situation and indicates the payoffs of the players.

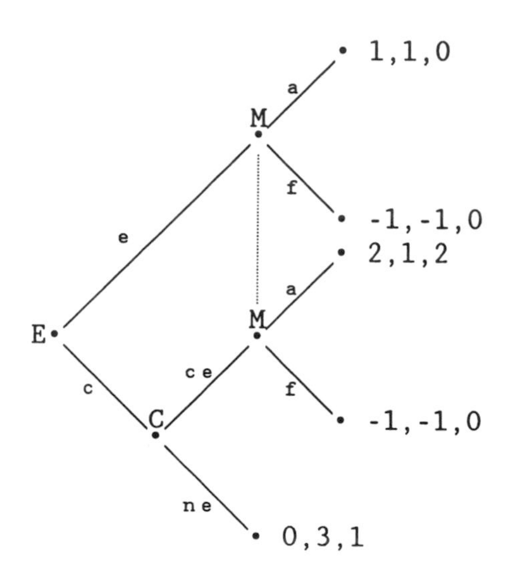

E's strategies: S_E = {enter, consult} = {e,c}

C's strategies: S_C = {enter, stay out} = {ce,ne}

M's strategies: S_M = {accommodate, fight} = {a,f}

Each payoff vector lists the payoffs in the order (r_E, r_M, r_C).

In this game there is no subgame except for the game itself. In particular, there is no subgame at the decision node of *C* since the information set of *M* contains nodes that do not follow *C*'s action. ■

Example 6.2 shows that there may be information sets containing a single node, like the information set of the consultant C, for which the set of successor nodes does not form a well-defined game in its own right. A subgame must be a game in extensive form that forms part of a larger game in extensive form but can be analyzed separately. Since the information set of the monopolist includes a node that is a successor of the decision node of C and a node that is not a successor of the decision node of C, the behavior of the monopolist cannot be determined within the set of nodes following the decision node of C. Hence, to analyze a subgame as a separate game, it must be possible to partition the set of information sets U such that any information set belongs either to the subgame or to its complement. In example 6.2, the monopolist's information set violates this condition for the part of the tree that follows the decision node of player C.

Given a game in extensive form $\Gamma = (I, (N, \sigma), (A, \alpha), (N_i)_{i \in I}, U, (A(u))_{u \in U}, r)$, for any node $x \in N$, denote by N_x the set of nodes following x, and by N_{-x} its complement $N \backslash N_x$.

Definition 6.1 (*subgame*). For any node x, the set of successor nodes of x, N_x, forms a subgame Γ_x (i) if $\{x\} \in U$ holds and (ii) if the set of information sets U can be partitioned into U_x and U_{-x} such that $u \in U_x$ implies $u \subseteq N_x$ and $u' \in U_{-x}$ implies $u' \subseteq N_{-x}$ for all $u, u' \in U$.

Condition (i) says a subgame must begin at an information set containing a single node, and condition (ii) requires that information sets lie either completely in the set of nodes following x, N_x, or in the complement N_{-x}. These conditions are trivially satisfied for games with perfect information because all information sets contain just one node.

Remark 6.2. One usually treats the game Γ as a subgame itself starting at the initial node 0, that is, $\Gamma_0 \equiv \Gamma$. With this convention, one knows that every game has at least one subgame. In this sense, there is only one subgame in example 6.2. ■

A subgame perfect equilibrium can now be defined as a behavior strategy combination such that the behavior strategy of each agent is a best response strategy against the behavior strategies of the other players in every subgame Γ_x.

Definition 6.2 (*subgame perfect equilibrium*). A behavior strategy combination $b^* = (b_1^*, \ldots, b_I^*)$ is a *subgame perfect equilibrium* if, for all players $i \in I$ and for all subgames Γ_x,

$$R_i(b^*|x) \geq R_i(b_i, b^*_{-i}|x)$$

for all behavior strategies $b_i \in B_i$ holds.

Note the similarity to the definition of a Nash equilibrium. In particular, every subgame perfect equilibrium is a Nash equilibrium because the equilibrium behavior strategy combination must be a Nash equilibrium, of the subgame Γ_0 which is identical to the game Γ itself. It follows from this observation that every game with a Nash equilibrium must also have a subgame perfect equilibrium. This justifies calling a subgame perfect equilibrium a refinement of the Nash equilibrium concept.

In games with perfect information in which all information sets contain a single node, this definition requires the equilibrium strategies to be optimal at all nodes. Example 6.1 illustrates this case.

EXAMPLE 6.1 (*resumed*). In this example, there are just two information sets, one for the entrant and one for the monopolist. Consider the following two Nash equilibria:

(i) $b^* = (b^*_E, b^*_M)$, with $b^*_E = (b^*_E(e), b^*_E(ne)) = (1, 0)$
and $b^*_M = (b^*_M(a), b^*_M(f)) = (1, 0)$;
(ii) $b' = (b'_E, b'_M)$, with $b'_E = (b'_E(e), b'_E(ne)) = (0, 1)$
and $b'_M = (b'_M(a), b'_M(f)) = (0, 1)$.

Let x be the node at which the monopolist has to move. b^* describes optimal behavior since $R_M(b^*|x) = 1 > R_M(b^*_E, b_M|x)$ holds for any b_M with $b_M(a) < 1$. Hence, b^* is a subgame perfect equilibrium. On the other hand, $R_M(b'|x) = -1 < R_M(b'_E, b_M|x)$ holds for any b_M with $b_M(a) > 0$. Thus, b' is not subgame perfect. ■

This example shows that subgame perfection can reduce the number of Nash equilibria. In finite games with perfect information in which no ties occur in players' payoffs, there is a unique subgame perfect equilibrium. On the other hand, as the following example illustrates, every Nash equilibrium may be subgame perfect.

EXAMPLE 6.2 (*resumed*). In this game the only subgame is the game itself. Hence, any Nash equilibrium of the game is a subgame perfect equilibrium. It is easy to check that there are three Nash equilibria in pure strategies:

(i) The entrant employs the consultant, who recommends staying out of the market, and the monopolist threatens to fight, (c, f, ne).
(ii) The entrant enters, the monopolist accommodates, and the consultant recommends staying out of the market (e, a, ne).

(iii) The consultant recommends entry, the entrant follows this recommendation, and the monopolist accommodates, (c, a, ce).

Notice that the first equilibrium can be interpreted as an equilibrium based on an incredible threat of the monopolist. Since this game has no subgames except for the game itself, subgame perfection will not eliminate such an equilibrium.

Equilibrium (ii) is somewhat strange as well. The only reason the consultant would recommend not entering the market is his belief, justified in equilibrium, that the entrant will not employ him. The consultant is therefore indifferent to the recommendation he would make if hired. Yet, it is the recommendation of the consultant to stay out of the market that makes it optimal for the entrant to enter without consultation. Thus, the outcome is worse than what could have been obtained with consultation and a recommendation to enter. Equilibrium (iii) is the only intuitively sensible one in this game. ■

In finite games with perfect information, the concept of subgame perfection coincides with backward induction. Notice, however, that subgame perfect equilibria are defined without reference to terminal nodes and that this concept can, therefore, be applied to infinite games. Chapter 7 on repeated games and Chapter 8 on bargaining will use this generalization. In games with imperfect information, however, subgame perfection loses much of its selection power. This section concludes with a remark on how to derive subgame perfect equilibria in finite games and an example illustrating this method.

Remark 6.3 (computation of subgame perfect equilibria). For games with a finite number of nodes, there is a simple procedure to derive subgame perfect equilibria. Since a subgame perfect equilibrium strategy combination b^* must be an equilibrium in each subgame, one can construct an equilibrium strategy combination working backward. First, one finds an equilibrium of the last subgame, which may imply just an optimal choice for one agent or a Nash equilibrium for some players. The equilibrium payoffs of this subgame can be used as the players' payoffs. This yields an induced shorter game to which one can apply the same procedure. Since the game is finite, working backward this way, one finally reaches the initial node. Any resulting sequence of mutually optimal behavior strategies defines a subgame perfect equilibrium. Since it is possible that there are multiple equilibria in each of these stages, one can combine them in various strings to subgame perfect equilibria. The following flow diagram summarizes this procedure.

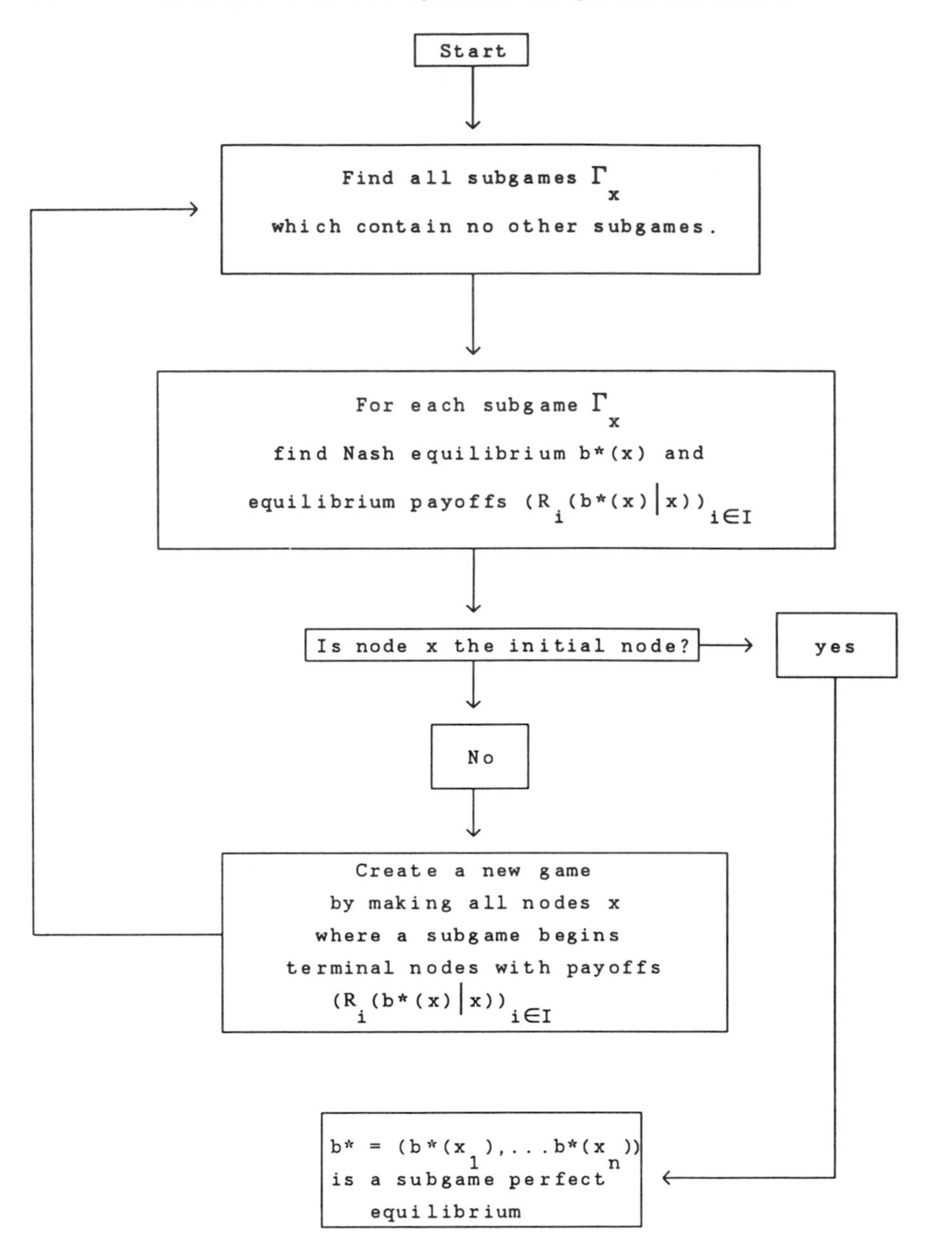
Start
Find all subgames Γ x
which contain no other subgames.
For each subgame Γ x
find Nash equilibrium b*(x) and
equilibrium payoffs (R i (b*(x)|x)) i∈I
Is node x the initial node?
yes
No
Create a new game
by making all nodes x
where a subgame begins
terminal nodes with payoffs
(R i (b*(x)|x)) i∈I
b* = (b*(x 1),...b*(x n))
is a subgame perfect
equilibrium

■

The following example shows how to apply the procedure described in the remark. In addition, it provides a further example for the fact that subgame perfection eliminates few Nash equilibria in games of imperfect information.

EXAMPLE 6.3. Consider the following modification of the entry game of example 6.1. The entrant first decides whether to enter or not, and then whether to enter with a high investment level, h, or a low one, l. With low investment, payoffs are as in example 6.1. With high investment, however, the entrant will face a smaller payoff if met by accommodating behavior but also a smaller loss in case of a fight. Note that the monopolist is indifferent between fighting and accommodating behavior in this case. The following game tree describes this situation.

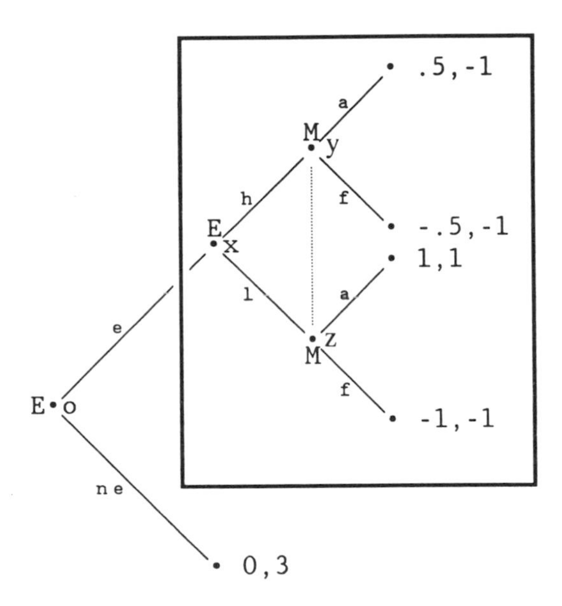

This game has two subgames, one starting at node x and the other at node o. The latter subgame is the game itself. It is not difficult to see that the subgame with origin at x has the following equilibria:

(i) $((b_E^*(h), b_E^*(l)), (b_M^*(a), b_M^*(f)) = ((0, 1), (1, 0))$, and
(ii) $((b'_E(h), b'_E(l)), (b'_M(a), b'_M(f)) = ((1, 0), (q, 1-q))$ for $0 \leq q \leq .5$.

Note that the monopolist is indifferent to any behavior strategy as long as the entrant chooses h, the high level of investment. On the other hand, as long as the monopolist chooses to fight with a probability greater than .5 ($q \leq .5$), the entrant will find it optimal to choose h. This explains the continuum of equilibria in case (ii). The expected payoffs in these equilibria are easily computed as follows:

(i) $(R_E(b^*|x), R_M(b^*|x)) = (1, 1)$, and
(ii) $(R_E(b'|x), R_M(b'|x)) = ((q - .5), -1)$ for $0 \leq q \leq .5$.

Taking these payoffs as the payoff combination at node x leads to the following reduced game.

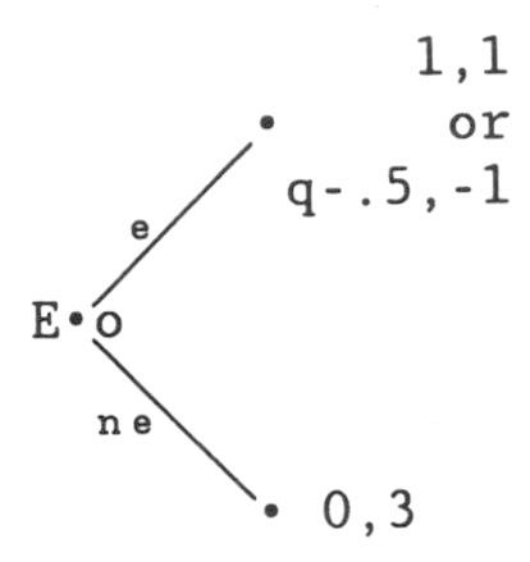

It is easy to see that the optimal strategy of the entrant depends on the equilibrium played in the second subgame. If the equilibrium with payoff $(1, 1)$ is played in the subgame starting at x, then it is optimal for the entrant to enter the market, $(b_E(e), b_E(ne)) = (1, 0)$. If one of the other equilibria follows in the subgame at x, then the entrant would not want to enter the market, $(b_E(e), b_E(ne)) = (0, 1)$.

Thus, there is a multiplicity of subgame perfect equilibria, namely

$$\begin{aligned} b^* &= ((b_E^*(e), b_E^*(ne)), (b_E^*(h), b_E^*(l)), (b_M^*(a), b_M^*(f)) \\ &= ((1,0), (0,1), (1,0)), \\ b' &= ((b_E'(e), b_E'(ne)), (b_E'(h), b_E'(l)), (b_M'(a), b_M'(f)) \\ &= ((0,1), (1,0), (q, 1-q)) \end{aligned}$$

with $0 \leq q \leq .5$. Note, however, that there are also Nash equilibria of the game that are not subgame perfect, namely $\tilde{b} = ((0, 1), (0, 1), (q, 1 - q))$ for $0 \leq q \leq .5$. In these Nash equilibria the entrant stays out of the market and the monopolist fights with high probability. The entrant, however, chooses the low investment level. Given that the monopolist will fight with high probability, this behavior is optimal only because the entrant does not enter the market. ■

Like example 6.2, the last example shows that subgame perfection is a selection criterion that loses much of its power in the context of games with imperfect information. Furthermore, in example 6.1, subgame perfectness eliminated the equilibrium based on the incredible threat of the monopolist to fight. With some imperfectness of information such an equilibrium could be sustained as a subgame perfect equilibrium in example 6.3.

6.2 Perfect Bayesian Equilibrium

The notion of a subgame perfect equilibrium is a natural generalization of the backward induction principle to infinite games and to games with imperfect information. For games with imperfect information, however, the concept proves to be far less successful as an equilibrium refinement. Particularly in games without proper subgames, every Nash equilibrium is subgame perfect. This chapter introduces another concept of refinement that is stronger than subgame perfection.

The basic idea of the backward induction concept consists in requiring equilibrium strategies to be optimal at each stage of the game that requires a player to move to eliminate equilibria based on incredible threats. In games with imperfect information, however, a player may have to move at an information set with several nodes. Which action is optimal at such an information set depends on the beliefs the player holds as to the particular node from which he moves. If a player could assign to each node of an information set a probability of being there, then he could determine the expected payoff of any behavior strategy combination, conditional on having reached this information set. This would allow him to determine the optimal strategy at this information set. The following example illustrates the problem.

EXAMPLE 6.3 (*resumed*). Consider the information set of the monopolist.

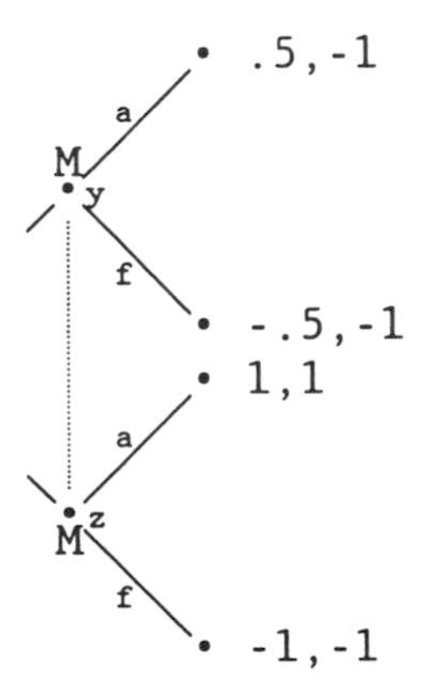

Whether the behavior strategy $(b^*_M(a), b^*_M(f)) = (1,0)$ is better or worse than a behavior strategy $(b'_M(a), b'_M(f)) = (q, 1-q)$ for some q, $0 \leq q \leq .5$, cannot be determined if the monopolist cannot assess the probability of being in node y or node z.

Suppose however that, for whatever reason, the monopolist believes he is in node y with probability $\mu(y)$ and in node z with probability $\mu(z) = 1 - \mu(y)$. Now one can calculate that the behavior strategy b_M^* yields an expected payoff $R_M(b_M^*|u) = \mu(y) \cdot (-1) + \mu(z) \cdot 1 = (\mu(z) - \mu(y))$. Similarly, the strategy b'_M achieves the expected payoff

$$\begin{aligned} R_M(b'_M|u) &= \mu(y) \cdot [q \cdot (-1) + (1-q) \cdot (-1)] \\ &\quad + \mu(z) \cdot [q \cdot 1 + (1-q) \cdot (-1)] \\ &= q \cdot (\mu(z) - \mu(y)) + (1-q) \cdot (-1). \end{aligned}$$

Hence, $R_M(b_M^*|u) \geq R_M(b'_M|u)$ holds for all $q \in [0, .5]$, with equality for $\mu(y) = 1$. Therefore, the monopolist must be certain that he is in node y or he will never choose a strategy b'_M. Even a small probability that he may be in node z would render the equilibrium strategy b'_M suboptimal at that information set. Thus, the credibility of b'_M depends on the beliefs of the monopolist. ■

This example shows how one can extend the notion that an equilibrium strategy is optimal at every stage at which a player has to move to games of imperfect information. For this extension, it is necessary to introduce beliefs at each information set. A *belief* at an information set is simply a probability distribution over the nodes of this information set. Clearly, information sets that contain just one node must assign probability one to this node.

Given such beliefs, it is possible to evaluate the optimality of behavior strategy combinations at every information set. What is optimal at an information set, however, depends on the particular beliefs held by the player. To avoid complete arbitrariness of these beliefs, certain consistency requirements have to be satisfied. In particular, beliefs must be consistent with the strategies played in equilibrium.

EXAMPLE 6.3 (*resumed*). Consider the equilibrium $((b_E^*(e), b_E^*(ne)), (b_E^*(h), b_E^*(l)), (b_M^*(a), b_M^*(f))) = ((1,0),(0,1),(1,0))$. Given that the entrant plays e and l with probability 1 according to this strategy, the monopolist should believe that he is with probability one at node z, $\mu(z) = 1$. In an equilibrium $((b'_E(e), b'_E(ne)), (b'_E(h), b'_E(l)), (b'_M(a), b'_M(f))) = ((0,1),(1,0),(q, 1-q))$, $(0 \leq q \leq .5)$, however, the monopolist will reach the information set $\{y, z\}$ with probability zero. Hence the equilibrium strategy provides no indication of the likelihood of being at y or z. Any belief of the monopolist will be consistent with the equilibrium strategies. Note, however, that only a belief $\mu(y) = 1$ will make the equilibrium strategy $(b'_M(a), b'_M(f)) = (q, 1-q)$ optimal for the monopolist. ■

This example shows that beliefs are determined by the equilibrium strategies for all information sets reached with positive probability, according to these strategies. Equilibrium strategies, however, put no constraints on beliefs at information sets with more than one node that have a zero probability of being reached during equilibrium play. Thus, beliefs are not completely specified by the requirement that beliefs be consistent with the equilibrium behavior strategies.

But what conditions should be imposed on beliefs at information sets not supposed to be reached, given the equilibrium strategies of the players? What is assumed about beliefs in those cases is often essential for the "rationality" of the equilibrium action. In this sense, one can say that out-of-equilibrium beliefs determine the equilibrium. In fact, most of the remainder of this chapter will be concerned with "reasonable" restrictions on beliefs at information sets off the equilibrium path. Before this discussion, however, the notion of beliefs and their relationship to equilibrium strategies will be introduced formally. The appropriate concept is a perfect Bayesian equilibrium.

DEFINITION 6.3 (*system of beliefs*). A *system of beliefs* is a probability distribution for each information set; formally, a mapping μ: $N \to [0,1]$ such that $\Sigma_{x \in u} \mu(x) = 1$ holds for all information sets $u \in U$.

Note that this definition implicitly assumes all players hold the same beliefs and know that other players hold the same beliefs. Thus, the system of beliefs has to be common knowledge.

A *perfect Bayesian equilibrium* is defined as a behavior strategy combination together with a belief system such that (i) behavior strategies are equilibrium strategies at each information set given the beliefs, and (ii) beliefs are consistent with equilibrium strategies. For the following formal definition of a perfect Bayesian equilibrium, recall that $\sigma(x)$ denotes the predecessor node and $\alpha(x)$ the predecessor action of node x, and $i(x)$ denotes the player who moves at node x.

DEFINITION 6.4 (*perfect Bayesian equilibrium*). A behavior strategy combination b^* and a system of beliefs μ form a perfect Bayesian equilibrium if the following two conditions are satisfied for all information sets $u \in U$ and all players $i \in I$:

(i) $$\sum_{x \in u} \mu(x) \cdot R_i(b_i^*, b_{-i}^* | x) \geq \sum_{x \in u} \mu(x) \cdot R_i(b_i, b_{-i}^* | x) \text{ for all } \quad b_i \in B_i,$$

and

(ii) $$\mu(x) = \frac{\mu(\sigma(x)) \cdot b_{i(\sigma(x))}^*(\alpha(x))}{\sum_{x' \in u} \mu(\sigma(x')) \cdot b_{i(\sigma(x'))}^*(\alpha(x'))}$$

$$\text{for } \quad \sum_{x' \in u} \mu(\sigma(x')) \cdot b_{i(\sigma(x'))}^*(\alpha(x')) \neq 0.$$

A perfect Bayesian equilibrium requires an equilibrium strategy to be optimal at each information set given the equilibrium strategy combination and given the belief system. Note that updating of beliefs occurs iteratively from beliefs at predecessor nodes to beliefs at successor nodes according to the equilibrium behavior strategy combination. This is equivalent to Bayesian updating for information sets reached with positive probability according to the equilibrium strategy combination. If an information set has probability zero of being reached from its predecessor nodes according to the equilibrium behavior strategy combination, then beliefs can be assigned arbitrarily. The following examples demonstrate how to apply this definition.

EXAMPLE 6.4. Consider the market for used cars. Cars are either of high quality s_H or of low quality s_L. Buyers value a car of quality s_i, $i = H, L$, at $a_B \cdot s_i$ and sellers at $a_S \cdot s_i$. It is assumed that $a_B > a_S$ holds to make a sale desirable whenever the quality of the car is known to the buyer and the seller. Assume further that a car is with probability q of high quality but that only sellers are informed about the quality. Sellers propose a price $p \in \{p_H, p_L\}$, with $p_H \equiv a_S \cdot s_H, p_L \equiv a_S \cdot s_L$, and buyers accept or reject the offer.[2] The following game tree represents this game.

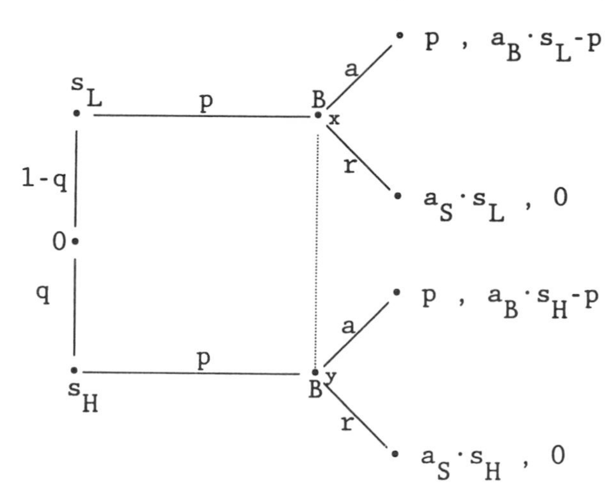

First, consider the buyer. Given a belief system $(\mu(x), \mu(y))$, he will buy a car if $\mu(x) \cdot [a_B \cdot s_L - p] + \mu(y) \cdot [a_B \cdot s_H - p] \geq 0$ holds and reject the offer otherwise. Given a pricing strategy $(p(s_H), p(s_L))$ of the seller, the buyer can update his beliefs in the following manner:

- if the two types of the seller charge different prices $(p(s_H), p(s_L)) = (p_H, p_L)$, then updating yields a belief

$$\mu(y) \equiv \text{Prob}\{s_i = s_H | p\} = \begin{cases} 1 & \text{for} \quad p = p_H \\ 0 & \text{for} \quad p = p_L \end{cases}, \quad \mu(x) = 1 - \mu(y);$$

[2]For simplicity it is assumed that buyers get all the surplus if the quality is known to them.

- if both types of the seller charge the same price $p(s_H) = p(s_L) = p$ then the buyer, knowing that a seller will find it optimal to sell a car only if $p \geq a_S \cdot s_i$ holds, will update his beliefs to

$$\mu(y) \equiv \text{Prob}\{s_i = s_H | p\} = \begin{cases} q & \text{for} \quad p \geq a_S \cdot s_H \\ 0 & \text{otherwise} \end{cases}, \quad \mu(x) = 1 - \mu(y).$$

Now consider the seller. When choosing a price, the seller must take into account that this choice will influence the payoff and also the beliefs of the buyer. Two cases have to be considered:

(i) Suppose that both types of the seller offer a different price, $(p(s_H), p(s_L)) = (p_H, p_L)$. In this case, the buyer can identify the seller's type and will accept the offer. Knowing that the buyer will accept the high price p_H, provides an incentive for the low-quality seller to charge a high price as well. Consequently, there can be no equilibrium with low-quality and high-quality sellers offering different prices.

(ii) Suppose that both types of the seller offer the same price. The common price must be p_H because otherwise the high-quality seller would not want to sell. Given that p_H is the price of both types of sellers, the buyer gains no information from the observed price and updates his beliefs to $\mu(y) = q$. If $a_B \cdot [q \cdot s_H + (1 - q) \cdot s_L] \geq p_H$ holds, then the buyer will accept the price in spite of uncertainty about the quality of the car. To ask for the high price is clearly optimal for both types of the seller in this case. On the other hand, if $a_B \cdot [q \cdot s_H + (1 - q) \cdot s_L] < p_H$ holds, the buyer will reject the offer. Charging the high price is again optimal for high-quality and for low-quality sellers, in this case because the buyer rejects the offer.

To summarize, the following pure strategies $(p^*(s_H), p^*(s_L), \alpha_B^*)$ and beliefs $(\mu(x), \mu(y))$ form a perfect Bayesian equilibrium:

$$p^*(s_H) = p^*(s_L) = p_H, \ \mu(y) = q, \ \mu(x) = 1 - q$$

$$\alpha_B^* = \begin{cases} a & \text{for} \quad a_B \cdot [q \cdot s_H + (1 - q) \cdot s_L] \geq p_H \\ r & \text{for} \quad a_B \cdot [q \cdot s_H + (1 - q) \cdot s_L] < p_H \end{cases}.$$

This example is a modified version of the "market for lemons" introduced by Akerloff (1970). Cars of high and of low quality will be sold for the price of the high-quality car if the average quality is high enough. Otherwise no cars will be traded. The informational asymmetry may lead to a break-down of the market. ■

By definition 6.3, $\mu(x) = 1$ must hold for any information set that contains just one node x. Therefore, it follows immediately from part (i) of definition 6.4 that every perfect Bayesian equilibrium will be subgame perfect. Thus, a perfect Bayesian equilibrium generalizes the concept of subgame perfect equilibrium. It is stronger, however, since it assigns arbitrary beliefs at information sets not reached during equilibrium play and requires equilibrium strategies to be optimal at all information sets given these beliefs.

In some cases, perfect Bayesian equilibrium is capable of reducing the set of equilibria in games of imperfect information as example 6.2 demonstrates. Recall that all equilibria are subgame perfect in this example.

EXAMPLE 6.2 (*resumed*). Denote the upper node in the monopolist's information set by x and the lower node by y. For any belief system $(\mu(x), \mu(y))$, the monopolist will find it optimal to choose to accommodate since this is a dominant strategy. This reduces the game to the following form with only the payoffs of the entrant and the consultant listed.

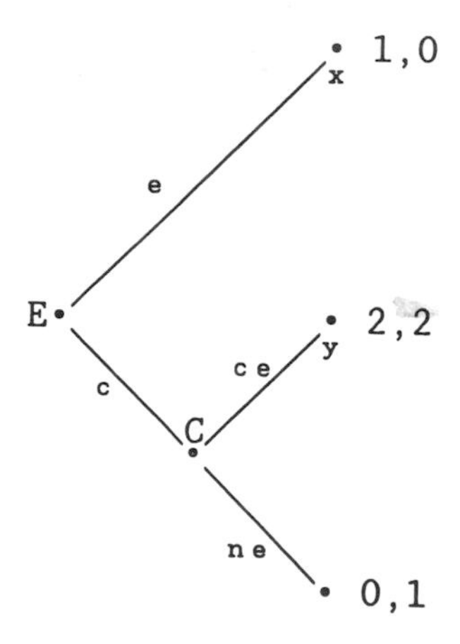

Given that the monopolist chooses to accommodate, the consultant will find it optimal to recommend entry and the entrant will find it optimal to consult before entering the market. Hence, there is a unique perfect Bayesian equilibrium (b^*, μ^*) with $(\mu^*(x), \mu^*(y)) = (0, 1)$ and $b^* = ((b_E^*(e), b_E^*(c)), (b_C^*(ce), b_C^*(ne)), (b_M^*(a), b_M^*(f))) = ((0, 1), (1, 0), (1, 0))$. Note that the unique perfect Bayesian equilibrium is the only reasonable equilibrium in this context. ■

The perfect Bayesian equilibrium concept, however, does not always reduce the Nash equilibrium set substantially. In example 6.3, where most Nash equilibria were subgame perfect, all subgame perfect equilibria are perfect Bayesian equilibria.

EXAMPLE 6.3 (*resumed*). For the equilibrium strategy b^* beliefs will be completely determined by the equilibrium strategies according to the updating rule in definition 6.4 (ii). For $q = 0$, it is easy to check that the following behavior strategy combination b' together with the belief system μ' forms a perfect Bayesian equilibrium:[3]

$$b' = ((b'_E(e), b'_E(ne)), (b'_E(h), b'_E(l)), (b'_M(a), b'_M(f)))$$
$$= ((0,1),(1,0),(0,1)),$$
$$\mu'(y) = 1.$$

In this equilibrium the monopolist fights entry that is rational only if he is absolutely certain that entry will occur with high investment. Since the entrant stays out under this threat, the resolve of the monopolist will not be tested. Note that in this case the belief μ' is supported by the behavior strategy $(b'_E(h), b'_E(l)) = (1,0)$ of the entrant, which is optimal only if the monopolist actually fights.

It is clear that any equilibrium that keeps the entrant out of the market must be supported by beliefs of the monopolist that the entrant will enter with a high investment level ($\mu(y) = 1$). The slightest doubt about the entrant's investment decision would induce the monopolist to accommodate. The equilibrium strategy "not to enter," however, does not preclude extreme beliefs like $\mu(y) = 1$. ■

Example 6.2 shows that the perfect Bayesian equilibrium concept is able to discriminate among multiple equilibria in cases in which the subgame perfection concept achieves no reduction in the number of equilibria. Example 6.3 on the other hand presents a game in which all subgame perfect equilibria may be supported by arbitrary specifications of out-of-equilibrium beliefs. The following section considers an approach that eliminates out-of-equilibrium beliefs that are not robust against perturbations of behavior strategies at each information set.

6.3 Sequential Equilibrium and Agent Normal Form Perfection

Perfect Bayesian equilibrium strengthens the subgame perfection concept by the requirement that equilibrium strategies be optimal at all information sets given a consistent system of beliefs. They are, however, not always effective in eliminating equilibria based on incredible threats because beliefs can be assigned arbitrarily at information sets not reached in equilibrium. This observation suggests imposing restrictions on out-of-equilibrium beliefs to obtain further refinements.

If all players choose completely mixed behavior strategies, then every information set will be reached with positive probability. In such a case

[3]For $q \in (0, .5)$ the same belief system supports the subgame perfect equilibrium strategy combination.

Bayesian updating will determine the beliefs at all information sets and condition (ii) of the definition of a perfect Bayesian equilibrium is sufficient for a unique system of beliefs. Recall that a behavior strategy b_i is called *completely mixed* if $b_i^u(a) > 0$ for all information sets u and all actions $a \in A(u)$ holds.

The concept of a sequential equilibrium restricts beliefs at information sets not reached in equilibrium by imposing the requirement that beliefs are determined by a sequence of completely mixed behavior strategy combinations converging to the equilibrium strategy combination. This implies that, in a sequential equilibrium, out-of-equilibrium beliefs have to be consistent with some small deviation from the equilibrium strategy combination. One can, therefore, interpret a sequential equilibrium as a perfect Bayesian equilibrium in which out-of-equilibrium beliefs are justified by a small deviation from the equilibrium strategy combination.

DEFINITION 6.5 (*sequential equilibrium*). A behavior strategy combination b^* and belief system μ^* form a *sequential equilibrium* if

(i) (b^*, μ^*) is a perfect Bayesian equilibrium

(ii) there is some sequence of completely mixed behavior strategy combinations $(b^\nu)_{\nu=1}^\infty$ such that $\lim_{\nu \to \infty} b^\nu = b^*$ and $\lim_{\nu \to \infty} \mu^\nu = \mu^*$ holds where $(\mu^\nu)_{\nu=1}^\infty$ is the sequence of beliefs derived from $(b^\nu)_{\nu=1}^\infty$ by Bayes' rule.

Note that a behavior strategy combination b^ν from a sequence $(b^\nu)_{\nu=1}^\infty$ need not be an equilibrium. The only requirement is that this sequence converge to the equilibrium strategy combination b^*. It follows immediately that a sequential equilibrium provides no extra restrictions on a perfect Bayesian equilibrium if there is just one information set in a game.

EXAMPLE 6.3 (*resumed*). The perfect Bayesian equilibrium in which the monopolist fights believing with probability one that the entrant has entered with a high investment level,

$$
\begin{aligned}
b' &= ((b'_E(e), b'_E(ne)), (b'_E(h), b'_E(l)), (b'_M(a), b'_M(f))) \\
&= ((0,1), (1,0), (0,1)), \\
&\qquad\qquad \mu'(y) = 1,
\end{aligned}
$$

is a sequential equilibrium. To see this, consider the following sequence of completely mixed behavior strategy combinations ($\nu \to \infty$):

$$
\begin{aligned}
b^\nu &= ((b_E^\nu(e), b_E^\nu(ne)), (b_E^\nu(h), b_E^\nu(l)), (b_M^\nu(a), b_M^\nu(f))) \\
&= \left(\left(\frac{1}{\nu}, 1 - \frac{1}{\nu}\right), \left(1 - \frac{1}{\nu}, \frac{1}{\nu}\right), \left(\frac{1}{\nu}, 1 - \frac{1}{\nu}\right)\right).
\end{aligned}
$$

Clearly, $\lim_{\nu \to \infty} b^{\nu} = b'$ holds for this sequence. Bayes' rule implies the following sequence of beliefs:

$$\mu^{\nu}(y) = \frac{b_E^{\nu}(e) \cdot b_E^{\nu}(h)}{b_E^{\nu}(e) \cdot (b_E^{\nu}(h) + b_E^{\nu}(l))} = \frac{b_E^{\nu}(h)}{(b_E^{\nu}(h) + b_E^{\nu}(l))} = 1 - \frac{1}{\nu}.$$

Since these beliefs converge to $\mu'(y) = 1$, the perfect Bayesian equilibrium (b', μ') is also a sequential equilibrium. ■

In example 6.3, the consistency requirement of sequential equilibrium puts no extra constraints on beliefs. This is different in the following example with two information sets following each other.

EXAMPLE 6.5. Consider the following modified version of the market entry game. The entrant enters with high investment, *eh*, or with low investment, *el*, or stays out of the market, *ne*. The monopolist cannot observe the investment decision of the entrant but has to react by accommodation, *a*, or fighting, *f*. A high investment level guarantees the entrant an advantage over the monopolist whether the monopolist fights or accommodates entry. If the monopolist accommodates entry, a regulator who is uninformed about the entrant's investment decision decides whether the market situation conforms to existing regulations (c) or does not (nc). Assume that an entrant's high investment strategy does not comply with regulations. The payoff of the regulator is related to a correct assessment of the investment decision. The following tree describes such a situation.

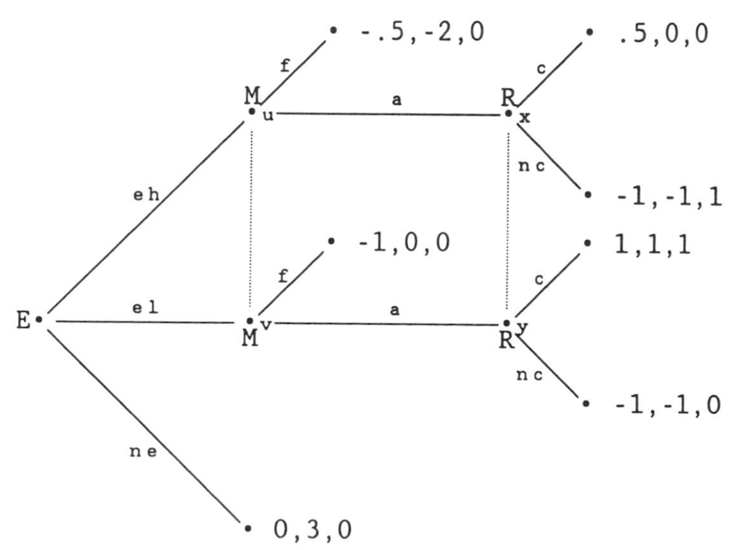

It is easy to check that the following behavior strategy combination b' and belief system μ' form a perfect Bayesian equilibrium:

$$\begin{aligned} b' &= \big((b'_E(eh), b'_E(el), b'_E(ne)), (b'_M(a), b'_M(f)), (b'_R(c), b'_R(nc))\big) \\ &= ((0,0,1),(0,1),(0,1)), \\ \mu' &= ((\mu'(u), \mu'(v)), (\mu'(x), \mu'(y))) = ((0,1),(1,0)). \end{aligned}$$

The out-of-equilibrium beliefs of the monopolist and the regulator are inconsistent because the monopolist believes that the entrant has chosen *el*, whereas the regulator believes *eh* to be the investment level of the entrant.

It is not difficult to see that the perfect Bayesian equilibrium (b', μ') does not constitute a sequential equilibrium. Any sequence of behavior strategies $(b^\nu)_{\nu=1}^\infty$ that justifies the belief $\mu'(v) = 1$ will justify $\mu'(y) = 1$ as well. This follows because

$$\mu^\nu(v) = \frac{b_E^\nu(el)}{(b_E^\nu(eh) + b_E^\nu(el))} = \frac{b_M^\nu(a) \cdot b_E^\nu(el)}{b_M^\nu(a) \cdot (b_E^\nu(eh) + b_E^\nu(el))} = \mu^\nu(y)$$

holds for all ν. For $\mu(v) = \mu(y) = 1$, however, b' is no longer a perfect Bayesian equilibrium. ■

The restrictions on out-of-equilibrium beliefs implied by the sequential equilibrium concept are sufficient to rule out the perfect Bayesian equilibrium of example 6.5 in which players hold beliefs that are inconsistent with each other. This consistency requirement of a sequential equilibrium does not preclude the unreasonable equilibrium of example 6.3. "Fighting" appears to be unreasonable there as the slightest doubt about the entrant's investment decision induces the monopolist to "accommodate." A sequential equilibrium restricts beliefs by forcing them to be justified by some deviation. The deviations themselves need not be Nash equilibria. Thus no mutual optimality is imposed for such a deviation. If the monopolist's choice of action has to be optimal given a small probability that the entrant deviates from the equilibrium strategy, then "fighting" is never chosen. Such a notion of robustness of an equilibrium against small deviations of the opponents is embodied in the concept of a perfect equilibrium introduced in Chapter 4.

It seems natural to apply this notion of robustness to games in extensive form as well. Since an equilibrium behavior strategy is supposed to be optimal at each information set, robustness against "trembles" has to be imposed at each information set of the extensive form. To apply the definition of a perfect equilibrium given in Chapter 4 at each information set, the extensive form is represented as a normal form game with an extended player set.

Each player of the extensive form is replaced by a group of agents in the extended strategic form. There is one agent for each information set at which this player has to move. A mixed strategy of such an agent corresponds to a behavior strategy of the original player at the information set the respective agent controls. All agents of the same player have the same payoff as this player and, therefore, the same incentives as this player. Since agents choose their moves independently in the normal form, these moves cannot be coordinated by the player. Hence, each choice of an agent must be optimal in its own right, which guarantees optimality of behavior strategies at each information set. The normal form arising from this expansion of the player set is called *agent normal form*.

DEFINITION 6.6 (*agent normal form*). The *agent normal form* of an extensive form game

$$\Gamma = \left(I, (N, \sigma), (A, \alpha), (N_i)_{i \in I}, U, (A(u))_{u \in U}, r\right)$$

is the normal form game

$$\Gamma^a = \left(U, (A(u))_{u \in U}, (p_u)_{u \in U}\right)$$

with U as the player set, $A(u)$ as the set of pure strategies for player $u \in U$, and the payoff function $p_u((a_u)_{u \in U}) = r_i(n(a_u)_{u \in U}))$ where i is the player who has to move at information set u, and $(a_u)_{u \in U}$ is a pure strategy combination with $a_u \in A(u)$.

Recall that a pure strategy of an extensive form game specifies an action at each information set and, therefore, determines a particular path through the game. The following example shows how to derive the agent normal form of a game in extensive form.

EXAMPLE 6.3 (*resumed*). In this game the entrant has to move at the information sets $\{o\}$ and $\{x\}$. Hence there are two agents for the entrant, one controlling the information set $\{o\}$, E_o, and one controlling the information set $\{x\}$, E_x. The monopolist has one information set only and is represented by a single agent M. The agent normal form of the game has three agents as players and can be summarized by the following payoff matrices.

E_o: ne

E_x \ M	a	f
h	0, 3	0, 3
l	0, 3	0, 3

E_o: e

E_x \ M	a	f
h	.5, −1	−.5, −1
l	1, 1	−1, −1

This agent normal form represents the extensive form game of example 6.3 as a three-player game in normal form in which player M chooses columns, player E_x rows, and player E_o one of the matrices. ■

The concept of the agent normal form makes it possible to apply the perfect equilibrium concept introduced in Chapter 4 to the extensive form of a game. Recall that a perfect equilibrium is a Nash equilibrium of the normal form game obtained as the limit of a sequence of Nash equilibria of perturbed games. A perturbed game is a game in which all players choose completely mixed strategies. As the constraints on the sets of mixed strategies that ruled out pure strategies are relaxed, the associated sequence of Nash equilibria of the perturbed games converges to a Nash equilibrium of the unperturbed game. All those Nash equilibria of the unperturbed game that can be obtained as a limit of some sequence of perturbed Nash equilibria are perfect equilibria.

According to theorem 4.6, every perfect equilibrium strategy is undominated. This property of a perfect equilibrium is useful for finding perfect equilibria in many games. Since every game with finite strategy sets has a perfect equilibrium (theorem 4.5), one of the undominated Nash equilibria of such a game must be perfect. If elimination of equilibria using dominated strategies leaves only one Nash equilibrium, this must be a perfect equilibrium.

Example 6.3 (*resumed*). It is easy to check that the agent normal form given earlier has the same set of equilibria as the extensive form game:

$$\text{(i)}\quad ((m^*_{Eo}(e), m^*_{Eo}(ne)), (m^*_{Ex}(h), m^*_{Ex}(l)), (m^*_M(a), m^*_M(f))) = ((1,0),(0,1),(1,0))$$

$$\text{(ii)}\quad ((m'_{Eo}(e), m'_{Eo}(ne)), (m'_{Ex}(h), m'_{Ex}(l)), (m'_M(a), m'_M(f))) = ((0,1),(1,0),(q,1-q)) \qquad \text{for } 0 \le q \le .5$$

Notice that "fighting" with a positive probability is a dominated strategy for the monopolist. Hence no equilibrium m' can be a perfect equilibrium, since $m'_M(f) > 0$ for all $q \in [0, .5]$ holds. Since every game has a perfect equilibrium, m^* must be the unique perfect equilibrium of this game. This perfect equilibrium is the equilibrium that does not use the incredible threat of the monopolist to fight. Thus, perfectness of an equilibrium applied to the agent normal form selects the reasonable equilibrium. ■

It is not difficult to see that every agent normal form perfect equilibrium is a sequential equilibrium. Notice that a mixed strategy combination of the agent normal form is a behavior strategy combination of the associated extensive form game. A sequence of Nash equilibria of the perturbed games forms a sequence of completely mixed behavior strategy combinations converging to the perfect equilibrium. Hence the following result can be stated without formal proof.

THEOREM 6.1. *Every agent normal form perfect equilibrium is a sequential equilibrium and, therefore, a perfect Bayesian equilibrium.*

The converse result is not true as example 6.3 demonstrates.[4] Hence perfectness of the agent normal form is a refinement of the sequential equilibrium concept.

One may be tempted to think that the concept of an agent normal form perfect equilibrium constitutes a completely satisfactory equilibrium refinement since it eliminates successfully all Nash equilibria that rely on incredible threats or promises. However, as the next chapter shows, there is a large class of games for which even an agent normal form perfect equilibrium will not restrict the beliefs at information sets not reached in equilibrium. This class includes signalling games which have many applications in economics.

6.4 Summary

A refinement of the Nash equilibrium concept was achieved in this chapter by eliminating equilibria with strategies that prescribe moves at some stage of the game which are suboptimal there. This appears to be a reasonable consistency requirement if actions are carried out sequentially. Such sequential rationality rules out Nash equilibria based on incredible threats.

In finite games with perfect information, sequential rationality could be captured by backward induction. For infinite games with imperfect information, however, this notion had to be extended. The concept of a subgame perfect equilibrium was an appropriate generalization for games with perfect information, but failed to refine the set of Nash equilibria in most games with imperfect information. A strengthening of subgame perfection was obtained by the perfect Bayesian equilibrium concept that introduces beliefs at information sets. These beliefs have to be consistent with the information contained in the equilibrium behavior strategies. This consistency requirement, however, does not determine beliefs at information sets not reached in equilibrium. Since these out-of-equilibrium beliefs support the optimality of equilibrium strategies at those information sets not reached during equilibrium play, they provide an element of arbitrari-

[4]It can be proved, however, that a sequential equilibrium is agent normal form perfect for "almost all" games (Kreps and Wilson, 1982).

ness that reduces the power of the perfect Bayesian equilibrium notion to guarantee sequential rationality.

Further constraints on out-of-equilibrium beliefs, and thus on equilibria, are imposed by two further refinements. Both of them rest on the fact that for completely mixed behavior strategies all information sets of a game will be reached with positive probability. For completely mixed behavior strategy combinations, beliefs at all information sets are determined by Bayesian updating. The notion of a sequential equilibrium requires, therefore, out-of-equilibrium beliefs to be supported by a sequence of completely mixed behavior strategies and associated beliefs converging to the perfect Bayesian equilibrium under consideration. For a sequential equilibrium, the completely mixed behavior strategies of the sequence were not required to be Nash equilibria. This distinguishes a sequential equilibrium from the related refinement of an agent normal form perfect equilibrium. In an agent normal form perfect equilibrium, the perfect Bayesian equilibrium had to be supported by a sequence of Nash equilibria of perturbed games that ruled out the choice of pure strategies.

6.5 Remarks on the Literature

The literature on refinements of the Nash equilibrium concept begins with a paper by Selten (1975) that introduced the concept of a perfect equilibrium in the normal and agent normal form of a game. This paper generalized the notion of subgame perfectness introduced before in Selten (1965). Kreps and Wilson (1982) introduce the notion of beliefs and of a sequential equilibrium. This paper also studies the relationship between the sequential equilibrium concept and the perfect equilibrium concept. Tirole (1988) and Fudenberg and Tirole (1989) use the concept of a perfect Bayesian equilibrium in numerous applications to industrial organization problems. In a recent paper Fudenberg and Tirole (1991a) relate this concept to other refinements.

The literature on equilibrium refinements has grown rapidly during the past decade and it is impossible to provide even a limited survey here. It is important to note there are many more refinements proposed and discussed in the literature not included in this chapter. The omissions are mainly motivated by an attempt to concentrate on those concepts most commonly applied to economic problems. Many of the omitted refinements are, however, of great importance in the context of a systematic treatment of game-theoretic ideas. In particular, the proper equilibrium concept of Myerson (1978) and the persistent equilibrium concept of Kalai and Samet (1984) should be mentioned here. A useful nontechnical survey over this field is provided by van Damme (1990).

Exercises

6.1 *Consider the following modification of the market entry game (example 6.1). Here the entrant has to choose an action x or y after the monopolist has accommodated. If action x is chosen the same outcome results as under accommodation before. If y is chosen the entrant receives* 0 *and the monopolist* -2.

   (a) Draw the game tree for this game and derive all Nash equilibria in pure strategies.
   (b) Use backward induction to select an equilibrium.
   (c) Indicate all subgames and show that the backward induction equilibrium is subgame-perfect.

6.2 *A seller of a car bargains with a potential buyer over the price. It is common knowledge that, at the beginning, the buyer's maximum willingness to pay is* 100 *and the seller's reservation value is* 90. *The seller proposes a price p that the buyer can accept or reject. If accepted the buyer buys the car for this price, if rejected the buyer makes a final take-it-or-leave-it offer of a price q. In the second round however, the seller's reservation price has risen to* $90 + \delta$, $\delta < 10$.

   (a) Describe actions, strategies, and payoffs of the two players.
   (b) Show that, for any price p, $p \in [90, 100]$, a take-it-or-leave-it offer of the seller that is accepted by the buyer forms a Nash equilibrium.
   (c) Show that the only subgame perfect equilibrium has the seller propose his second-stage reservation price and the buyer accept this offer.

6.3 *Consider the seller of a used car and a potential buyer. Quality of cars is measured by a number s between* 0 *and* 1. *The seller knows the quality of the car, whereas the buyer knows only the probability distribution for the quality of used cars* $G(s) = s$. *Assume the seller values a car of quality s according to the function* $V_1(s) = \theta_1 \cdot s$, *and the buyer values it according to* $V_2(s) = \theta_2 \cdot s$ *with* $\theta_1 < \theta_2$.

   (a) Derive the payoff functions of the buyer and seller and determine the demand and supply functions of these agents under the assumption that both know that the quality of the car is s_o. What is the equilibrium price?
   (b) Suppose that the buyer does not know the quality s and obtains an offer to get the car at a price p. How does this affect the quality assessment of the buyer?
   (c) What equilibrium price will be charged in the market under asymmetric information? Does an equilibrium exist?

6.4 *Consider a monopolistic chain store that faces a threat of entry from two potential entrants. The entry game proceedes as in example 6.1, but* *the potential entrants are uncertain as to the payoff of the chain store if it fights entry (all other payoffs are as in example 6.1). Suppose there is a probability q that the monopolistic chain store has a payoff of -1; with probability $(1-q)$, its payoff is 2. Suppose the entrants make their decision to enter or not to enter sequentially. If the first entrant enters, then the second observes the chain store's reaction before making its own decision. The payoff of the monopolist at the end of stage 2 is the sum of its payoffs in each stage of the game.*

(a) Assume that $q < .5$ holds. Derive the perfect Bayesian equilibrium strategies and beliefs for the chain store and the two entrants. Can it be a perfect Bayesian equilibrium strategy for the type of chain store with payoff -1 to fight in stage 1, or is this an incredible threat?

(b) How does the result change if $q \geq .5$ holds?

6.5 *Consider an insurance company that insures against property damage D. It is known there are two types of potential customers who differ in their loss probability $q_H > q_L$ but are otherwise identical. Suppose that both types of customers have a wealth W before damage occurs and a strictly concave utility function $u(\cdot)$ strictly increasing in wealth. The proportion of high-risk customers, with q_H, is μ. The type of a customer is private information.*

(a) Insurance contracts $\gamma = (x, y)$ specify the payment of the insurer to the insured in case of damage, $x \geq 0$, and the premium payment of the insured to the insurer, $y \geq 0$. Draw a diagram with the indifference curves over (x, y)-combinations for both types of customers.

(b) Suppose the insurer is a monopolist. Show that, for any μ, there is a perfect Bayesian equilibrium in which the insurer proposes two types of contracts $\gamma_i = (x_i, y_i)$, $i = H, L$, and the customer selects the contract. Show the equilibrium contracts in the (x, y)-diagram.

(c) Is the equilibrium a pooling or a separating equilibrium?

7

Refinements of the Nash Equilibrium Concept: Forward Induction

For games with perfect information, sequential rationality provides a powerful refinement of the Nash equilibrium concept. In its strongest form, as agent normal form perfect equilibrium, it eliminates equilibria based on incredible threats or promises even in games with imperfect information. There are however many games, particularly games of incomplete information, in which the sequential rationality requirement does not reduce the number of equilibria. Of special importance for economics are signalling games in which credibility of threats or promises is not an issue. Sequential rationality, therefore, does not eliminate any equilibrium in this class of games. The following example will illustrate this point.

Example 7.1. Consider bank B and applicant A, who applies for a loan to finance an investment project. It is known that there are applicants with projects that yield a low return and applicants with projects yielding a high return, but only the applicant knows the type. The *ex-ante* probability of facing an applicant with the high-return project is $1/4$. Applications for low-return projects are standard business and can be processed without extra costs. Applications for a high-return project, however, require a larger loan and, therefore, an approval process costly for both parties. A_l (A_h) denotes an applicant with a low-return (high-return) project and a_l (a_h) an application for a low-return (high-return) loan. The bank may either approve the loan (a) or reject it (r). The following game tree represents this model and indicates the payoffs.

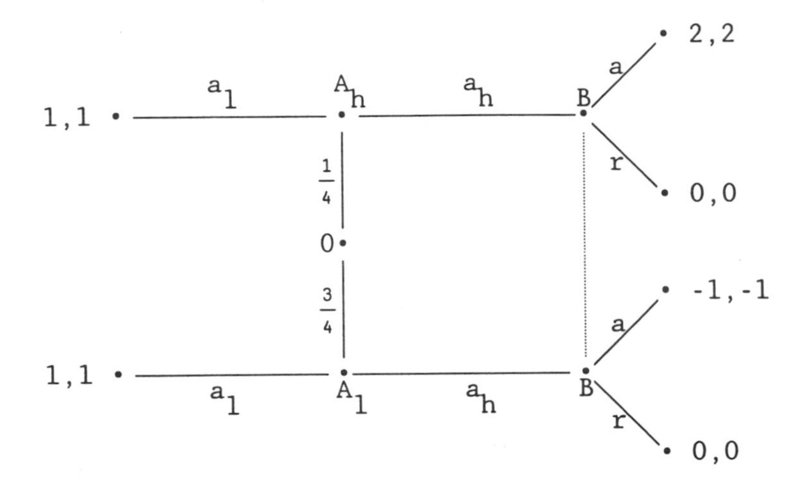

It is not difficult to check that the following behavior strategy combinations, $b_A^h = (b_A^h(a_h), b_A^h(a_l))$, $b_A^l = (b_A^l(a_h), b_A^l(a_l))$, and $b_B = (b_B(a), b_B(r))$ and beliefs, $\mu = (\mu(A_h), \mu(A_l))$, form perfect Bayesian equilibria of this game:

(i) separating equilibrium:

$$b_A^h = (1,0), \qquad b_A^l = (0,1), \qquad b_B = (1,0), \qquad \mu = (1,0);$$

(ii) pooling equilibria: for $0 \leq q \leq \frac{1}{3}$

$$b_A^h = (0,1), \qquad b_A^l = (0,1), \qquad b_B = (0,1), \qquad \mu = (q, 1-q).$$

Note that the bank's information set contains one node for each type of applicant. A belief about the type of the applicant is, therefore, the same as a belief about the node where the bank moves.

All of these equilibria are perfect in the agent normal form and are, therefore, sequential equilibria. To see this, consider the following perturbations on behavior strategies: $0 < \eta_i \leq b_i(\cdot)$ for $i = A_h, A_l, B$. Notice that both actions of a player have the same constraint η_i. Since each type of the applicant is treated as a separate player, perturbing the behavior strategies is equivalent to perturbing the mixed strategies of the agent normal form. For sufficiently small η_i, the perturbed game has the following equilibria:

(i) separating equilibrium:

$$b_A^h = (1-\eta_{Ah}, \eta_{Ah}), \qquad b_A^l = (\eta_{Al}, 1-\eta_{Al}) \qquad b_B = (1-\eta_B, \eta_B),$$

$$\mu(A_h) = \frac{.25 \cdot b_A^h(a_h)}{.25 \cdot b_A^h(a_h) + .75 \cdot b_A^l(a_h)} = \frac{1-\eta_{Ah}}{1-\eta_{Ah} + 3 \cdot \eta_{Al}},$$

$$\mu(A_l) = 1 - \mu(A_h);$$

(ii) pooling equilibrium:

$$b_A^h = (\eta_{Ah}, 1 - \eta_{Ah}), \qquad b_A^l = (\eta_{Al}, 1 - \eta_{Al}), \qquad b_B = (\eta_B, 1 - \eta_B),$$

$$\mu(A_h) = \frac{.25 \cdot b_A^h(a_h)}{.25 \cdot b_A^h(a_h) + .75 \cdot b_A^l(a_h)} = \frac{\eta_{Ah}}{\eta_{Ah} + 3 \cdot \eta_{Al}},$$

$$\mu(A_l) = 1 - \mu(A_h),$$

with $\eta_{Ah} = (3 \cdot q/1 - q) \cdot \eta_{Al}$ for any $q \in [0, \frac{1}{3}]$.

Notice that $\mu(A_h) = q$ holds in this case for all (η_{Ah}, η_{Al}). These behavior strategy combinations form perfect Bayesian equilibria given a perturbation $\eta = (\eta_B, \eta_{Ah}, \eta_{Al})$. As η converges to zero, these equilibria converge to the unperturbed equilibria of the game. It is necessary, however, to perturb the pooling equilibria in a particular way, $\eta_{Ah} = (3 \cdot q/1 - q) \cdot \eta_{Al}$; but for a perfect equilibrium it suffices to find *some* sequence of perturbed equilibria that converges to the equilibrium under consideration. Thus, one can choose an appropriate perturbation for each equilibrium. Incidentally, this example shows that agent normal form perfect equilibria are not robust against arbitrary perturbations. ■

In example 7.1, all perfect Bayesian equilibria are perfect in the agent normal form. Thus no selection is achieved by the refinements based on sequential rationality. Yet the pooling equilibria require the bank to believe that there is a high probability that it faces a low-return applicant if it gets a high-return application. This is possible in the pooling equilibria because there is zero probability of observing a_h according to the equilibrium strategy. This leaves the beliefs of the bank unconstrained if a_h actually occurs. Nevertheless, the beliefs in this contingency determine the optimal action of the bank, and the bank's behavior in turn influences the behavior of the applicant. Thus, in a pooling equilibrium, an applicant with the high-return project will not apply for a high-return project loan because the bank would misinterpret this move.

This example makes it clear why the backward induction logic embodied in the agent normal form perfectness concept does not distinguish between these equilibria. Whether a threat is credible may be tested by playing the strategy that induces the threatened behavior with some positive probability. Whether an action reveals some information requires analyzing the motivation of the agents who chose the action. Creating some risk through perturbations that each player may have to carry out all equilibrium actions does not provide such information.

The pooling equilibria in example 7.1 appear unreasonable because it seems implausible for the bank, faced with an application a_h, to believe that the applicant was the type with the smallest incentive to submit such an application. The low-return applicant has a dominant strategy to choose a_l, but the high-return applicant could find it optimal to choose a_h. Thus, the bank should put a higher probability on type A_h because of the stronger incentives of this type to make such an application.

In this argument, an analysis of the incentives of the different types of players reveals which types are more likely to choose the action leading to the respective information set. Such reasoning about the incentives of players is called forward induction. The following example illustrates this line of argument in a different context.

EXAMPLE 7.2. Two players work together to produce some output. If both choose a high level of effort e_h they can achieve a high payoff, 20 for each of them, whereas coordination on a low effort level produces a payoff of 10 for each of them. Lack of coordination is very costly for the high-effort player. Assume that player 1 has the additional option of participating in this joint venture (p) or not (n). In the latter case, player 1 obtains a payoff of 15. The following game tree and payoff matrix represent this game. p/e_h denotes the strategy of player 1 first to participate and then to choose the high-effort level. An analogous notation obtains for p/e_l, n/e_h, and n/e_l.

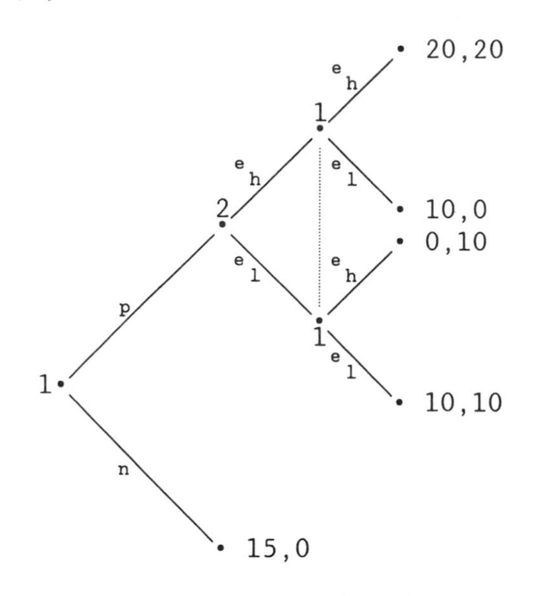

		Player 2	
		e_h	e_l
Player 1	p/e_h	20, 20	0, 10
	p/e_l	10, 0	10, 10
	n/e_h	15, 0	15, 0
	n/e_l	15, 0	15, 0

The subgame beginning at the information set of player 2 has two equilibria, (e_h, e_h) and (e_l, e_l). Hence, there are two subgame perfect equilibria, $(n/e_l, e_l)$ and $(p/e_h, e_h)$, which are both perfect in the agent normal form. The equilibrium $(n/e_l, e_l)$ is strange because the low effort level is optimal for player 2 only if she believes that player 1 will also choose the low effort level. But if player 1 participates in the joint venture, why would she choose a low effort level yielding not more than 10, since she can obtain a certain payoff of 15? Thus, given that player 1 may refuse to participate, it seems unreasonable to assume that she would participate with a low effort level. If player 2 expects player 1 to enter with a high effort level, she will find it optimal to choose the high effort level e_h, which provides an incentive for player 1 to participate with a high effort level. Thus, only the equilibrium $(p/e_h, e_h)$ seems sensible. ■

Examples 7.1 and 7.2 suggest as a new principle for belief formation that beliefs should take into account the incentives of the players for actions that led to the information set in question. The new principle requires a player to assume that opponents do not choose actions that necessarily lead to a payoff that is inferior to some outcome they could have obtained earlier with another action. Since this principle reasons from past behavior forward it is often referred to as forward induction. Arguments about belief formation based on the forward induction principle are particularly important in the context of signalling games.

7.1 Signalling Games

Signalling games have a simple structure. There are two players. Player 1 has private information about some characteristic and moves first. Player 2, moving second, observes the action of the first player but does not know the type of player 1. The move of player 1 is a signal for player 2 that may reveal the type of player 1 (separating equilibrium) or may not reveal it (pooling equilibrium).

Definition 7.1 (*signalling game*). A *signalling game* is a two-player game of incomplete information about player 1's characteristics in which player 1 moves first and each player moves just once.

In signalling games, there are as many information sets for player 2 as there are actions for player 1 and each information set of player 2 contains as many nodes as there are types of player 1. Thus, player 1's strategy is type-contingent and player 2's strategy specifies an action for each signal of player 1, that is, for each information set. Furthermore, one can identify beliefs about the node in an information set with beliefs about the type of player 1.

The special features of a signalling game suggest the following modified notation. To simplify the exposition of the analysis of signalling games, only pure strategies are considered in this chapter. It should be noted, however, that most games also have equilibria in strictly mixed behavior strategies and some games have only equilibria in strictly mixed behavior strategies.[1] Let T denote the set of types of player 1 with initial type distribution μ. A_i is the set of actions available to player i, and for given actions of player 1 and player 2, a_1 and a_2, $p_i(a_1, a_2; t)$ is the payoff to player i if player 1's type is t. A strategy of player 1 is a type-contingent choice of actions $s_1 = (\alpha(t))_{t \in T}$ with $\alpha(t) \in A_1$ for all $t \in T$. A strategy for player 2, $s_2 = (\beta(a))_{a \in A_1}$, consists of a choice of action for each information set, or equivalently for each action of player 1.

Furthermore, it is convenient to write $\mu(t|a)$ for the beliefs of player 2 about the type of player 1 after observing player 1's action a. Given a strategy of player 1, $s_1 = ((\alpha(t))_{t \in T}$, let $T(a) = \{t \in T | \alpha(t) = a\}$ be the set of types choosing action a. If $T(a) \neq \emptyset$ and therefore $[\Sigma_{t \in T(a)} \mu(t)] > 0$ holds, that is, if there is a type choosing action a according to strategy s_1, then player 2's beliefs can be derived according to Bayes' rule as $\mu(t|a) = \mu(t)/[\Sigma_{t \in T(a)} \mu(t)]$; otherwise any belief $\mu(\cdot|a)$ that satisfies $\mu(t|a) \geq 0$ for all $t \in T$ and $\Sigma_{t \in T} \mu(t|a) = 1$ is adequate. Adapting definition 6.4 to the notation of a signalling game, a perfect Bayesian equilibrium is a strategy combination $(s_1^*, s_2^*) = ((\alpha^*(t))_{t \in T}, (\beta^*(a))_{a \in A_1})$ and a belief system $(\mu(\cdot|a))_{a \in A_1}$ obtained according to Bayes' rule wherever possible.

The following game tree, with two types for player 1, $T = \{t_1, t_2\}$, and two actions for each player, $A_1 = \{a_1, b_1\}$ and $A_2 = \{a_2, b_2\}$, represents the prototype of a signalling game.

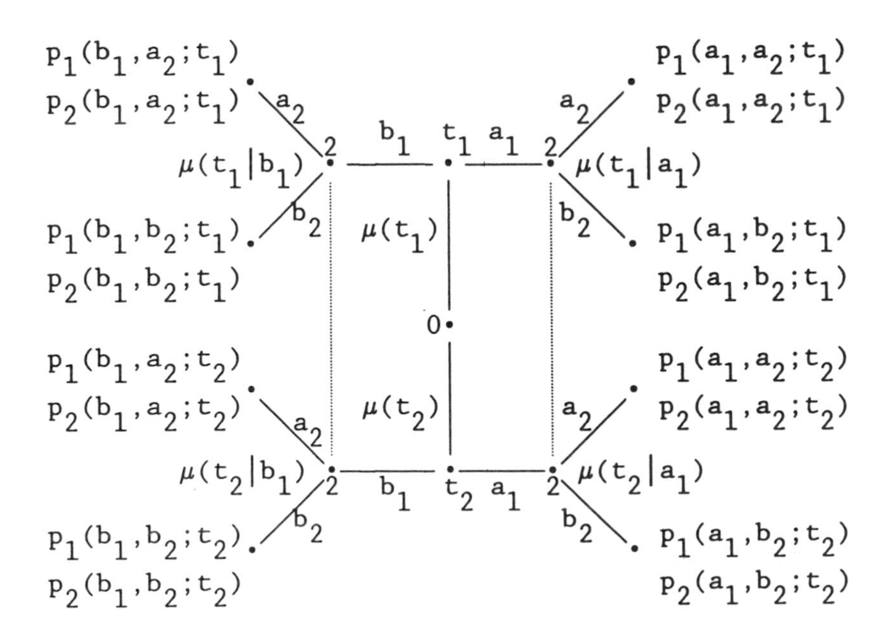

[1]Dasgupta and Maskin (1986) provide some examples for the latter case.

Two types of equilibria are of particular interest in signalling games:

- separating equilibria at which each type of player 1 chooses a different action, thus revealing her type
- pooling equilibria at which each type of player 1 chooses the same action, thus providing no information about her type[2]

In the prototype game here, $(s_1^*, s_2^*) = ((\alpha^*(t_1), \alpha^*(t_2)), (\beta^*(a_1), \beta^*(b_1))$ is a separating equilibrium if the following hold:

(i) $\alpha^*(t_1) \neq \alpha^*(t_2)$ and therefore $\mu(t_i|\alpha^*(t_i)) = 1$
(ii) for each type t_i, $i = 1, 2$

$$p_1(\alpha^*(t_i), \beta^*(\alpha^*(t_i)); t_i) \geq p_1(a, \beta^*(\alpha^*(t_i)); t_i) \quad \text{for all} \quad a \in A_1,$$
$$p_2(\alpha^*(t_i), \beta^*(*\alpha(t_i)); t_i) \geq p_2(\alpha^*(t_i), a; t_i) \quad \text{for all} \quad a \in A_2$$

A separating equilibrium is revealing, condition (i), and both players must choose maximizing actions given the type of player 1, condition (ii). If player 1 has just two actions available, as in the prototype game, then beliefs are completely determined in a perfect Bayesian equilibrium. However, if player 1 has more actions than types, then there are necessarily actions that will not be played in equilibrium. After such out-of-equilibrium actions, beliefs are not determined by Bayesian updating because $T(a) = \emptyset$.

Similarly, a strategy combination $(s_1', s_2') = ((\alpha'(t_1), \alpha'(t_2)), (\beta'(a_1), \beta'(b_1)))$ is a pooling equilibrium if the following conditions hold:

(i) $\alpha'(t_1) = \alpha'(t_2) \equiv \alpha$ and therefore $\mu(t_i|\alpha'(t_i)) = \mu(t_i)$, $\beta'(\alpha) \equiv \beta$
(ii) for each type t_i, $i = 1, 2$ $p_1(\alpha, \beta; t_i) \geq p_1(a, \beta; t_i)$ for all $a \in A_1$,

$$\mu(t_1) \cdot p_2(\alpha, \beta; t_1) + \mu(t_2) \cdot p_2(\alpha, \beta; t_2)$$
$$\geq \mu(t_1) \cdot p_2(\alpha, a; t_1) + \mu(t_2) \cdot p_2(\alpha, a; t_2) \quad \text{for all} \quad a \in A_2$$

The pooling equilibrium leaves the action and the beliefs of player 2 indeterminate at the information set at which player 1 has chosen the non-equilibrium action $a \neq \alpha$. Although action a does not occur during equilibrium play, it nevertheless influences which actions α and β form equilibrium strategies. As in the case of the separating equilibrium, there will be more information sets off the equilibrium path if there are more actions for player 1 than there are types of this player.

Most economic applications consider signalling games with few possible types and many possible actions for player 1. Thus typically, there are many perfect Bayesian equilibria. The forward induction principle provides a means to reduce the number of possible beliefs following the

[2]There are of course equilibria at which some types choose different actions and others the same, but such semi-revealing equilibria will be disregarded in this chapter.

out-of-equilibrium signals. Suppose that player 2 observes an action of player 1 that is not an equilibrium action. How should she assess the situation? According to the forward induction logic, player 2 should not assume that player 1 chose an action that under no circumstances will yield her a better payoff than the equilibrium action. The basic idea can be described as follows:

(i) For any out-of-equilibrium move of player 1 restrict beliefs μ to those types who would not be worse off with this move, for some associated best response of the second player, than they are in equilibrium.

(ii) If there are beliefs satisfying this restriction and an associated best response of player 2 such that no type of player 1 wants to choose this out-of-equilibrium move, then the equilibrium satisfies the "intuitive criterion."

To formalize the intuitive criterion,[3] for any action a of player 1 denote by $BR_2(a, \mu(\cdot|a))$ the set of best response strategies of player 2 to action a and beliefs $\mu(\cdot|a)$ on T. Let $\hat{p}_1(a,t) = \max\{p_1(a,b;t)|b \in BR_2(a, \mu(\cdot|a)$ for some $\mu(\cdot|a)\}$ be the maximum payoff that player 1 of type t may achieve from action a and a best response of player 2 to some belief $\mu(\cdot|a)$ that is consistent with Bayesian updating. Note that for out-of-equilibrium actions a, any belief is consistent with Bayesian updating. Finally,

$$\hat{T}(a) = \{t \in T | p_1(\alpha^*(t), \beta^*(\alpha^*(t);t) > \hat{p}_1(a,t)\}$$

is the set of types for which action a would necessarily lead to a worse payoff than the equilibrium strategy (s_1^*, s_2^*).

Definition 7.2. A perfect Bayesian equilibrium of a signalling game $((s_1^*, s_2^*), \mu)$ satisfies the *intuitive criterion* if the following condition holds: For any out-of-equilibrium action of player 1 a, $a \neq \alpha^*(t)$ for all $t \in T$, with $\hat{T}(a) \neq T$, there exists a $\mu'(\cdot|a)$ with support $T/\hat{T}(a)$ and a best response of player 2 $\mathscr{b}$, $\mathscr{b} \in BR_2(a, \mu'(\cdot|a))$, such that

$$((\alpha^*(t))_{t\in T}, ((\beta^*(a))_{a \neq a}, \mathscr{b}), ((\mu^*(\cdot|a))_{a\neq a}, \mu'(\cdot|a)))$$

is also a perfect Bayesian equilibrium.

According to this definition, a perfect Bayesian equilibrium satisfies the intuitive criterion if it is not based on beliefs that an out-of-equilibrium signal comes from a type that does not gain from sending this signal. The following example illustrates the power of the intuitive criterion to reduce the number of perfect Bayesian equilibria in signalling games.

Example 7.1 (*resumed*). Recall that there are two types of applicants for a bank loan, A_h and A_l, which have two actions: either to apply

[3]Cho and Kreps (1987) define the intuitive criterion as a negative test.

for a high-return loan a_h or for a low-return loan a_l. The only difference from the prototype signalling game lies in the assumption that the bank will only have to decide about acceptance, a, or rejection, r, after a high-return application comes in. Low-return loan applications will be granted in any case. In the notation of a signalling game, the perfect Bayesian equilibrium strategies $(s_1^*, s_2^*) = ((\alpha^*(A_h), \alpha^*(A_l)), (\beta^*(a_h))$ and beliefs $(\mu(A_h|a_h), \mu(A_l|a_h))$ can be written as follows:

(i) separating equilibrium:

$$\alpha^*(A_h) = a_h \qquad \alpha^*(A_l) = a_1, \qquad \beta^*(a_h) = a,$$
$$\mu(A_h|a_h) = 1 \qquad \mu(A_l|a_h) = 0$$

(ii) pooling equilibria: for $0 \le q \le \frac{1}{3}$,

$$\alpha^*(A_h) = \alpha^*(A_l) = a_l, \qquad \beta^*(a_h) = r$$
$$\mu(A_h|a_h) = q, \qquad \mu(A_l|a_h) = 1 - q$$

Notice that there is a unique separating equilibrium. Since the applicant has only two actions, both played in equilibrium, no information set off the equilibrium path exists in a separating equilibrium. There are however, pooling equilibria as well. In these equilibria, an application for a high-return loan should not be observed because it would be rejected based on the bank's beliefs that a high-return applicant is unlikely, $q \le 1/3$ to send such an application.

In a pooling equilibrium, the payoff of each type of applicant is $p_l(a_l, r; A_i) = 1$, $i = h, l$. If the bank observes an application a_h and forms out-of-equilibrium beliefs $(\mu(A_h|a_h), \mu(A_l|a_h)) = (m, 1 - m)$, $m \in [0, 1]$, the best response of the bank would be

$$BR_B(a_h, (m, 1 - m)) = \begin{cases} \{a\} & \text{for } m > 1/3 \\ \{a, r\} & \text{for } m = 1/3 \\ \{r\} & \text{for } m < 1/3 \end{cases}.$$

Hence, the best possible payoff for the two types from applying for the high-return loan is

$$\hat{p}_1(a_h, t) = \begin{cases} 2 & \text{for } t = A_h \\ 0 & \text{for } t = A_l \end{cases}.$$

For the low-return applicant, the best possible payoff from applying for the high-return loan is clearly worse than the equilibrium payoff $p_l(a_l, r; A_1) = 1$. Hence according to the intuitive criterion, the bank should not believe that the applicant with a low-return project applies for the high-return loan, $\hat{T}(a_h) = \{A_l\}$. If the low-return applicant is ruled out as a possible sender of a high-return application, then the beliefs of the bank should be concentrated on $T \setminus \hat{T}(a_h) = \{A_h\}$. Thus, the bank should conclude from this argument that the application a_h comes with probability one from the agent with the high-return project, $\mu(A_h|a_h) = 1$. Given this belief, the bank's optimal strategy is to accept the application,

$\beta(a_h) = a$. But given such a strategy, it is suboptimal for the applicant with the high-return project to apply for a low-return loan. Thus, no pooling equilibrium passes according to the intuitive criterion. The separating equilibrium is the unique perfect Bayesian equilibrium of this game that satisfies the intuitive criterion. ■

In example 7.1, all pooling equilibria failed the intuitive criterion and only the separating equilibrium passed the test. The education-signalling game of Spence (1973) provides an example for a game in which not all separating equilibria satisfy the intuitive criterion.

EXAMPLE 7.3 (*Spence's signalling game*).[4] A firm offers employment to workers with different productivity. Assume that there are just two types of workers, one with high productivity π_h and another with low productivity π_l, $\pi_h > \pi_l$. Productivity of a worker is private information, but the firm knows that the probability of hiring a high-productivity worker is q, $0 < q < 1$. Furthermore, it is known that high-productivity workers have a comparative advantage in acquiring an education level e reflected in different cost functions: $C_t(e) = c_t \cdot e$, $t = h, l$, with $c_l > c_h$.

Assume that each worker chooses a level of education e and a wage rate w and proposes the contract $\gamma = (w, e)$ to the firm. The education level has no productive effect but provides a signal for the firm that may help to distinguish workers of different types. The firm accepts, a, or rejects, r, the proposed contract depending on the profit it expects to make. The payoffs of the players are given in the following game tree that represents this game for a proposed contract $\gamma = (w, e)$.

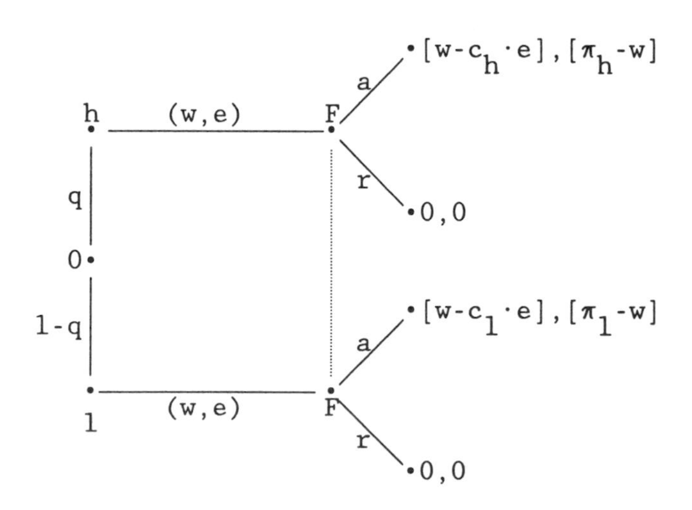

[4] This is the model of Spence (1973), modified as in Tirole (1988, pp. 448–450) to conform to the standard structure of a signalling game.

Implicit in these payoffs is the assumption that the education level is not a sunk cost for the worker. Furthermore, it is immediately clear from this tree that the firm will reject any offer yielding a negative expected profit $[\mu(h|\gamma)\cdot\pi_h + \mu(l|\gamma)\cdot\pi_l - w]$. Notice that the expected profit of the firm depends on the updated beliefs of the firm about the worker's type, $\mu(h|\gamma)$ and $\mu(l|\gamma)$. The education level serves as a signal for this updating process because the more productive worker has a lower cost of obtaining a given level of education.

There is a continuum of contracts in this game, $\gamma = (w, e) \in \mathbb{R}^2_+$. Thus, there are many more actions than types of player 1 and hence many information sets not reached in equilibrium. Beliefs at information sets off the equilibrium path will, therefore, support a multiplicity of perfect Bayesian equilibria.

- *Separating equilibria:* Any two contracts $\gamma' = (w', e')$ and $\gamma'' = (w'', 0)$, that satisfy
  (i) $[\pi_h - w'] \geq 0, \qquad w'' = \pi_l$
  (ii) $[w' - c_h\cdot e'] \geq w'' \geq [w' - c_l\cdot e']$
  will support a separating equilibrium

$$((s_1^*, s_2^*), \mu) = \big(\alpha^*(h), \alpha^*(l), (\beta^*(\gamma))_{\gamma\in\mathbb{R}^2},$$
$$(\mu(h|\gamma), \mu(l|\gamma))_{\gamma\in\mathbb{R}^2}\big)$$

with $\alpha^*(h) = \gamma'$, $\alpha^*(l) = \gamma''$,

$$\beta^*(\gamma) = \begin{cases} a & \text{for } \gamma \in \{(w,e)|w < \pi_l\} \cup \{\gamma', \gamma''\} \\ r & \text{otherwise} \end{cases},$$

$$\mu(h|\gamma) = \begin{cases} 1 & \text{for } \gamma = \gamma' \\ 0 & \text{otherwise} \end{cases}.$$

It is not difficult to check that the strategies and beliefs $((s_1^*, s_2^*), \mu)$ given here indeed form a perfect Bayesian equilibrium. If a high-productivity worker proposes γ', the contract will be accepted. By condition (ii), this yields a better payoff for her than either suggesting the contract γ'' or a contract with a wage below π_l which would be accepted, or any other contract γ which would be rejected. By a similar argument, γ'' is an optimal action for the low-productivity worker. Notice that the separating contracts γ' and γ'' satisfy condition (ii), which is called *self-selection constraint* or *incentive constraint*. Since both types of worker offer a different contract, the firm will learn the type of the worker from her offer. It is optimal for the firm to accept the contract γ', since it comes from the high-productivity worker, and the contract γ'' suggested by the low-productivity worker. If any other contract γ is proposed, Bayesian updating does not restrict the beliefs of the firm. Given the assumption that the firm believes that any contract other than γ' comes from a low-productivity worker, it must reject any contract with a wage rate above the low productivity level π_l and accept any contract with a wage rate below π_l.

The following diagram shows a separating contract pair (γ', γ'') with $\gamma'' = (\pi_l, 0)$. Any contract γ' that satisfies conditions (i) and (ii) must lie in the area between the lines $w = w'' + c_l \cdot e$, $w = w'' + c_h \cdot e$, and below π_h.

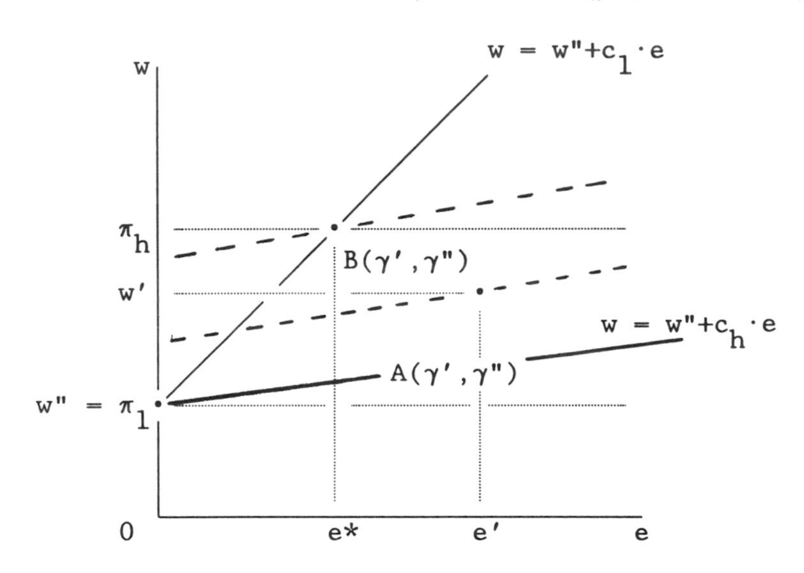

In the separating equilibria, a low-productivity worker will choose the lowest level of education, $e'' = 0$, and will be paid a wage equal to her marginal product, $w'' = \pi_l$. The firm believes that, except for γ', any contract specifying a wage higher than π_l comes from the low-productivity worker. Thus it will reject such an offer. This raises the question of whether such beliefs are compatible with the intuitive criterion.

Consider any out-of-equilibrium contract γ, then $\mu(h|\gamma) = m$ and $\mu(l|\gamma) = 1 - m$ is compatible with Bayesian updating for any $m \in [0, 1]$. The best response of the firm to any contract proposal depends on these beliefs as follows:

$$BR_F(\gamma, m) = \begin{cases} \{a\} & \\ \{a, r\} & \text{for } \left[m \cdot \pi_h + (1-m) \cdot \pi_l\right] - w \begin{matrix} > 0 \\ = 0 \\ < 0 \end{matrix} . \\ \{r\} & \end{cases}$$

If the firm believes that the worker who proposes the contract has a high productivity, then the firm will accept any proposal with a wage below π_h. Thus, a worker of type t who proposes the contract $\gamma = (w, e)$ with $w \leq \pi_h$ can hope to obtain a payoff $\hat{p}_W(\gamma, t) = w - c_t \cdot e$. Comparing the payoff $\hat{p}_W(\gamma, t)$ to the equilibrium payoff of this worker, one can identify those types of the worker who are worse off with the contract γ than with the equilibrium contract. Denote the set of contracts that are worse than the equilibrium contract for both types of worker

$$A(\gamma', \gamma'') = \{(w, e) | w' - c_h \cdot e' > w - c_h \cdot e,\ w'' > w - c_l \cdot e\},$$

and for the low-productivity worker alone

$$B(\gamma', \gamma'') = \{(w, e) | w' - c_h \cdot e' \leq w - c_h \cdot e,\ w'' > w - c_l \cdot e\}.$$

Both sets are indicated in the diagram. The types of workers that cannot gain from a contract $\gamma = (w, e)$ are in $\hat{T}((w, e))$,

$$\hat{T}((w, e)) = \begin{cases} \{h, l\} & \text{for } (w, e) \in A(\gamma', \gamma'') \\ \{l\} & \text{for } (w, e) \in B(\gamma', \gamma'') \end{cases}.$$

For out-of-equilibrium contracts in $A(\gamma', \gamma'')$, $T = \hat{T}((w, e))$ holds and no constraints are imposed on beliefs. For contracts γ in $B(\gamma', \gamma'')$, however, $T/\hat{T}(\gamma) = \{h\}$, the firm should believe with probability one that the high-productivity worker makes the proposal. If the beliefs of the firm are modified to $\mu'(h|\gamma) = 1$ for $\gamma \in B(\gamma', \gamma'')$, then it will accept any contract in $B(\gamma', \gamma'')$ with $w \leq \pi_h$. The perfect Bayesian equilibrium contract pair (γ', γ'') in the diagram does not satisfy the intuitive criterion, because the worker with high productivity prefers a contract in $B(\gamma', \gamma'')$ with $\pi_h \geq w > w'$ and such a contract would be accepted.

It is not difficult to see that the only contract pair that yields a separating perfect Bayesian equilibrium that satisfies the intuitive criterion is the pair (γ^*, γ'') in the diagram. In this case, $\gamma \in B(\gamma^*, \gamma'')$ implies $w > \pi_h$. Thus, there are no out-of-equilibrium beliefs that satisfy the intuitive criterion and are profitable for the firm.

The equilibrium allocation in the unique separating equilibrium remaining after applying the intuitive criterion can be summarized as follows: Both types of worker get a wage equal to their marginal product. The worker with low productivity chooses no education, while the high-productivity worker chooses just enough education to be distinguished from the other type. The firm makes no profits.

- *Pooling equilibria:* There are also pooling equilibria in this game. Given any contract $\bar{\gamma}$ with the following properties
  (i) $[q \cdot \pi_h + (1 - q) \cdot \pi_l] - \bar{w} \geq 0$
  (ii) $[\bar{w} - c_l \cdot \bar{e}] \geq \pi_l$
  then the following strategies and beliefs form a perfect Bayesian equilibrium:

$$((\bar{s}_1, \bar{s}_2), \mu) = \left(\bar{\alpha}(h), \bar{\alpha}(l), (\bar{\beta}(\gamma))_{\gamma \in \mathbb{R}^2}, (\mu(h|\gamma), \mu(l|\gamma))_{\gamma \in \mathbb{R}^2}\right)$$

with $\bar{\alpha}(h) = \bar{\alpha}(l) = \bar{\gamma}$,

$$\bar{\beta}(\gamma) = \begin{cases} a & \text{for } \gamma \in \{(w,e) | w < \pi_l\} \cup \{\bar{\gamma}\} \\ r & \text{otherwise} \end{cases},$$

$$\mu(h|\gamma) = \begin{cases} q & \text{for } \gamma = \bar{\gamma} \\ 0 & \text{otherwise} \end{cases}.$$

It is not difficult to see that these strategies and beliefs form a perfect Bayesian equilibrium. Since both types of worker offer the same contract, the firm obtains no new information from the equilibrium strategies and its updated beliefs will equal the prior beliefs. Given a contract proposal $\bar{\gamma}$ satisfying condition (i), the firm will optimally accept the contract. Without extra constraints on out-of-equilibrium beliefs, it is possible to assume the firm believes the worker to be of low productivity for any other contract proposal. This justifies a rejection of any contract with a wage rate above π_l. Given this strategy of the firm, based on these beliefs, both types of worker will find it optimal to propose the contract $\bar{\gamma}$.

The following diagram shows a pooling contract $\bar{\gamma}$. Note that any contract in the area between π_l and $q \cdot \pi_h + (1-q) \cdot \pi_l$ and above the broken line $w = \pi_l + c_l \cdot e$ satisfies conditions (i) and (ii).

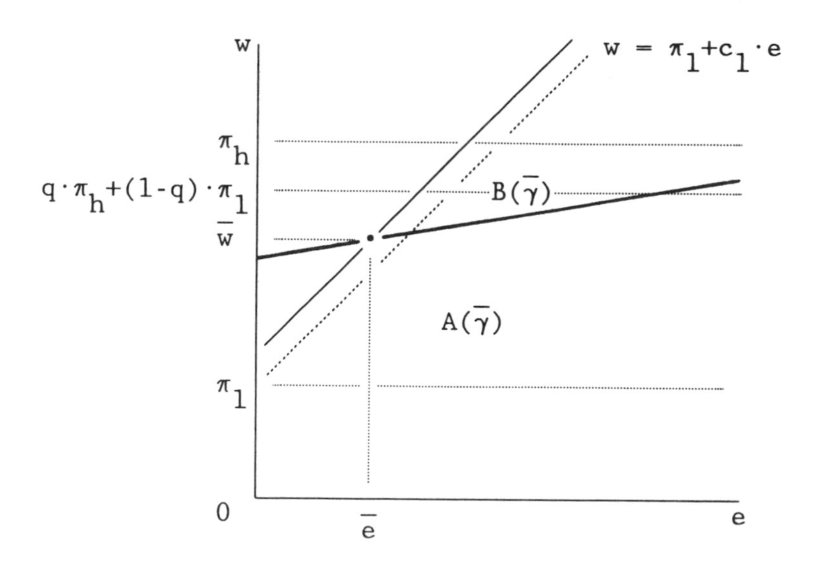

Again there are many equilibria. They are supported by out-of-equilibrium beliefs that put probability one on the low-productivity worker whenever a contract $\gamma \neq \bar{\gamma}$ is proposed. As in the case of the separating equilibria, one finds that contracts in $B(\bar{\gamma})$ would not benefit a low-productivity worker.

According to the intuitive criterion, such contract proposals should induce a belief of the firm that the proposal comes from a high-productivity worker. Therefore, the firm should accept any proposed contract in $B(\bar{\gamma})$ with $w \leq \pi_h$. There are, however, contracts $\gamma = (w, e)$ in $B(\bar{\gamma})$ with $\bar{w} < w \leq \pi_h$. Any such contract would be preferred by the high-productivity worker to the equilibrium contract. Thus the perfect Bayesian equilibrium pooling contract $\bar{\gamma}$ shown in the diagram fails to satisfy the intuitive criterion.

In fact, no perfect Bayesian equilibrium pooling contract γ will satisfy the intuitive criterion if the set of contracts $B(\gamma)$ with $w \leq \pi_h$ is non-empty. It is straightforward to check in the diagram that every pooling equilibrium contract γ has a non-empty $B(\gamma)$. It follows that no pooling equilibrium will satisfy the intuitive criterion. The unique perfect Bayesian equilibrium consistent with forward induction as represented by the intuitive criterion is the separating contract pair (γ^*, γ'').[5] ■

For signalling games, the forward induction idea is captured by the intuitive criterion. This test, however, is defined with reference to the structure of signalling games. Because of this context-specific definition, it is impossible to apply the intuitive criterion directly to other types of games. This can be checked easily by trying to apply the intuitive criterion to example 7.2 used to motivate the forward induction principle.

Therefore, alternative criteria have been suggested in the literature[6] to capture the forward induction idea:

(i) Equilibrium strategies shall be undominated in the normal form (admissibility).
(ii) Deletion of dominated strategies from a game does not change the equilibrium (iterated dominance).
(iii) Deletion of strategies that are not a best response to any equilibrium strategy does not affect the equilibrium (equilibrium domination).

All three additional requirements for an equilibrium are based on the idea that an equilibrium should not depend on dominated strategies. Admissibility requires that the equilibrium strategies be undominated. Iterated dominance rules out that equilibria may be sustained by beliefs that some player chooses a dominated strategy off the equilibrium path. The last requirements captures best the notion of the intuitive criterion: no player shall believe that another player would choose a strategy that cannot yield

[5]Cho and Kreps (1987) show that uniqueness of equilibrium under the intuitive criterion holds for models with two types of workers only.

[6]Kohlberg and Mertens (1986) suggest all of these criteria though reserve the name "forward induction" for the last. Van Damme (1990) compares various formalizations of the forward induction idea.

a higher payoff for her than the equilibrium strategy. As formalized in (iii), however, the intuitive criterion logic can be applied to general games.

EXAMPLE 7.2 (*resumed*). Reconsider the normal form of this game.

Player 2

Player 1		e_h	e_l
	p/e_h	20, 20	0, 10
	p/e_l	10, 0	10, 10
	n/e_h	15, 0	15, 0
	n/e_l	15, 0	15, 0

There are three equilibria, $(n/e_h, e_l)$, $(n/e_l, e_l)$ and $(p/e_h, e_h)$, but only $(p/e_h, e_h)$ satisfies the forward induction principle.

Since none of these equilibria uses a dominated strategy in the normal form, all equilibria satisfy admissibility. To check the equilibrium domination criterion (iii), note that p/e_l is the only strategy not used in any equilibrium. In fact, p/e_l is not a best response to any equilibrium strategy. Deleting this strategy leaves the following game.

Player 2

Player 1		e_h	e_l
	p/e_h	20, 20	0, 10
	n/e_h	15, 0	15, 0
	n/e_l	15, 0	15, 0

It is easy to check that all equilibria of the original game remain equilibria of the game obtained after deleting this strategy. Thus, criterion (iii) does not discriminate in this example either.

Iterated dominance requires the equilibrium to be robust against iterated deletion of dominated strategies. It is easy to see that p/e_l is dominated by n/e_l. Deleting p/e_l leads to the same reduced game as before. In this new game player 2's strategy e_l is dominated by e_h. Deleting e_l in the second round of the elimination process leaves the first column of the previous game matrix. Thus, after iterated deletion of dominated strategies, player 2 is left with strategy e_h only, and player 1 will choose p/e_h as the best response to e_h. The remaining strategy combination $(p/e_h, e_h)$ is the Nash equilibrium that satisfies the forward induction logic. ■

Unfortunately, there does not seem to be a unique criterion that captures the forward induction notion in all types of games. For signalling games, the intuitive criterion, a variant of equilibrium domination proved to be most suitable, whereas for example 7.2 the iterated dominance criterion appears to be appropriate.

7.2 Stable Equilibria

Two different approaches to restrict out-of-equilibrium beliefs have been considered so far:

(i) the backward induction (or sequential rationality) criterion that imposes consistency with the players' incentives in later stages of the game
(ii) the forward induction principle that requires beliefs to be consistent with the players' incentives in previous moves of the game

A natural question concerns the existence of equilibria satisfying both criteria. An attempt to answer this was made by Kohlberg and Mertens (1986) with the concept of a stable set of equilibria.

The difficulties with such a search for the "ideal" refinement of the Nash equilibrium concept were emerging already at the end of the section on forward induction. No simple criterion could be found which would capture this notion satisfactorily. Kohlberg and Mertens (1986) demonstrate with many examples that, in general, it is impossible to find an equilibrium strategy combination that satisfies sequential rationality and forward induction simultaneously. It is possible, however, to find sets of equilibria that have the same equilibrium payoff distribution and the same equilibrium path that contain equilibria some of which may satisfy the forward induction criteria and others the backward induction principle.

To understand that multiple equilibria can have the same payoff and the same equilibrium path, it is useful to consider the set of all Nash equilibria in a simple game. Typically the set of Nash equilibria consists of several connected sets of strategy combinations called *components*. The following example illustrates the structure of the equilibrium set for a finite game.

EXAMPLE 7.4 (*example* 6.1 *resumed*). Reconsider example 6.1 in its normal form.

		Player 2	
		a	f
Player 1	e	$1, 1$	$-1, -1$
	ne	$0, 3$	$0, 3$

Denote the mixed strategies of both players by $(m_1(e), m_1(ne)) = (p, 1-p)$ and $(m_2(a), m_2(f)) = (q, 1-q)$ respectively. Note that there is no distinction between mixed and behavior strategies in this case, since both players move only once. The following diagram shows the best response correspondences as solid lines.

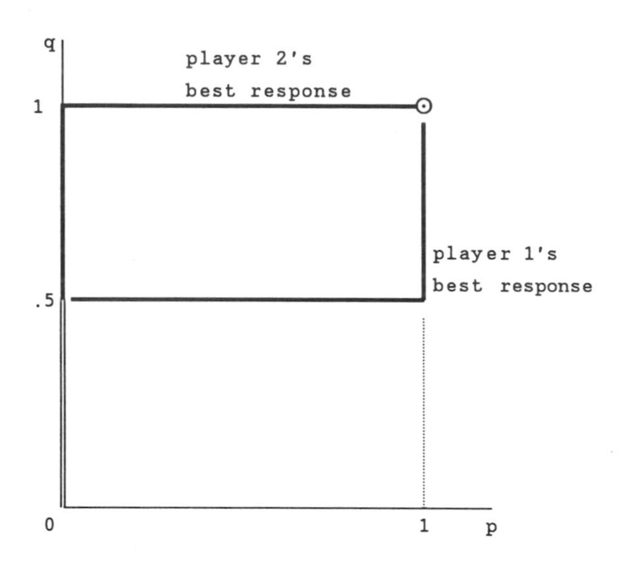

The equilibrium set consists of two components: $A \equiv \{(p,q) | p = 1,\ q = 1\}$ indicated by $\odot$ and $B \equiv \{(p,q) | p = 0,\ 0 \leq q \leq .5\}$ indicated by ═══. Notice that the equilibrium payoffs and the equilibrium actions, that is, the equilibrium path, are the same on each component:

- component A contains a single equilibrium with payoff combination $(1,1)$ and equilibrium path "*e* followed by *a*"
- all equilibria in component B yield the payoff combination $(0,3)$ and have the equilibrium path "*ne*"

What distinguishes the equilibria in component B is a different probability of player 2 choosing to fight. Although this part of the equilibrium strategy combination is important for the incentives of the players, it will not affect equilibrium play or the payoffs achieved in equilibrium. ■

The structure of the equilibrium set in example 7.4 is in fact typical for games with finite strategy sets in general. To guarantee that all stable equilibria have a common payoff, stable equilibria are required to lie in one component. Kohlberg and Mertens (1986) suggest calling a set of equilibria stable if it is a subset of a single component and if it contains equilibria that satisfy the backward induction principle and equilibria that satisfy forward induction in the form of admissibility, iterated dominance, and equilibrium domination. Because these equilibria are required to lie in the same component of the Nash equilibrium set, generally they will all have the same equilibrium payoff distribution and the same equilibrium path.

Remark 7.1. Kohlberg and Mertens (1986) actually require equilibria in a stable set to satisfy further criteria, particularly that they be invariant to transformations of the game tree they consider strategically irrelevant. In this context they present a strong argument for considering games in normal form only. ■

A *stable set of equilibria* is defined as a set of equilibria in the normal form of a game that is stable in the following sense: There is a class of perturbations of the normal form of a game such that for every perturbation in this class, there is an equilibrium close to this stable set of equilibria. A formal definition is not given here, because this definition is difficult to apply directly. Instead some properties of equilibria in a stable set are noted that can be checked easily and that allow one in many games to identify stable sets of equilibria.

THEOREM 7.1.[7]

(i) *For every game, there exists a stable set of equilibria.*
(ii) *Every equilibrium in a stable set is a perfect equilibrium in the normal form.*
(iii) *A stable set of equilibria contains a stable set of equilibria of any game derived by deletion of dominated strategies.*
(iv) *A stable set of equilibria contains a stable set of equilibria of any game derived by deletion of strategies that are not a best response to any equilibrium strategy in the set.*
(v) *A stable set of equilibria contains a sequential equilibrium.*

Theorem 7.1 establishes that sets of stable equilibria exist and that they contain equilibria that satisfy forward and backward induction. Theorem 4.6 showed that every perfect equilibrium strategy in the normal form is undominated. Thus, by (ii), all equilibria in a stable set satisfy admissibility. Furthermore, according to (iii) and (iv), a stable set contains equilibria that satisfy iterated dominance and equilibrium domination. The notion of forward induction is, therefore, captured by properties (ii), (iii), and (iv). Finally, property (v) guarantees that a stable set contains an equilibrium that satisfies the backward induction property.

[7]A proof of these claims (and of further properties) can be found in Mertens, (1989, theorems 1 to 6).

On the other hand, though existence of a stable set of equilibria is guaranteed for all games, the theorem does not claim that there is always an equilibrium in this set that satisfies properties (iii), (iv), and (v) simultaneously. Thus, there may not be a single equilibrium compatible with forward and backward induction. Furthermore, there may be more than one stable set of equilibria in a game.

The section concludes with two examples that illustrate how one can find stable sets of equilibria using the properties given in theorem 7.1.

EXAMPLE 7.4 (*resumed*). It is easy to check that all equilibria in component B use a dominated strategy for player 2 ($q = 1$ dominates $q \in [0, \frac{1}{2}]$). Hence, the set of equilibria in B does not satisfy property (ii) of theorem 7.1. Equilibria in component B are, therefore, not stable. By theorem 7.1 (i), existence of a stable set of equilibria is guaranteed. Thus, A must be the stable set. Note that the stable set of equilibria contains a single equilibrium. ■

EXAMPLE 7.5 (*example* 6.3 *resumed*). Reconsider the entry game of example 6.3 For convenience, the strategic form of this game is repeated.

		Monopolist	
		a	f
Entrant	e/h	$.5, -1$	$-.5, -1$
	e/l	$1, 1$	$-1, -1$
	ne/h	$0, 3$	$0, 3$
	ne/l	$0, 3$	$0, 3$

It is not difficult to check that the set of Nash equilibria consists of the following two components:

$$A = \{(m_E, m_M) \in \Delta^4 \times \Delta^2 | m_E(e/l) = 1, m_M(a) = 1\},$$
$$B = \{(m_E, m_M) \in \Delta^4 \times \Delta^2 | m_E(e/h) = m_E(e/l) = 0, m_M(a) \leq .5\}.$$

No equilibrium in component B is perfect since a dominated strategy, f, is played with positive probability. Hence, by property (ii) of theorem 7.1, no equilibrium in B is admissible and B does not contain a stable set. By property (i), existence of a stable set is guaranteed. Thus A must be a stable set. ■

7.3 Summary

In many games, particularly those with incomplete information, refinements based on the backward induction principle do not constrain beliefs if a non-equilibrium action occurs. Often in signalling games for example, no perfect Bayesian equilibrium can be ruled out in spite of a large number of equilibria based on various beliefs about the type of the player sending the signal. The forward induction principle, however, as formalized by the intuitive criterion, is a very effective refinement in signalling games.

It is difficult to find an adequate formalization of the forward induction notion in more general games. Several concepts seem to capture aspects of this idea: equilibria should not use strategies that are dominated in the normal form of the game (admissibility); equilibria should be robust against deletion of dominated strategies (iterated dominance); and equilibria should be independent of strategies that are not a best response to any equilibrium strategy (equilibrium domination). Unfortunately, there is no equilibrium concept satisfying all three versions of this principle for a single strategy combination.

Attempts to provide an equilibrium refinement that satisfies the sequential rationality principle as well as the forward induction principle challenge the traditional notion of an equilibrium as a single strategy combination. Instead a stable set of equilibria is suggested that, in general, contains multiple equilibria with common equilibrium path and common equilibrium payoffs. In a stable set of equilibria, there are equilibria satisfying the three versions of the forward induction principle and also at least one sequential equilibrium.

7.4 Remarks on the Literature

Kohlberg and Mertens (1986) suggest the forward induction principle as an additional refinement to the backward induction notion. Signalling games were studied in Cho and Kreps (1987) where, among other criteria, the intuitive criterion was proposed. This class of games is of particular importance in economics because it provides a link between the substantial literature on equilibrium concepts in models with asymmetric information (Riley, 1987) and game-theoretic equilibrium concepts. Cho and Kreps (1987) analyze the relationship between forward induction and the concept of a stable equilibrium. In the context of signalling games, the concept of a divine equilibrium by Banks and Sobel (1987) deserves special mention. Finally, signalling games in which the action of the sender is costless have been studied by Crawford and Sobel (1982).

Attempting to unify the growing number of refinements and to provide an equilibrium concept that would satisfy additional desirable properties, Kohlberg and Mertens (1986) noted that it was impossible to maintain the notion of an equilibrium as a single "strategy combination" if one wants to satisfy the forward induction and backward induction principles at the same time. On the other hand, their study of the Nash equilibrium correspondence revealed that the set of Nash equilibria consists of a finite number of components. Together with a result by Kreps and Wilson (1982) that finite games have finitely many equilibrium payoff distributions, this implies that finite games generically must have constant payoff distributions on each component. Hence, considering sets of strategy combinations that lie in the same component, instead of single strategy combinations, appears acceptable and satisfies all desired properties. The concept of a stable set of equilibria fulfills these requirements. The definition of a stable set of equilibria in Kohlberg and Mertens (1986) was not completely satisfactory and a modified version was suggested in Mertens (1989). Theorem 7.1 is based on this latter paper. Van Damme (1990) provides an excellent survey, and van Damme (1991) a comprehensive treatment of the literature on refinements of the Nash equilibrium concept.

Exercises

7.1 *Consider the Spence signalling model of example 7.3.*

(a) For the case of the pooling equilibrium, formally define the set of contracts that are preferred to the equilibrium contract by the high-productivity worker but not by the low-productivity worker, $B(\bar{\gamma})$. Derive the best response correspondence of the firm if it believes that contract proposals from the set $B(\bar{\gamma})$ come from a high-productivity worker.

(b) Show that the pooling equilibrium strategies of the two types of worker together with these modified beliefs and the associated best response correspondence do not form a perfect Bayesian equilibrium. Which condition of a perfect Bayesian equilibrium is violated?

7.2 *Consider an incumbent monopolist who may have either low or high costs of production. A low-cost monopolist, M_l, maximizes profit by choosing to supply a quantity x_l yielding a profit of 4. A high-cost monopolist, M_h, will choose a quantity x_h earning a profit of 3. The quantity of the high-cost monopolist x_h would result in a profit of 3 for the low-cost monopolist, whereas the high-cost monopolist would obtain a profit of 1 with the quantity x_l.*

Suppose an entrant E, not knowing the cost situation of the monopolist, considers entering the market. Prior beliefs indicate there is a 25% chance of facing the high-cost monopolist. The entrant monitors the behavior of the monopolist for one period before making a decision. If the entrant enters the market and meets the monopolist with high costs M_h, the duopoly profit will be 1 for the entrant and 1 for the *monopolist. If she enters and faces the low-cost monopolist M_l, payoffs are -1 for the entrant and 2 for the monopolist. Thus, the entrant wants to enter if and only if the high-cost monopolist is in the market.*

The payoff for the monopolist is the sum of the profits in both periods. If no entry occurs in the second period, the monopolist chooses the profit-maximizing quantity.

(a) Draw the game tree of this game and show that there is a separating perfect Bayesian equilibrium. Are there any out-of-equilibrium beliefs?

(b) Show that there is a pooling equilibrium, based on the out-of-equilibrium belief that the monopolist has low production costs, that does not satisfy the intuitive criterion.

(c) Show that there is another pooling equilibrium that satisfies the intuitive criterion.

7.3 *Consider a version of the entry game in which, after entry has occurred, the entrant and the monopolist choose simultaneously an output quantity, u or v in the case of the monopolist and x or y in the case of the entrant. The profits from the four possible quantity choices are given in the game tree.*

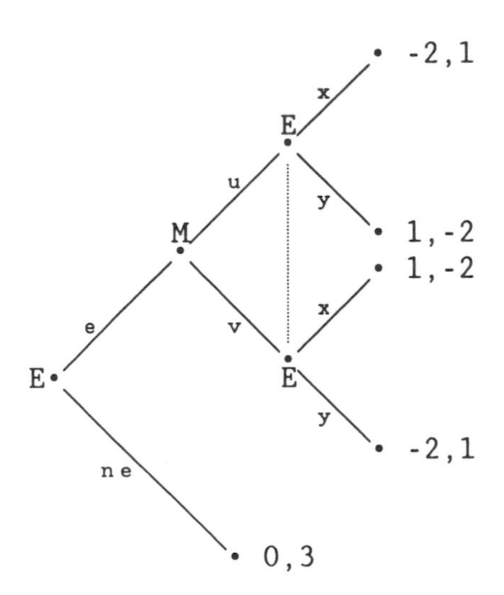

(a) Derive the unique subgame perfect equilibrium of this game.

(b) Derive the set of all Nash equilibria of this game. How many components does the equilibrium set have?

(c) Show that the set of all Nash equilibria is a stable set of equilibria.

7.4 *Consider two firms, a monopolist and a potential entrant. The entrant can enter the market in the second period, after observing the behavior of the monopolist in the first period. After the entrant enters the cost structure of the monopolist is revealed and both firms engage in Cournot-type competition. Because the entrant does not know the cost situation of the monopolist when the entry decision has to be made, the behavior of the monopolist in the first period may provide crucial information about the monopolist's cost.*

Assume that both firms have constant unit costs, that the entrant faces an entrance fee of $K = 100$, *and that the inverse demand function of the market is given as*

$$p(x) = \max\{0, 24 - (1/3) \cdot x\}.$$

Let us assume that, according to the entrant's assessment, the monopolist's unit cost is $c'' = 9$ *with probability* q *and* $c' = 3$ *with probability* $(1 - q)$. *The entrant has unit costs* $c = 6$.

(a) Derive the profit function for both types of the monopolist and determine the profit-maximizing output and profit level for each of them.
(b) Derive the Cournot equilibrium quantity and profit levels for the entrant and the monopolist of either type.
(c) Show that there is a perfect Bayesian equilibrium at which each type of the monopolist behaves like the low-cost monopolist provided that the *ex-ante* expected profit from entering is negative.
(d) Show that there is a separating perfect Bayesian equilibrium at which the low-cost monopolist chooses a higher than optimal quantity in stage one in order to keep the entrant out of the market.
(e) Comment on the beliefs in case (c) and (d). Do they satisfy the intuitive criterion?

7.5 *Reconsider the game of example 4.1.*

		Player 2	
		L	R
Player 1	T	2, 1	.5, .5
	B	0, 0	1, 2

Suppose that player 1 *has an option not to participate in this game that will yield her a payoff of 1.5.*

(a) Draw a game tree of the extensive form of this game.
(b) Derive all subgame perfect equilibria.
(c) Which equilibria satisfy the forward induction principle?

8

Repeated Games and Folk Theorems

In many social situations one can observe individuals appearing to act against their self-interest. Sellers of a product, for example, may exchange a defective good even if there is no contractual requirement to do so; workers and managers keep agreements not legally binding; or illegal cartel arrangements are respected by the participants although individually each has an incentive to break away. People behave cooperatively and forgo immediate advantages although they could exploit others. One possible explanation for such behavior rests on the assumption that people interact repeatedly rather than just once and therefore have an incentive to maintain the goodwill of their partners. Thus interest in long-term cooperation may overcome the incentive to take an immediate advantage. This explanation of cooperative behavior is based on self-interest. In this debate the prisoners' dilemma has become one of the most influential and thought-provoking games.

EXAMPLE 8.1 (*prisoners' dilemma*). Example 3.1 presents the "story" of this game. Its payoff matrix is repeated.

		Player 2	
		N	C
Player 1	N	$1, 1$	$-1, 3$
	C	$3, -1$	$0, 0$

..e challenge this game poses results from the fact that there is a unique dominant strategy equilibrium (C,C) yielding a payoff strictly worse for each player than the payoff from (N,N). On the other hand, no agreement of the players to play (N,N) that is not enforced by some outside agency is stable since each player has an incentive to deviate unilaterally. But if both players follow this incentive the outcome will be the "bad" equilibrium (C,C). This reasoning leads to predicting (C,C) as the outcome of the game, in stark contrast to numerous experimental studies in which players are consistently observed playing cooperative strategies in prisoners' dilemma-type games.[1] ■

Early in this debate, the point was made that the noncooperative outcome in games like the prisoners' dilemma results from assuming the game is played just once. If the game were repeated, it was argued, the long-term benefit of cooperation would overcome the short-term incentive to act noncooperatively. For instance, to achieve the cooperative outcome, a player may choose to play cooperatively as long as the other player does the same and to respond to noncooperative behavior by also acting noncooperatively. If both players used this strategy, they would cooperate forever. Notice that repeating a game provides the opportunity to condition actions on past behavior. In recent years, the question of whether repetition of a game would allow outcomes other than Nash equilibria of the basic game has been studied extensively. A substantial amount of conceptual and formal modeling is required for this analysis. The results vindicate some of the intuitions about "cooperation in long-term relationships," but raise questions regarding other aspects. Before a detailed exposition of the formal structure of a repeated game, however, an economic example sharing some features of the prisoner's dilemma will be discussed.

EXAMPLE 8.2 (*Cournot duopoly*). Consider two firms producing a homogeneous product with constant marginal costs c. The maximum price obtained in the market depends on the total quantity $q \equiv q_1 + q_2$ sold in the market. Hence the profit of each firm depends on its own output level as well as on that of its competitor. Suppose that the inverse demand function is linear, $p(q_1, q_2) = a - b \cdot (q_1 + q_2)$ for some positive constants $a > c$ and b. The profit of firm 1 and 2 can be written as follows: $\pi_i(q_1, q_2) = (a - b \cdot (q_1 + q_2)) \cdot q_i - c \cdot q_i$, $i = 1, 2$.

Profit maximization of firm i, given firm j's quantity choice, yields the best response function $r_i(q_j) = [(a-c)/b - q_j]/2$, $i \neq j$. The following diagram shows some isoprofit lines and the best response functions for these firms. The shaded area indicates quantity combinations that yield

[1] Rapoport (1989) provides a short description of experimental results on the prisoners' dilemma and contains further literature on this topic.

negative profits for both firms. The Nash equilibrium strategy combination (q_1^*, q_2^*) is indicated by $\circ$. The set of Pareto optimal strategy combinations is given by the locus of tangencies of isoprofit lines that run from $(a-c)/(2b)$ on the vertical axis to $(a-c)/(2b)$ on the horizontal axis.

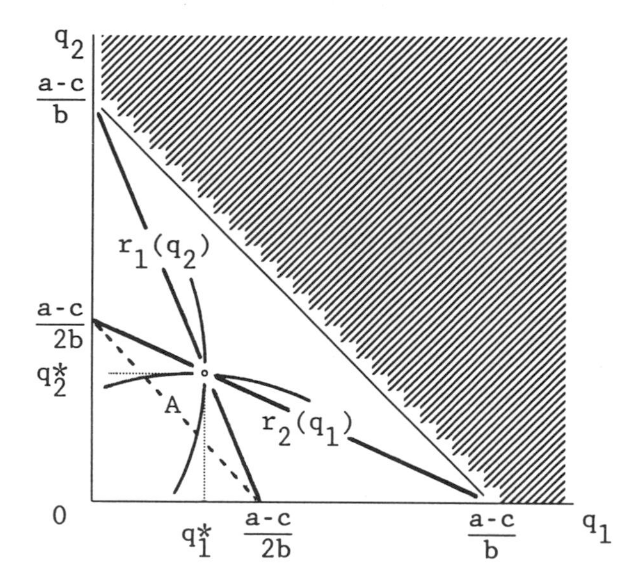

If this duopoly situation arose only once, it would be difficult to predict any outcome other than the Nash equilibrium. Indeed, this solution was suggested as equilibrium by Cournot (1838) based on an argument about dynamic adjustments. Similar to the prisoners' dilemma, however, the equilibrium outcome is not Pareto-optimal. This means that there are quantity combinations (in region A of the diagram) that yield both firms a larger profit than the equilibrium quantities. Such a situation makes cooperation of the two firms profitable. With an agreement to produce not more than the monopoly quantity $(a-c)/2b$ in total, they would maximize joint profits. The joint profit, achieved by cooperation, can be shared in a manner that yields each of them a higher profit than they obtain in the Nash equilibrium. Such a cartel agreement, however, cannot be expected to last, because both firms could increase their profits by unilaterally deviating from the agreement. According to this argument, cooperative behavior, that is, cartels, should not be observed in duopoly markets for which no outside enforcement possibilities[2] exist. This conclusion changes dramatically if firms make their quantity decisions repeat-

[2]Shapiro (1989) surveys the literature on oligopoly theory. Section 2.4 studies conditions that facilitate cartel formation.

edly. New strategic possibilities arise and collusion may be sustained by the prospect of future gains from cooperation. ■

8.1 The Formal Description of a Repeated Game

A repeated game is the special case of a game in extensive form in which the game tree consists of a sequence of strategic form games. In principle, it is possible to analyze repeated games with the notation of extensive form games introduced in Chapter 1. The special structure of these games allows, however, for an alternative simpler notation in this context. Consider a game in strategic form $\Gamma = (I, S, p)$, the constituent game (or stage game) repeated T times. In this section, T may be a finite number or infinity. The following example shows the extensive form of a twice repeated version of the prisoners' dilemma game.

EXAMPLE 8.1 (*Prisoners' dilemma, resumed*). The following game tree represents a twice-repeated version of the prisoners' dilemma ($T = 2$).

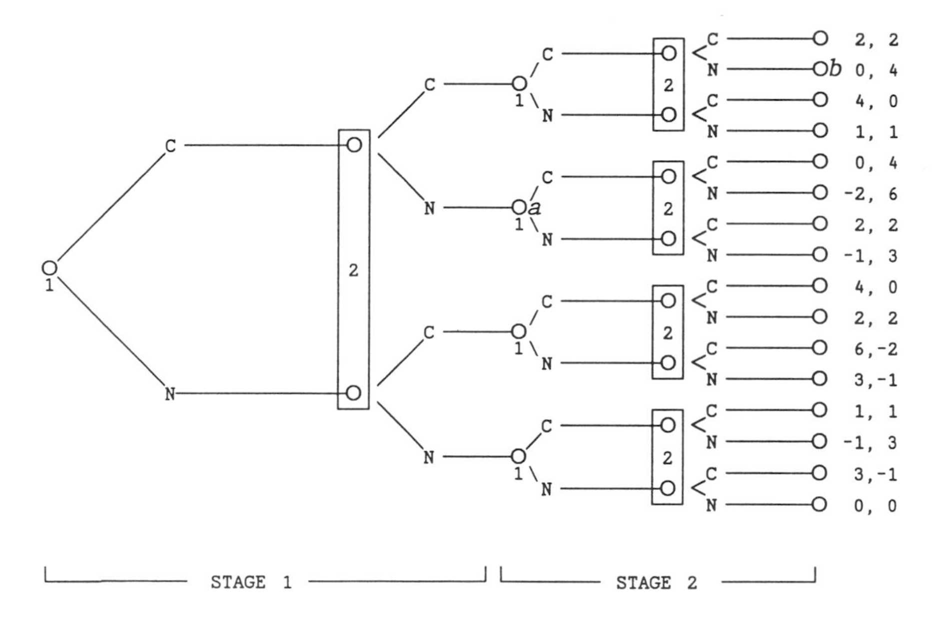

This example shows that, even with just two strategies per player in the stage game, the number of nodes is rapidly expanding. ■

In any extensive form game, a strategy for a player $i \in I$ must specify the choice of action at each node at which this player has to move. As noted in Chapter 1, it is possible to identify each decision node with the unique path leading to this node. The special nature of a repeated game makes it easy to describe the unique path leading to a particular decision node in the tth repetition of the game by the sequence of strategy combinations played in the stage games up to period t.

Each decision node at which a player has to move is identified by the history that led to this node. A *history up to period* t, h^t, (or a t-*history*) is a sequence of strategy combinations played up to this stage $(s^1, s^2, \ldots, s^{t-1})$. In this example of the twice-repeated prisoners' dilemma, decision node a can be identified with history $h^2 = ((C, N))$, and terminal node b with history $h^3 = ((C, C), (C, N))$. The set of possible histories up to period t is the $(t-1)$-fold Cartesian product of the set of all strategy combinations in the stage game, that is, $S^{t-1} = S \times S \times \cdots \times S$ $(t-1$ times). Since a normal form game is played in each stage, the information of all players is the same at the beginning of each stage. Notice that there may be imperfect information within a stage game. The strategic implications of imperfect information within a stage game however, are completely captured by the strategy combination played in the stage game. Thus only sequences of strategy combinations, one for each stage game played, need to be recorded.

Remark 8.1. In this chapter, only pure strategy combinations are considered. Indeed, all results of this chapter remain unchanged if agents play mixed strategy combinations, provided that mixed strategy combinations can be observed. If mixed strategies played cannot be observed directly, then the problem becomes much more difficult. In this case, each agent has to infer from the observed pure strategy whether a particular mixed strategy was used. This statistical decision problem complicates the strategic analysis considerably. ■

The strategies $s_i \in S_i$ of an agent in the constituent game Γ are called *actions* in the repeated game to distinguish them from the strategy of the repeated game itself denoted by σ_i. A strategy σ_i for player i in the repeated game consists of history-dependent action choices in each stage of the game. Let $a_i^t(h^t) \in S_i$ be the action chosen by agent i in stage t after observing history h^t, then one can write a strategy for the repeated game as

$$\sigma_i = \left(a_i^1(h^1), a_i^2(h^2), \ldots, a_i^T(h^T)\right)$$

for a finite horizon game and

$$\sigma_i = \left(a_i^1(h^1), a_i^2(h^2), \ldots\right)$$

for an infinite horizon game.

A strategy in the repeated game is a sequence of action choices that depend on the history that led to this stage. An action in stage $t, a_i^t(\cdot)$, is a function that associates with each history in S^{t-1} an action from S_i. The set of possible actions in stage t is, therefore, the set of functions $A_i^t = \{a_i^t | a_i^t \colon S^{t-1} \to S_i\}$. Since period 1 has no history, one defines $S^0 = \{0\}$ as the set containing only the initial node. The initial node is the unique history of period 1, $h^1 = 0$, and an action choice in period 1 selects a unique action $a_i^1(h^1) \in S_i$ for the first stage game.

Choosing a strategy in a repeated game means choosing a sequence of functions that specify for any possible history what the agent will do. In a repeated game, the strategy set of agent i, Σ_i, is therefore, the T-fold Cartesian product of action function sets A_i^t, that is, $\Sigma_i = A_i^1 \times A_i^2 \times \cdots \times A_i^T$. Finally, denote by $\Sigma = \Sigma_1 \times \cdots \times \Sigma_I$ the set of strategy combinations in a repeated game with typical elements $\sigma = (\sigma_1, \sigma_2, \ldots \sigma_I)$. A strategy combination σ contains a plan for each player that specifies the action of this player after any possible history. The set of plans contained in a strategy combination σ determines uniquely the sequence of action combinations that is actually played. Such a sequence of actually played action combinations is called *path of play* in the repeated game and is denoted by $\pi(\sigma) = (\pi^1(\sigma), \pi^2(\sigma), \pi^3(\sigma), \ldots.)$. For a given strategy combination σ, the action combination played in period t, $\pi^t(\sigma)$, can be defined recursively as follows:

$$\pi^1(\sigma) = \left(a_1^1(h^1), a_2^1(h^1), \ldots, a_I^1(h^1)\right),$$

$$\pi^t(\sigma) = \left(a_1^t\left(\pi^1(\sigma), \ldots, \pi^{t-1}(\sigma)\right), a_2^t\left(\pi^1(\sigma), \ldots, \pi^{t-1}(\sigma)\right), \ldots, \right.$$
$$\left. a_I^t\left(\pi^1(\sigma), \ldots, \pi^{t-1}(\sigma)\right)\right).$$

The path of play $\pi(\sigma)$ is the actual history of the game if strategy combination σ is used.

It remains to describe the payoffs of the players in a repeated game. A natural notion of payoff in a repeated game is the sum of the discounted or undiscounted payoffs of the stage games. In addition, to make comparisons between payoffs in a stage game and payoffs in the repeated game meaningful, one considers the average, rather than the total, payoff of the repeated game. Maximizing the average payoff over the number of stages leads to the same choice as maximizing the sum of payoffs because the number of rounds in the repeated game is not a choice variable.

Remark 8.2. Throughout this chapter, repeated games with identical discount factors $\delta \in [0, 1]$ for all players will be considered. This is without loss of generality for all results in this chapter and avoids unnecessarily complicated notation. ■

Summing the stage payoffs is a valid procedure if the game has a finite horizon, $T < \infty$. In a finite-horizon game in which players discount the future with a common discount factor δ, the average payoff is given as

$$\mathscr{P}_i^T(\sigma) = \left(\sum_{t=1}^{T} \delta^{t-1} \right)^{-1} \cdot \left(\sum_{t=1}^{T} \delta^{t-1} \cdot p_i(\pi^t(\sigma)) \right).$$

The discount factor[3] $\delta \in [0,1]$ covers the extreme cases in which players care for the present only, $\delta = 0$, and $\delta = 1$ cases in which there is no discounting of the future at all. Note that for $\delta = 1$ the payoff of the repeated game is simply the average payoff $T^{-1} \cdot (\sum_{t=1}^{T} p_i(\pi^t(\sigma)))$. The normalizing factor $(\sum_{t=1}^{T} \delta^{t-1})^{-1}$ makes the sum of the stage-game payoffs a weighted average that, for $\delta = 1$, coincides with the average payoff. These average payoffs of the repeated game can be compared with payoffs of the stage game. In the finite-horizon case, this normalization has no influence on the strategic choice of agents, since it multiplies the sum of the payoffs by a constant number $(\sum_{t=1}^{T} \delta^{t-1})^{-1}$.

For infinite-horizon games, $T = \infty$, the sum of stage-game payoffs is defined as the limit of the sequence of payoffs in T-fold repetitions of the stage game as T diverges to infinity. This limit may not be well-defined. If payoffs of the stage game are bounded, then the sequence of average payoffs will converge as T goes to infinity provided that the discount factor is strictly less than one, $\delta < 1$. For $\delta = 1$, bounded payoffs of the stage game guarantee in general only that the sequence of average payoffs will not diverge and it remains a possibility that average payoffs cycle in a bounded range. In this case, the average payoff of the infinitely repeated game is defined to be the limit point of the smallest converging subsequence that one can select. This limit point is denoted by "lim inf" of the sequence. In summary, for infinitely repeated games, one uses the following average payoffs[4] that are always well-defined if the stage-game payoffs remain bounded:

$$\mathscr{P}_i^\infty(\sigma) = (1-\delta) \cdot \left(\sum_{t=1}^{\infty} \delta^{t-1} \cdot p_i(\pi^t(\sigma)) \right) \quad \text{for} \quad \delta < 1,$$

$$\mathscr{P}_i^\infty(\sigma) = \liminf_{T \to \infty} \frac{1}{T} \cdot \left(\sum_{t=1}^{T} p_i(\pi^t(\sigma)) \right) \quad \text{for} \quad \delta = 1.$$

For any game in strategic form Γ and any common discount factor $\delta \in [0,1]$, one can now analyze the T-times repeated game as the normal form game

$$\Gamma^T(\delta) = (I, \Sigma, \mathscr{P})$$

with strategy sets Σ and payoff functions $\mathscr{P} = (\mathscr{P}_1^T, \mathscr{P}_2^T, \ldots, \mathscr{P}_I^T)$.

[3]A discount factor δ corresponds to a discount rate $r \equiv (1-\delta)/\delta$.
[4]Note that $\lim_{T \to \infty} \sum_{t=1}^{T} \delta^{t-1} = 1/(1-\delta)$ holds.

An immediate question concerns the existence of Nash equilibria in repeated games. Intuition suggests that playing a Nash equilibrium strategy combination of the constituent game in each stage of a repeated game will yield a Nash equilibrium of the repeated game. Indeed, the following result that holds whether T is finite or infinite confirms this conjecture.

THEOREM 8.1. *Let s^* be a Nash equilibrium of the constituent game Γ and define the repeated game strategy combination σ^* by $a_i^t(h^t) = s_i^*$ for all $t = 1, 2, \ldots, T$ and all $i \in I$, then, for all $\delta \in [0, 1]$, σ^* is a Nash equilibrium of $\Gamma^T(\delta)$.*

Proof. Suppose that player i chooses an alternative strategy whereas all other players follow the strategy combination σ_{-i}^*. Since s^* is a Nash equilibrium of the stage game, one knows that $p_i(s^*) \geq p_i(s_i, s_{-i}^*)$ for all $s_i \in S_i$ holds. Hence,

$$\mathscr{P}_i(\sigma^*) = \left(\sum_{t=1}^{T} \delta^{t-1} \right)^{-1} \cdot \sum_{t=1}^{T} \delta^{t-1} \cdot p_i(s^*)$$

$$\geq \left(\sum_{t=1}^{T} \delta^{t-1} \right)^{-1} \cdot \sum_{t=1}^{T} \delta^{t-1} \cdot p_i(s_i^t, s_{-i}^*)$$

must hold for any strategy sequence $(s_i^1, s_i^2, \ldots, s_i^T)$. Since any strategy of the repeated game $\sigma_i = (a_i^1(h^1), \ldots, a_i^T(h^T))$ together with σ_{-i}^* induces some sequence $(s_i^1, s_i^2, \ldots, s_i^T)$, one has $\mathscr{P}_i(\sigma^*) \geq \mathscr{P}_i(\sigma_i, \sigma_{-i}^*)$ for all $\sigma_i \in \Sigma_i$. ■

Theorem 8.1 establishes that repeated play of a Nash equilibrium forms a Nash equilibrium of the repeated game. Hence there is no existence problem for Nash equilibria in repeated games provided that the stage game has an equilibrium. The average payoff from repeated Nash equilibrium play is, of course, exactly the payoff of the Nash equilibrium in the stage game. The following subsections will study whether other payoff combinations of the constituent game can be achieved as equilibria in the repeated game. Two interesting results will emerge from this analysis:

- every equilibrium strategy has the form that the game follows a prescribed sequence of actions until a deviation occurs followed by a punishment phase
- every outcome for which a "sufficient" punishment exists can be implemented in a repeated game

Making precise what punishment possibilities exist in a given game constitutes most of the remainder of this chapter.

8.2 Repeated Games with an Infinite Horizon

In this section infinitely repeated games are studied. There are at least two reasons one should consider infinite repetitions of games in a world in which players have finite lives. Firstly, to assume finiteness of a repeated game means to specify precisely (or with very high probability) in which round T the game will end. If the application under consideration provides such a clear time limit, then it is not appropriate to use an infinite-horizon model. But there are many applications in which it is impossible to determine the final date of an interaction with certainty or high probability. If this is the case, then an infinite-horizon model makes sense, even if there is a positive probability in each stage that the game will be terminated. The probability of continuation at each stage of the repeated game works like a discount factor.

A second reason one may want to study infinite repetitions of a game is to investigate the consistency of the model as the number of repetitions increases. If a result derived in a repeated game remains correct for every finite repetition of the game, irrespective of the number of repetitions, but becomes incorrect for an infinite horizon, then this observation may indicate a problem with the underlying model. In fact, the following sections of this chapter will reveal a marked difference in the results between finitely and infinitely repeated games. This raises some questions about the model of repetition of a stage game itself.

Repeating a game substantially increases the set of possible strategies because actions, that is, strategies of the stage game, can be made conditional on the observed behavior of other players in previous stages of the game. This makes possible strategies that enforce a particular behavior in the stage game by threatening to choose some other action if the opponents do not follow the prescribed behavior. In this manner, strategy combinations of the stage game that do not form a Nash equilibrium may be implemented as a Nash equilibrium in the repeated game. Implementing payoff combinations that are not equilibrium payoff combinations of the stage game relies on threats that prevent players from deviating to a best response strategy. Hence, it is clear that no payoff combination can be obtained in a repeated game that will give some agent less than the worst punishment that the other players can inflict on him.

Suppose that all players except player i cooperate in punishing player i. Player i can always choose a best response to the strategy combination that the opponents play. Thus, the highest punishment that the opponents can inflict on player i in the stage game is

$$w_i = \min_{s_{-i}} \max_{s_i} p_i(s_i, s_{-i}).$$

Denote by r^i_{-i} the strategy combination that player i's opponents have to choose for this punishment,

$$r_{-i}^{i} \in \arg\min_{s_{-i}} \max_{s_i} p_i(s_i, s_{-i}).$$

The punishment w_i is the worst payoff that the other players can inflict on player i by choosing strategy combination r_{-i}^{i}. Thus, under no circumstances can the average payoff of a repeated game yield less to a player than w_i.

Remark 8.3. The worst punishment that opponents can impose on a player, w_i, usually exceeds the security level of this player, v_i. The security level or maximin value, discussed in Chapter 2 in the context of zero-sum games, is the payoff level a player can guarantee himself, v_i. For example, in the pure coordination game,

		Player 2	
		L	R
Player 1	T	1, 1	0, 0
	B	0, 0	1, 1

$$v_1 = \max_{s_1} \min_{s_2} p_1(s_1, s_2) = 0 < 1 = \min_{s_2} \max_{s_1} p_1(s_1, s_2) = w_1$$

holds. ■

By choosing a best response strategy to the punishment strategy combination r_{-i}^{i} in each stage of the game, player i will obtain an average payoff of w_i in the repeated game. It is, therefore, impossible to achieve a payoff combination in the repeated game that does not give each player an average payoff of at least w_i. In fact, any payoff combination that is supposed to be achieved by a threat to punish the player with w_i must have a payoff strictly better than w_i; otherwise punishment with w_i cannot be a threat for this player. This leaves the set of all feasible payoff combinations strictly greater than $w = (w_1, w_2, \ldots, w_I)$ as potential outcomes in a repeated game. Denote by $\mathbb{P}(\Gamma)$ the set of all feasible and individually rational payoff vectors, formally,

$$\mathbb{P}(\Gamma) = \{(p_1(s), p_2(s), \ldots p_I(s)) | s \in S,\ p_i(s) > w_i \quad \text{for all } i \in I\}.$$

The following examples illustrate this set.

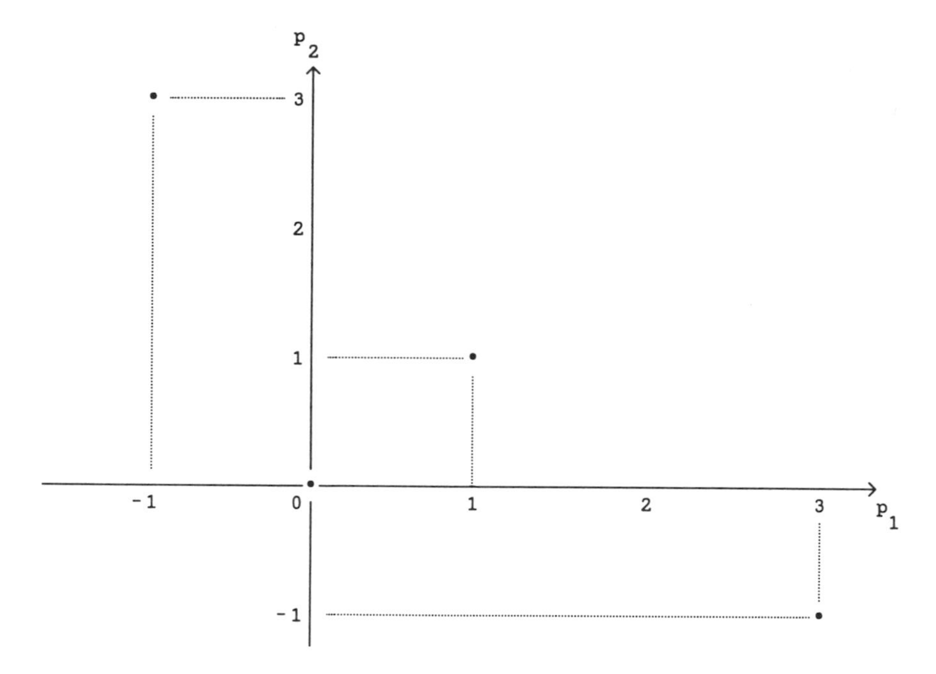

EXAMPLE 8.1 (*resumed*). Reconsider the prisoners' dilemma game. The four feasible payoff combinations of this game are indicated in the diagram by ○. In this game, the Nash equilibrium payoff coincides with the worst punishment, $w_i = 0$. Thus, the set of feasible and individually rational payoffs consists of just one element, $\mathbb{P}(\Gamma) = \{(1, 1)\}$. ■

The prisoners' dilemma example is special because the Nash equilibrium payoffs yield the worst outcome for both players and the only feasible and individually rational payoff combination is Pareto-optimal. The duopoly model of example 8.2 illustrates that a Nash equilibrium of a game may have a payoff that is strictly preferred to the punishment payoff without being Pareto-optimal.

EXAMPLE 8.2 (*resumed*). Reconsider the Cournot duopoly model: The maximal joint profit is equal to the monopoly profit π^m. Producing the monopoly output in different proportions allows the duopolists to share this profit in any desired proportion without affecting the total profit. Other feasible output combinations yield profit combinations below the line from $(0, \pi^m)$ to $(\pi^m, 0)$. The Nash equilibrium gives both firms the same positive profit π^c, but the total profit of the duopolists is less than the monopoly profit, $\pi^c + \pi^c < \pi^m$. By increasing its output to the quantity where price equals marginal cost, $q^o \equiv (a - c)/b$, each firm can force the other firm's profits to zero. Since each firm can choose a best response to the other firm's quantity choice, however, a punishment strategy cannot inflict a loss on a firm. Thus, $w_i = 0$ is the worst punishment for each firm. The diagram below shows the set $\mathbb{P}(\Gamma)$ for the duopoly game.

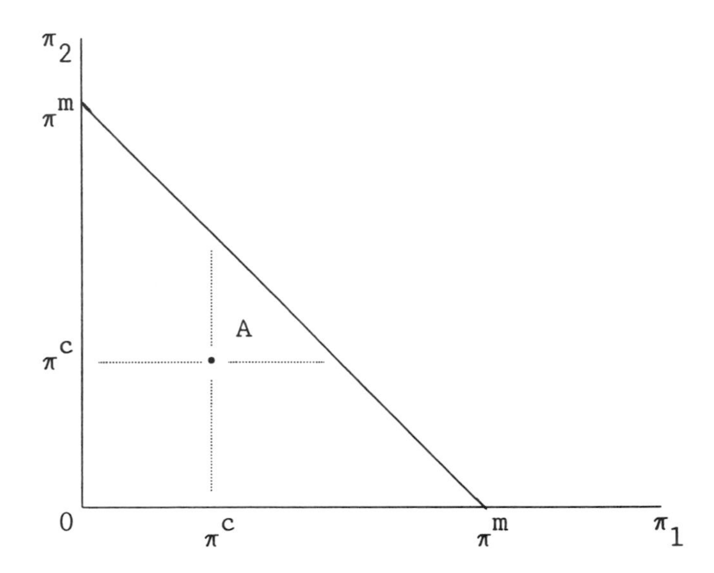

Note that there are profit combinations in the triangle A with a higher profit for both firms than the Nash equilibrium. ■

The next sections will investigate which payoff combinations from the set of feasible and individually rational payoffs $\mathbb{P}(\Gamma)$ can be obtained as Nash equilibria in a repeated game.

8.2.1 Nash Equilibria

Infinitely repeated games allow players to enforce agreements with a strategy of the following type: *I will play the strategy on which we agreed as long as you follow our agreement*; *if you deviate from this behavior I will play my punishment strategy against you forever*. The Cournot duopoly example illustrates that this kind of strategy can enforce any Pareto-optimal profit combination.

EXAMPLE 8.2 (*resumed*). Reconsider the Cournot duopoly. Suppose that the two firms agreed to produce the quantities $(\bar{q}_1, \bar{q}_2)$, $\bar{q} + \bar{q}_2 = q^m \equiv (a-c)/2b$. This yields a Pareto-optimal profit combination $(\bar{\pi}_1, \bar{\pi}_2)$, with $\bar{\pi}_i = (p^m - c) \cdot \bar{q}_i$, $i = 1,2$, and $p^m \equiv (a+c)/2$. Note that p^m and q^m denote the monopoly price and the monopoly quantity respectively. Recall that q^o denotes the quantity which forces the competing firm to zero profits.

In any stage of this game, both players will choose a quantity combination (q_1, q_2). A history of the game is therefore a sequence of such quantity combinations, $h^t = ((q_1^1, q_2^1), (q_1^2, q_2^2), \ldots, (q_1^{t-1}, q_2^{t-1}))$. Denote by $h^t(t-1) = (q_1^{t-1}, q_2^{t-1})$ the last quantity combination observed in history h^t.

Consider the following claim: the strategy combination

$$(\hat{\sigma}_1, \hat{\sigma}_2) = \big((\hat{a}_1^1(h^1), \hat{a}_1^2(h^2), \hat{a}_1^3(h^3), \ldots),\ (\hat{a}_2^1(h^1), \hat{a}_2^2(h^2), \hat{a}_2^3(h^3), \ldots)\big) \text{ with}$$

$$\hat{a}_i^t(h^t) = \begin{cases} q^o & \text{for } h^t(t-1) = (q^o, q_j) \text{ and } h^t(t-1) = (\bar{q}_i, q_j') \\ \bar{q}_i & \text{otherwise,} \end{cases}$$

$q_j' \neq \bar{q}_j$, and $i = 1, 2$, $j \neq i$, forms a Nash equilibrium of the infinitely repeated game $\Gamma^\infty(\delta)$ if δ is close to one. Notice that this strategy requires firm i to play the punishment strategy q^o when a deviation of firm j occurs and whenever it played the punishment strategy before. Thus, a deviation of firm j will be punished by producing the quantity q^o forever, irrespective of firm j's action choices after the deviation. In all other cases, firm i follows the cooperative strategy $\bar{q}_i$, in particular in period 1 when $h^1 = 0$ obtains.

The strategy combination $(\hat{\sigma}_1, \hat{\sigma}_2)$ is a Nash equilibrium if $\mathscr{P}_1^\infty(\hat{\sigma}_1, \hat{\sigma}_2) \geq \mathscr{P}_1^\infty(\sigma_1, \hat{\sigma}_2)$ for all $\sigma_1 \in \Sigma_1$ and $\mathscr{P}_2^\infty(\hat{\sigma}_1, \hat{\sigma}_2) \geq \mathscr{P}_2^\infty(\hat{\sigma}_1, \sigma_2)$ for all $\sigma_2 \in \Sigma_2$ hold. Consider firm 1 and assume that firm 2 behaves according to the equilibrium strategy $\hat{\sigma}_2$. If firm 1 follows the equilibrium strategy, $(\bar{q}_1, \bar{q}_2)$ will be produced in each period yielding a profit of $\bar{\pi}_1$ per period. If it chooses any other strategy σ_1, the following three phases of play have to be considered:

- an initial phase in which σ_1 prescribes the agreed behavior $\bar{q}_1$: during this phase firm 1 will earn the profit $\bar{\pi}_1$
- the period when the first deviation occurs, say period τ: in period τ firm 1 will earn a higher profit by deviating from $\bar{q}_1$; this profit cannot exceed the profit obtained by playing the best response to $\bar{q}_2$, say $\tilde{\pi}_1$
- the phase after the deviation, $t > \tau$: in this phase firm 2 will produce the quantity q^o to punish firm 1 for the deviation; the best that firm 1 can do then is to produce nothing and to earn zero profits; any positive output yields a loss for firm 1

If there is an initial phase, $\tau > 1$, then firm 1 will earn the same profit during this phase as with the equilibrium strategy. Without loss of generality, one can therefore assume, that the deviation occurs in period 1. A deviation is unprofitable if the higher profit made in period 1, $\tilde{\pi}_1$, does not justify the guaranteed loss in the future that is equal to the discounted value of the infinite profit stream, $\bar{\pi}_1/(1-\delta)$, that firm 1 would have earned by choosing its equilibrium quantity. The equilibrium strategy of firm 1, $\hat{\sigma}_1$, is therefore a best response to $\hat{\sigma}_2$ if $\bar{\pi}_1/(1-\delta) \geq \tilde{\pi}_1$ holds. This inequality is satisfied for all discount factors $\delta \geq \delta_1^o \equiv (\tilde{\pi}_1 - \bar{\pi}_1)/\tilde{\pi}_1$.

An analogous argument for firm 2 confirms that there is a critical discount factor δ_2^o such that firm 2 will not deviate if the discount factor is higher than δ_2^o. Let $\delta^o = \max\{\delta_1^o, \delta_2^o\}$ be the larger of these two critical discount factors. Then one can claim that the strategy combination $(\hat{\sigma}_1, \hat{\sigma}_2)$ is a Nash equilibrium of the repeated game that implements the agreement to produce $(\bar{q}_1, \bar{q}_2)$ if the discount rate is high enough, $\delta \geq \delta^o$. Notice that δ^o depends on the agreement that is to be implemented.

This proves the claim made above for any agreement $(\bar{q}_1, \bar{q}_2)$. In fact, the argument that established that $(\hat{\sigma}_1, \hat{\sigma}_2)$ forms a Nash equilibrium for high enough discount factors did not depend on the agreement being implemented as long as the best deviations from the agreement $(\tilde{\pi}_1, \tilde{\pi}_2)$ exceeded the payoffs from the agreement $(\bar{\pi}_1, \bar{\pi}_2)$. In particular, the implemented profits need not be Pareto-optimal or even better than the Nash equilibrium profits. ■

The Cournot duopoly example illustrates a case in which every payoff combination in $\mathbb{P}(\Gamma)$ can be implemented if the game is repeated infinitely often. The main theorem of this section extends this result to general stage games. It is called *Folk Theorem* because it has been well-known among game theorists for a long time without a printed reference. This theorem demonstrates that virtually any payoff vector that gives each player more than his worst punishment payoff can be implemented as a Nash equilibrium of an infinitely repeated game provided that agents are sufficiently patient, that is, that their discount factor exceeds a critical value δ^o. Note, however, that this critical discount factor δ^o varies with the payoff vector $p = (p_1, p_2, \ldots, p_I)$ that is implemented.

THEOREM 8.2 (*Folk theorem*). *For any stage game Γ and any individually rational payoff vector $p \in \mathbb{P}(\Gamma)$, there exists $\delta^o \in (0,1)$ such that, for any $\delta \geq \delta^o$, $\Gamma^\infty(\delta)$ has a Nash equilibrium σ^* such that $\mathscr{P}_i^\infty(\sigma^*) = p_i$ for all $i \in I$ holds.*

Proof. Denote by $h^t(t-1)$ the last strategy combination in the history $h^t = (s^1, s^2, s^3, \ldots, s^{t-1})$, $h^t(t-1) = s^{t-1}$, and let r_i^j be the part of the punishment strategy against player j,

$$r_{-j}^j = \left(r_1^j, r_2^j, \ldots, r_{j-1}^j, r_{j+1}^j, \ldots, r_I^j\right)$$

that player i has to play.

For any payoff vector $p \in \mathbb{P}(\Gamma)$, let s be the strategy combination of the stage game that yields this payoff vector, $p = (p_1(s), p_2(s), \ldots, p_I(s))$. Consider the following strategy of the infinitely repeated game: $\sigma_i^* = (a_i^*(h^1), a_i^*(h^2), a_i^*(h^3), \ldots.)$ with

$$a_i^*(h^t) = \begin{cases} r_i^j & \text{if } h^t(t-1) = (s_j', s_{-j}) \text{ or } h^t(t-1) = \left(s_j'', r_{-j}^j\right) \\ s_i & \text{otherwise,} \end{cases}$$

and $s_j', s_j'', s_j' \neq s_j$, arbitrary strategies of some player $j \neq i$.

If player i follows this strategy, he will switch to the punishment strategy against the first player who deviates from strategy combination s and play this punishment strategy forever. For any other history, however, player i chooses s_i. This implies in particular that player i begins with s_i and does not play a punishment strategy if simultaneous deviations occur. The latter assumption is without consequence for the equilibrium play because a Nash equilibrium does not require optimality against coordinated deviations of other players.

If all players follow the strategy combination σ^*, the path $\pi(\sigma^*) = (s, s, s, \ldots.)$ results which yields an average payoff

$$\begin{aligned} \mathscr{P}_i^\infty(\sigma^*) &= (1-\delta) \cdot \sum_{t=1}^{\infty} \delta^{t-1} \cdot p_i(\pi^t(\sigma^*)) \\ &= (1-\delta) \cdot \sum_{t=1}^{\infty} \delta^{t-1} \cdot p_i(s) = p_i(s) \end{aligned}$$

for each player $i \in I$. It remains to be shown that σ^* is a Nash equilibrium of the repeated game $\Gamma^\infty(\delta)$ for δ larger than some critical value δ^o.

Consider an arbitrary player $i \in I$, and assume all other players follow the strategy combination σ_{-i}^*. Assume further that there is some period $\tau \geq 1$ when player i decides to deviate from the strategy σ_i^*. Note that up to period τ the strategy combination s will be played in each period. The best payoff player i can get in the period of deviation, τ, is $p_i(\rho_i(s_{-i}), s_{-i})$ where $\rho_i(\cdot)$ denotes the best response mapping of player i. If $\rho_i(s_{-i}) \neq s_i$ holds, then according to the strategy σ_{-i}^* played by the other players, punishment actions r_{-i}^i by all other players will follow in any period after τ. Again $\rho_i(r_{-i}^i)$ is the best player i can do in these punishment periods yielding $p_i(\rho_i(r_{-i}^i), r_{-i}^i) = w_i$.

It is, therefore, not optimal for player i to deviate in period τ if the payoff from following σ_i^*, namely $p_i(s)$, is greater than the best payoff from deviating and being punished thereafter, that is, if

$$\begin{aligned} p_i(s) &\geq (1-\delta) \cdot \left[\sum_{t=1}^{\tau-1} \delta^{t-1} \cdot p_i(s) + \delta^{\tau-1} \cdot p_i(\rho_i(s_{-i}), s_{-i})) + \sum_{t=\tau+1}^{\infty} \delta^{t-1} \cdot w_i \right] \\ &= (1-\delta^{\tau-1}) \cdot p_i(s) + (1-\delta) \cdot \delta^{\tau-1} \cdot p_i(\rho_i(s_{-i}), s_{-i})) + \delta^{\tau} \cdot w_i \end{aligned}$$

or equivalently

$$p_i(s) \geq (1-\delta) \cdot p_i(\rho_i(s_{-i}), s_{-i})) + \delta \cdot w_i$$

holds. This condition is satisfied for any

$$\delta \geq \frac{(p_i(\rho_i(s_{-i}), s_{-i}) - p_i(s))}{(p_i(\rho_i(s_{-i}), s_{-i}) - w_i)} := \delta_i^o.$$

The same argument holds for any player $i \in I$ and for any round τ of the game. Let $\delta^o = \max\{\delta_1^o, \delta_2^o, \ldots, \delta_I^o\}$, then no player has an incentive to deviate from the equilibrium strategy σ_i^* in any repeated game $\Gamma^\infty(\delta)$ with $\delta \geq \delta^o$ as long as all other players follow the equilibrium strategy. This proves that σ^* is a Nash equilibrium of $\Gamma^\infty(\delta)$ for $\delta \geq \delta^o$. ■

Notice that $\delta_i^o = 0$ follows if s is a Nash equilibrium strategy combination of the stage game Γ. An infinite repetition of a Nash equilibrium of the stage game is, therefore, a Nash equilibrium of the infinitely repeated game for any discount factor δ. This confirms the result of theorem 8.1. If the implemented strategy combination does not form a Nash equilibrium of the stage game however, then the discount factor becomes important. The greater the difference between the payoff to be implemented and the payoff from deviation to a best response, the more patient a player has to be.

Remark 8.4. The result of theorem 8.2 is more general than stated here. Example 8.1 shows that the set of feasible and individually rational payoff combinations $\mathbb{P}(\Gamma)$ need not be convex. However, any payoff vector in the convex hull of $\mathbb{P}(\Gamma)$ can be achieved as a Nash equilibrium of an infinitely repeated game. ■

8.2.2 Subgame Perfect Equilibria

In repeated games, cooperation between players is maintained by the threat to punish any deviation. For sufficiently patient players, with δ close to one, it is possible to obtain any feasible payoff profile that exceeds the punishment payoffs as the average payoff of a Nash equilibrium in an infinitely repeated game. By threatening to punish a deviation with a joint effort of the other players, r_{-i}^i, the range of equilibria payoffs can be extended to include every individually rational payoff combination. Playing such a punishment strategy however may be very costly for the players who have to carry out the punishment. To maintain the punishment, they must be willing to forgo the opportunity to use a best response to the offender's action forever. Play during the punishment phase, therefore, may not be a Nash equilibrium. The reason such behavior can be part of an equilibrium strategy is that no deviation occurs along the equilibrium path precisely because of the threat. In equilibrium, punishment will be unnecessary. This raises the question of the credibility of such a threat.

EXAMPLE 8.2 (*resumed*). Reconsider the Cournot duopoly. To enforce the equilibrium quantity combination $(\bar{q}_1, \bar{q}_2)$ in the infinitely repeated game, players threatened to punish any deviation by playing q^o, the quantity that yields zero profits for the firm that deviates.

The profit of the firm that produces q^o to punish the other firm will be zero as well. Thus, to maintain the infinite punishment the firm would have to forgo all future profits. How credible is such a threat given that the Nash equilibrium in the stage game yields positive profits for both firms? ■

In Chapter 6, it was shown that the backward induction principle rules out equilibria built on incredible threats and promises. Repeated games have perfect information at the beginning of each stage when an action, that is, a strategy of the stage game, has to be chosen. As argued in Chapter 6, this suggests applying the concept of a subgame perfect equilibrium to infinitely repeated games to eliminate equilibria that rely on incredible threats.

In repeated games, a subgame begins at the beginning of each stage game. This is easy to see in the diagram of the twice-repeated prisoners' dilemma. Hence, a subgame begins after each history h^t. It is therefore possible to define a subgame and strategies in a subgame in terms of histories. Recall that a history $h^t = (s^1, s^2, s^3, \ldots, s^{t-1})$ describes a path of play up to period t. Given a particular history, $h^t \in S^{t-1}$, a *continuation history* of h^t, denoted $h^{t'}(h^t)$, is any history $h^{t'} \in S^{t'-1}$, $t' > t$, that has h^t as the first $(t-1)$ strategy combinations. A continuation strategy for player $i \in I$ after history h^t is defined as

$$\sigma_i(h^t) = \left(a_i^t(h^{t+1}(h^t)), a_i^{t+1}(h^{t+2}(h^t)), \ldots\right)$$

Note that $\sigma_i(h^t)$ still has an infinite sequence of action functions. Let $\sigma(h^t) = (\sigma_1(h^t), \sigma_2(h^t), \ldots, \sigma_I(h^t))$ be a strategy combination after history h^t, then $\mathscr{P}^\infty(\sigma(h^t))$ is well-defined.

DEFINITION 8.1. *A* subgame perfect equilibrium *of the repeated game* $\Gamma^\infty(\delta)$ *is a strategy combination* σ^* *such that* $\sigma^*(h^t) = (\sigma_1^*(h^t), \sigma_2^*(h^t), \ldots, \sigma_I^*(h^t))$ *is a Nash equilibrium for all* $h^t \in S^{t-1}$ *and all* $t \geq 1$.

Requiring an equilibrium strategy combination σ^* to be subgame perfect instead of being a Nash equilibrium only, rules out incredible threats because, after any history h^t, $\sigma^*(h^t)$ must still represent optimal behavior for all players. In particular, after any deviation the following punishment strategy must be optimal for each player, given that everyone else follows this punishment strategy. Hence punishment strategy combinations themselves have to form subgame perfect equilibria. It is easy to see that a repeated game strategy combination that has each agent in each stage game play a Nash equilibrium strategy of the stage game will be subgame perfect. Hence, there is no existence problem for subgame perfect equilibria in infinitely repeated games.

The following result will show that any payoff vector yielding a strictly higher payoff for each player than some Nash equilibrium of the constituent game can be implemented as a subgame perfect equilibrium of the infinitely repeated game provided agents are sufficiently patient. This can be achieved by using the threat to revert to the Nash equilibrium in case of deviation of any player. In contrast to the threat to play the minimax strategy, this is a credible threat since it leads to a Nash equilibrium in each stage of the punishment phase.

THEOREM 8.3 (*Friedman, 1971*). *Let s^* be a Nash equilibrium of the constituent game Γ. Then for any payoff vector p of Γ with $p_i > p_i(s^*)$ for all $i \in I$, there is a discount factor $\delta^o \in (0,1)$ such that, for all $\delta \geq \delta^o$, $\Gamma^\infty(\delta)$ has a subgame perfect equilibrium σ^* with $\mathscr{P}_i^\infty(\sigma^*) = p_i$ for all $i \in I$.*

Proof. Let s be the strategy combination of the constituent game that yields the payoff vector $p = (p_1(s), p_2(s), \ldots, p_I(s))$, and consider the following repeated game strategy $\sigma^* = (\sigma_1^*, \sigma_2^*, \ldots, \sigma_I^*)$:

$$\sigma_i^* = \left(a_i^{*1}(h^1), a_i^{*2}(h^2), \ldots.\right) \text{ with}$$

$$a_i^{*t}(h^t) = \begin{cases} s_i & \text{if } h^t(t-1) = s \text{ or } h^t(t-1) = 0 \text{ holds} \\ s_i^* & \text{otherwise.} \end{cases}$$

As in theorem 8.2, one checks that σ^* is a Nash equilibrium of $\Gamma^\infty(\delta)$ for all $\delta \geq \delta^o = \max\{\delta_1^o, \delta_2^o, \ldots, \delta_I^o\}$, with

$$\delta_i^o = \frac{\left(p_i(\rho_i(s_{-i}), s_{-i}) - p_i(s)\right)}{\left(p_i(\rho_i(s_{-i}), s_{-i}) - p_i(s^*)\right)},$$

and that σ^* induces an equilibrium path $\pi(\sigma^*)$ that yields a payoff $\mathscr{P}_i^\infty(\sigma^*) = p_i$ for each player.

To check that σ^* is a subgame perfect equilibrium, consider an arbitrary history h^t.

(i) For $h^t(t-1) \neq s$, all other players will play s_{-i}^* in any future round. Hence, playing s_i^* is a best response for player i. Therefore, $\sigma^*(h^t)$ is a Nash equilibrium if a deviation has occurred.
(ii) If $h^t(t-1) = s$ holds, then $\sigma_i^*(h^t)$ prescribes playing $a_i^*(h^{t'}(h^t))$ for all $t' > t$ which is a best response in any game $\Gamma^\infty(\delta)$ with $\delta \geq \delta^o$ if no deviation has occurred. ■

In an immediate application of theorem 8.3 to the Cournot duopoly game, one can conclude that any output combination in region A of the Cournot duopoly diagram can be achieved as a subgame perfect equilibrium in an infinite repetition of the game, if both firms have high enough discount factors. This illustrates that the subgame perfection criterion together with a threat to play the Nash equilibrium of the stage game forever has the power to restrict substantially the set of possible outcomes in the infinite repetition of the game.

The crucial element for obtaining a particular payoff combination of the stage game as a subgame perfect equilibrium in the infinitely repeated game is the capability to punish deviations with a strategy combination that is a subgame perfect equilibrium itself. Repeating a Nash equilibrium of a stage game is always a subgame perfect equilibrium in the repeated game. Thus, a threat of punishment with repeated play of a stage game Nash equilibrium can induce any play that yields for all players an average payoff greater than the equilibrium payoff of the stage game.

In general, however, repetition of a stage game equilibrium is not the most severe subgame perfect punishment. It is possible to construct a harsher subgame perfect punishment sequence of plays against each player according to the following "stick and carrot" principle:

- for T periods all players play r^i_{-i}, the minimax strategy combination against player i
- for all following periods all players except player i receive a reward that is high enough to make it optimal to go along with the punishment phase

Note that the "reward" induces players to pursue the punishment of another player even if doing so is harmful to the punisher. Such rewards can be constructed provided that there are as many linearly independent payoff combinations as there are players in the game. Since punishment phases are finite but reward phases are infinite, such a punishment behavior will be subgame perfect.

Given a subgame perfect punishment strategy combination against each player, a subgame perfect equilibrium strategy that implements any payoff combination $p \in \mathbb{P}(\Gamma)$ can be constructed according to the following simple rule:

> Each player begins with the cooperative action sequence that implements the payoff p; if a player deviates from this strategy, all players switch to the punishment strategy against this player; if some other player deviates from this punishment strategy, all other players turn to the subgame perfect punishment strategy against this player, etcetera.

It can be shown that every payoff combination in $\mathbb{P}(\Gamma)$ can be obtained as a subgame perfect equilibrium of an infinitely repeated game if δ is close to one and if the necessary rewards can be constructed. Thus with minor qualifications, the folk theorem holds for subgame perfect equilibria as well. The following section will show that, in finitely repeated games, this result fails to hold.

8.3 Repeated Games with a Finite Horizon

Any finite or infinite repetition of a stage game equilibrium forms a Nash equilibrium; in fact a subgame perfect equilibrium, of the repeated game, according to theorem 8.1. In contrast to infinitely repeated games, however, this is often the only equilibrium of a finitely repeated game. For example, a straightforward backward induction argument demonstrates that the prisoners' dilemma (example 8.1) has no other equilibrium in any finite repetition of the game. In fact, one can prove the following result.

THEOREM 8.4.

(i) *If there is a unique Nash equilibrium* s^* *of the stage game* Γ *such that* $p_i(s^*) = w_i$ *for all* $i \in I$ *holds, then the only payoff vector that can be obtained as a Nash equilibrium in any finite repetition of* Γ *is* $(p_1(s^*), p_2(s^*), \ldots, p_I(s^*))$.

(ii) *If there is a Nash equilibrium* s^* *of the stage game* Γ *such that* $p_i(s^*) > w_i$ *for all* $i \in I$ *holds, then any payoff vector in* $\mathbb{P}(\Gamma)$ *can be obtained as a Nash equilibrium of the finitely repeated game* $\Gamma^T(\delta)$, *for* δ *close to one and* T *sufficiently high.*

Instead of proving theorem 8.4,[5] the following example will illustrate the result. In the prisoners' dilemma (example 8.1), $s^* = (C, C)$ and $p_i(C, C) = 0 = w_i$ holds. Hence, part (i) of theorem 8.4 can be directly applied to this example. Modifying the original prisoners' dilemma by adding a strictly dominated strategy for each player leaves the equilibrium set unchanged, but both players' minimax payoffs are now smaller than the equilibrium payoffs of the stage game, $w_i < p_i(s^*)$, $i = 1, 2$.

EXAMPLE 8.3. Consider the prisoners' dilemma with additional dominated strategies.

		Player 2		
		N	C	A
Player 1	N	$1, 1$	$-1, 3$	$-4, -4$
	C	$3, -1$	$0, 0$	$-3, -4$
	A	$-4, -4$	$-4, -3$	$-4, -4$

[5]Proofs for these results can be found in van Damme (1991, pp. 195–198).

The original prisoners' dilemma is contained in the upper left-hand corner of the matrix. The strategy combination (C,C) remains the unique equilibrium in dominant strategies of the game. Yet, the minimax strategies and values have changed to $r_2^1 = r_1^2 = A$ and $w_1 = w_2 = -3$. Hence, for both players,

$$p_i(C,C) = 0 > -3 = w_i.$$

Suppose, for simplicity of exposition, that there is no discounting, $\delta = 1$. If the game is repeated twice, then cooperative behavior can be implemented in the first period as a Nash equilibrium of $\Gamma^2(1)$. Consider the following strategy combination (σ_1^*, σ_2^*) for $\Gamma^2(1)$: for $i = 1,2$, $\sigma_i^* = (a_i^1(h^1), a_i^2(h^2))$ with

$$a_i^1(h^1) = N, \quad \text{and} \quad a_i^2(h^2) = \begin{cases} C & \text{if} \quad h^2 = (C,C) \\ A & \text{otherwise} \end{cases}.$$

To check that $\sigma^* = (\sigma_1^*, \sigma_2^*)$ forms a Nash equilibrium for $\Gamma^2(1)$, suppose player 2 follows the equilibrium strategy σ_2^*. Could player 1 do better than playing σ_1^* as well? Following the equilibrium strategy yields the average payoff $\mathscr{P}_1^2(\sigma^*) = (1/2) \cdot [p_1(C,C) + p_1(N,N)] = (1/2)$. If player 1 deviates in the best possible way in period 1 and in period 2 and, hence, plays C in the first stage and C in the second stage, this would yield an average payoff $\mathscr{P}_1^2(\cdot, \sigma_2^*) = (1/2) \cdot [p_1(C,N) + p_1(C,A)] = (1/2) \cdot [3 + (-3)] = 0$. It is easy to check that no other deviation would achieve a better result for player 1. Thus, σ^* is indeed a Nash equilibrium of $\Gamma^2(1)$. ■

Example 8.3 indicates the basic idea used in the proof of theorem 8.4. (ii). Using backward induction, it is clear that a Nash equilibrium of the stage game has to be played in the last stage of the repeated game. Otherwise, the action of some player could be changed to achieve a higher payoff for this player in the last period. To induce the desired behavior in an earlier period ("cooperation in period 1" in the example), one must find a way to threaten the opponents with a worse outcome than the Nash equilibrium payoff of the stage game. This was impossible in the original version of the prisoners' dilemma game (example 8.1), but is possible in the modified version of example 8.3. The proof of theorem 8.4 generalizes this argument. Note, however, that the length of the repeated game T and the discount factor matter for the result. For $\delta < 1$, the threat to play the strategy A in example 8.3 may have to be upheld for more than two periods to implement the cooperative actions (N,N) in the first stage.

Example 8.3 may also serve as an illustration for the fact that the Nash equilibrium strategy, which implements cooperation by threatening to play the minimax strategy against the deviating player, usually fails to be subgame perfect. In example 8.3, to punish player 1 for a deviation, player

2 has to play A in the second round which is not optimal for him and yields him a stage-payoff of -4. In contrast to infinitely repeated games, however, it is impossible to construct "stick-and-carrot"-type subgame perfect punishments in a finitely repeated game. This raises the question of whether there are subgame perfect equilibria in a finitely repeated game that implement nonequilibrium payoffs of the stage game. The following theorem provides the answer to this question.

THEOREM 8.5. *If the constituent game Γ has a unique Nash equilibrium s^*, then any finite repetition of Γ has a unique subgame perfect equilibrium σ^* with $\mathscr{P}_i^T(\sigma^*) = p_i(s^*)$ for all $i \in I$.*

Proof. The theorem can be proved by backward induction.

(i) In period T, the last period, each player must play a best response to the other players' actions; otherwise, there would be some player who could improve his payoff by moving to a best response in the last period. Since there is a unique equilibrium of the stage game, s^*, it follows that the equilibrium strategy of the repeated game σ^* must have $a_i^T(h^T) = s_i^*$ for any history $h^T \in S^{T-1}$ and all $i \in I$.

(ii) Consider an arbitrary period τ, $0 \leq \tau < T$. Assume that the Nash equilibrium of the constituent game is played in all periods after τ, $a_i^t(h^t) = s_i^*$ for all $h^t \in S^{t-1}$, all $i \in I$, and all $t > \tau$. This implies that there is no subgame perfect threat possible for periods after τ and a deviation in τ cannot be punished afterwards. Thus each player $i \in I$ must play a best response to the stage game strategy of his opponents in period τ. Hence, in period τ, a Nash equilibrium of the stage game will be played again. Since there is a unique equilibrium in the stage game, $a_i^\tau(h^\tau) = s_i^*$ for all $h^\tau \in S^{\tau-1}$ follows. The claim of the proposition follows therefore by induction. ■

As an immediate implication of this theorem, all games with a unique Nash equilibrium do not provide an opportunity to threaten the opponents in a credible way if the game is finitely repeated. The only payoff combination that can be supported by a subgame perfect equilibrium is, therefore, the unique Nash equilibrium outcome of the stage game. This is a consequence of the backward induction argument that requires that in the last period a Nash equilibrium of the stage game be played. If there is a unique Nash equilibrium of the constituent game, then there is no possibility of making a credible threat to induce any other play for the second but last period, etcetera. Note the essential difference to infinitely repeated games in which there is no last period and therefore no constraint to play a Nash equilibrium in the last period.

The fact that a Nash equilibrium of the stage game has to be played in the last round of a repeated game does not rule out all possibilities of credible threats if the constituent game has more than one Nash equilibrium. If the payoffs in one Nash equilibrium are strictly worse for all players than in the other, then a player may credibly threaten to revert to the bad equilibrium in the final stages.

THEOREM 8.6.[6] *If a game* Γ *has two Nash equilibria* s^* *and* s^{**} *with* $p_i(s^*) > p_i(s^{**})$ *for all* $i \in I$, *then there exists* $\delta^o \in (0,1)$ *and* $T(t')$ *such that any payoff vector* p *with* $p_i > p_i(s^*)$ *for all* $i \in I$ *can be implemented for the first* t' *periods as a subgame perfect equilibrium of any finitely repeated game* $\Gamma^T(\delta)$ *with* $T \geq T(t')$ *and* $\delta \geq \delta^o$.

This section concludes with an example illustrating the structure of the equilibrium strategies.

EXAMPLE 8.4. Consider the following two-player game with three strategies for each player.

		Player 2		
		T	M	B
	T	4, 4	1, 5	0, 2
Player 1	M	5, 1	3, 3	0, 0
	B	2, 0	0, 0	1, 1

This game has two equilibria in pure strategies, namely (M, M) and (B, B). The equilibrium (M, M) dominates the other equilibrium (B, B), $p_i(M, M) = 3 > p_i(B, B) = 1$, but the payoffs from (T, T), which is not a Nash equilibrium, are strictly superior to the equilibrium payoffs. For $\delta = 1$, the following strategy σ^* will implement the payoff vector (4, 4) in period 1 if the game is repeated twice. Consider

$$\sigma^* = (\sigma_1^*, \sigma_2^*), \qquad \sigma_i^* = \left(a_i^1(h^1), a_i^2(h^2)\right),$$

with

$$a_i^1(h^1) = T \quad \text{and} \quad a_i^2(h^2) = \begin{cases} M & \text{if } h^2 = (T, T) \\ B & \text{otherwise,} \end{cases}$$

for $i = 1, 2$.

[6] A proof of a more general version of this result is contained in van Damme (1991, pp. 198–206).

One checks that the best possible deviation for a player consists in a switch to the best response to T in the first stage and to the "bad" equilibrium strategy in the second stage. For player 1, this yields

$$\mathscr{P}_1(\cdot, \sigma_2^*) = (1/2) \cdot [p_1(M,T) + p_1(B,B)] = (1/2) \cdot [5+1] = 3,$$

whereas the equilibrium strategy yields

$$\mathscr{P}_1(\sigma^*) = (1/2) \cdot [p_1(T,T) + p_1(M,M)] = (1/2) \cdot [4+3] = 3.5.$$

This confirms that σ_1^* is a best response to σ_2^* in the twice-repeated stage game.

By symmetry, the same argument holds for player 2. Hence, in this game it is possible to implement the nonequilibrium strategy combination (T,T) for one period as the first stage of a subgame perfect equilibrium path. This is achieved by credibly threatening to play the "bad" Nash equilibrium of the stage game in the last stage, instead of the "good" one, if a deviation occurs in round 1. ■

8.4 Summary

The study of repeated games is concerned with the question of to what extent it is possible to obtain nonequilibrium payoff vectors of a strategic form game as equilibrium average payoffs, if the game is played repeatedly by the same players. The answer depends mainly on the threat possibilities contained in the constituent game. In addition, one has to distinguish between finite and infinite repetitions of the game. The main difference between these two approaches rests on the fact that for finitely repeated games a Nash equilibrium of the stage game has to be played in the last round. This puts a constraint on possible threats in the case of finite repetitions that makes implementation of nonequilibrium strategy combinations of the stage game more difficult.

The second general observation is that implementation of nonequilibrium strategy combinations of the stage game through repetition may be impossible if agents are too impatient, and in the case of finite repetitions, if the horizon T is to short. What "too low" for the discount factor and "too short" for the time horizon mean depends largely on the difference between the equilibrium average payoff and the payoff from the best deviation. If players are patient enough and if the number of repetitions is high enough, then one obtains the following results:

(i) Every payoff combination of the constituent game yielding each player a higher payoff than the minimax payoff can be implemented as a Nash equilibrium of an infinite repetition of this game.

(ii) Every payoff combination of the constituent game yielding each player a higher payoff than the Nash equilibrium payoff in the stage game can be implemented as a Nash equilibrium of a finite repetition of this game, provided the Nash equilibrium payoff in the stage game is higher than the minimax payoff for all players.
(iii) Every payoff combination of the constituent game yielding each player a higher payoff than the minimax payoff can be implemented as a subgame perfect equilibrium of an infinite repetition of this game, provided that appropriate reward structures can be constructed.
(iv) Every payoff combination of the constituent game yielding each player a higher payoff than some Nash equilibrium payoff can be implemented as a subgame perfect equilibrium of a finite repetition of this game, provided there is another Nash equilibrium yielding a smaller payoff to all agents.

8.5 Remarks on the Literature

Friedman (1971) proved that a payoff vector of the constituent game that has a higher payoff for each player than the Nash equilibrium can be obtained as a Nash equilibrium; in fact a subgame perfect equilibrium, of the infinitely repeated game if the discount factor is high enough. The folk theorem for infinitely repeated games with no discounting ($\delta = 1$) was studied by Aumann and Shapley (1976) and Rubinstein (1979). A complete characterization of the set of payoff combinations that can be obtained by repeating a game was achieved by Abreu (1983), Benoit and Krishna (1985), and Fudenberg and Maskin (1986). Excellent surveys of these results can be found in van Damme (1991, Chapter 8) and Sabourian (1989). A recent development in this field not covered in this chapter is renegotiation-proofness. Van Damme (1991, pp. 207–211) summarizes some of the results on renegotiation-proofness.

Exercises

8.1 *Consider the prisoners' dilemma game in example 8.1.*
(a) Show that the payoff combination $(1, 1)$ can be achieved as a subgame perfect equilibrium of an infinite repetition of the game if both players have discount factors larger than .75.
(b) Use backward induction to show that there is no Nash equilibrium of the T-times repeated game that achieves the payoff $(1, 1)$ even in one stage of the repeated game.

8.2 *Consider the following payoff matrix of a two-player game with three strategies for each player.*

		Player 2		
		A	S	L
Player 1	A	3, 3	1, 4	0, 0
	S	4, 1	0, 0	−1, −1
	L	0, 0	−1, −1	−2, −2

(a) Derive the Nash equilibria and the worst punishment strategies and payoffs for this game.
(b) Derive a Nash equilibrium of the infinitely repeated game that has an equilibrium path $((A, A), (A, A), (A, A), \ldots)$.
(c) Is the Nash equilibrium of the repeated game that was derived in (b) subgame perfect?

8.3 *Consider two producers of a homogeneous good who have the same constant average costs of production, c, and face the demand function $x = \max\{0, A - b \cdot p\}$, where p denotes the price of the product and A, b, and c are parameters with $A > c > 0$, $b > 0$. Assume that both producers simultaneously set a price for their product. The producer who sets the lower price gets the whole market. If both producers charge the same price, then they share the demand equally.*

(a) Derive the Nash equilibrium of this game and draw a diagram that shows the set of feasible profit combinations. Determine the worst punishment that a producer can inflict on the opponent.
(b) Suppose this game is repeated infinitely. Derive a Nash equilibrium of the repeated game that enforces an equilibrium path in which both firms charge the monopoly price and share the profit equally. Does the equilibrium depend on the discount factor?
(c) Is this equilibrium derived in (b) unique? Is it subgame perfect?
(d) To what extent does the answer in part (b) depend on the assumption that the game is repeated infinitely often?

8.4 *Consider the game of example 4.1.*

		Player 2	
		L	R
Player 1	T	2, 1	.5, .5
	B	0, 0	1, 2

Show that, for high enough discount factors, there is a subgame perfect equilibrium of the infinitely repeated game that implements alternative play of (T, L) and (B, R). What punishment strategies have to be used?

8.5 *Consider two firms producing different products that are close substitutes. Both firms are monopolists regarding their own product market. Suppose that the demand functions for the two products have the following form:*

$$d_i(p_1, p_2) = \max\{0, 100 + 0.5 \cdot p_j - p_i\}, \qquad i, j = 1, 2, i \neq j,$$

where p_1 and p_2 denote the prices of the two firms. The production costs of the two firms are $c_i(x_i) = 20 \cdot x_i$, $i = 1, 2$.

(a) Determine the set of feasible and individually rational payoff combinations and indicate this set in a $(p_1 - p_2)$-diagram.

(b) Derive the smallest discount factor for which there is an infinitely repeated game that implements the symmetric Pareto-optimal profit combination as a subgame perfect equilibrium.

9

Bargaining Theory

Exchange is one of the most fundamental activities in economics. Though economic theory has been quite successful in explaining and analyzing trade in markets with many traders, it turns out to be surprisingly difficult to predict the outcome of negotiations about goods between two parties. In a large group of similar traders, no single trader can successfully deviate from the general terms of trade, because there are many others willing to trade under those conditions. The anonymity of large markets and the negligible influence of a single trader in such a market makes it possible to model each individual's behavior separately. In small groups, however, agents have to realize that the bargaining outcome depends on all individuals' behavior jointly. This interdependence creates immense strategic possibilities in negotiations among a small group of agents and accounts for the difficulties in analyzing these interactions.

Bargaining theory has a long tradition in economics. Early studies can be traced back to Edgeworth who analyzed this problem and discovered an indeterminacy of the bargaining process.[1] Though this problem remained unresolved for many decades, analysis of the bargaining problem provided a major stimulus for economics and some of the most recent developments in game theory.

[1]Newman (1987) provides a survey of Edgeworth's writing on contracts and exchange.

EXAMPLE 9.1 (*trade in the Edgeworth box*). Consider two consumers 1 and 2 who consume two goods in quantities x and y. Assume that each consumer has a non-negative endowment of the two goods $(\bar{x}_i, \bar{y}_i)$, $i = 1, 2$, and that preferences of these agents are given by strictly increasing and quasi-concave utility functions $u_i(x_i, y_i)$, $i = 1, 2$. The Edgeworth box diagram represents such an exchange situation.

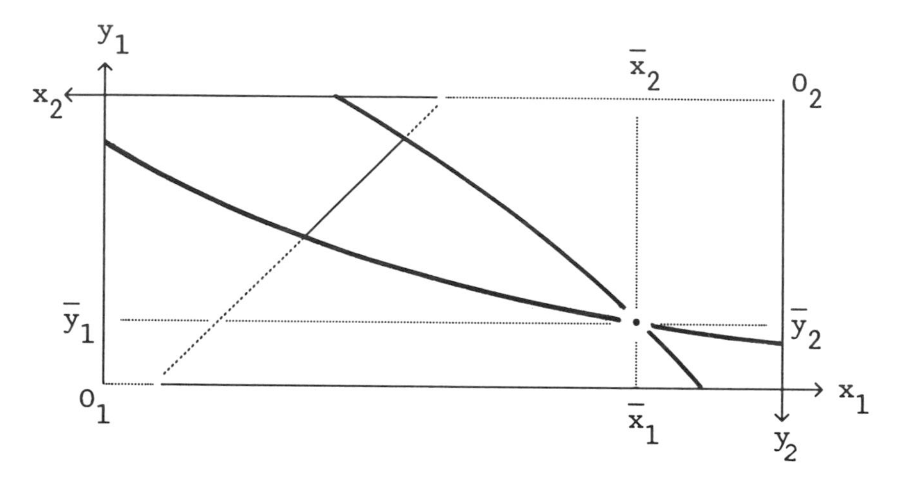

Traditional analysis does not predict a single allocation as the outcome of the bargaining process. Instead, a set of possible outcomes is identified as the rational result of any negotiation. This prediction is not based on a study of the bargaining process itself, however, but rests on the following two assumptions about the outcome of the bargaining process:

(i) Rational players will not accept allocations that make them worse off than their endowment (individual rationality).
(ii) Rational players will exhaust all potential gains from trade (efficiency or Pareto optimality).

The first of these principles assumes that agents are free to refuse to trade and, therefore, will only accept improving allocations. The second assumes that consumers, knowing the preferences of their bargaining partners, will not waste mutually beneficial opportunities. These principles are sufficient to predict a bargaining outcome on the part of the contract curve drawn solidly in the diagram. These allocations lie between the indifference curves passing through the endowment point (individual rationality) and on the locus of allocations where one agent maximizes utility given a certain level of utility of the other agent (efficiency). ■

A second important and much studied economic example is concerned with a prediction of the outcome of wage negotiations between trade unions representing workers on the one hand and firms as employers on the other.

EXAMPLE 9.2 (*firm-trade union bargaining*). Consider a trade union bargaining over wages and employment with a firm. An agreement specifies a wage rate w and an employment level n. Assuming that the trade-off for the union between different wage-employment pairs can be represented by a utility function $\nu(w,n) = w \cdot n - C(n)$, and that the firm's objective is profit maximization given a revenue function $R(n)$ that relates labor input n to the maximal revenue the firm can achieve with this input, one obtains the following payoff functions of the players:

$$\nu(w,n) = w \cdot n - C(n)$$
$$\pi(w,n) = R(n) - w \cdot n.$$

The cost function $C(n)$ of the trade union is supposed to represent the opportunity costs of its members.[2]

What can be said about the possible outcome of this interaction? Since both players have an interest in making the surplus they can share as big as possible, n will be chosen such that marginal revenue equals the marginal opportunity cost of working, $R'(n^*) = C'(n^*)$. Hence the appropriate employment level can be found agreeably. Note that n^* maximizes the joint surplus $\nu(w,n) + \pi(w,n)$ which does not depend on the wage rate. The employment level is therefore efficient. The level of wages w, however, remains contentious and forms the main object of bargaining. The following diagram represents a typical situation for an increasing and concave revenue function $R(n)$ and an increasing and convex opportunity cost function $C(n)$.

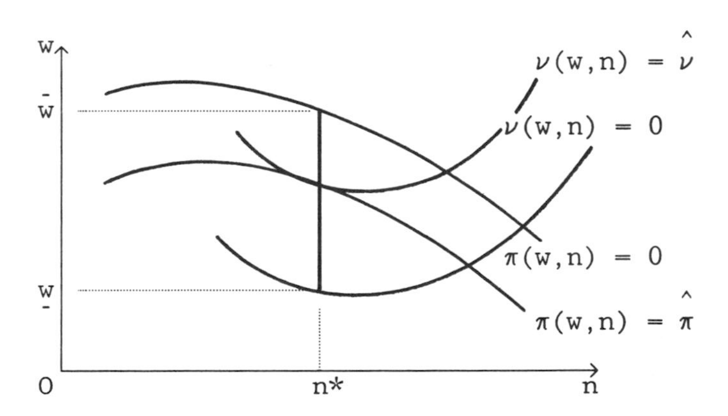

Assuming the firm will not accept a loss and the union wants to be compensated for the opportunity costs of its members, $C(n) \leq w \cdot n \leq R(n)$, one obtains the range $[\underline{w}, \overline{w}]$ of possible wage agreements. The set of contracts $\{(w,n) | w \in [\underline{w}, \overline{w}],\ n = n^*\}$ is individually rational and Pareto-

[2]McDonald and Solow (1981) consider more sophisticated versions of trade union objectives.

optimal. In general, there is little more that one can say about the outcome of this wage negotiation process. ■

In the bargaining problem between the trade union and the firm, payoffs can be transferred between the players directly through wage payments because the payoff functions of the firm $\pi(w, n)$ and the trade union $\nu(w, n)$ are both linear in the wage rate w. Payoff functions that are linear in one good are called *quasi-linear*, and if different players have payoff functions that are quasi-linear in the same good, then utility can be transferred between these players. Assuming transferable utility for a game means, therefore, that all players have utility functions that are quasi-linear in some good.[3] Note that transferable utility is not assumed for the Edgeworth box example.

In many applications, bargaining can be viewed as a decision about how to allocate interrelated payoffs between two players. This is obvious in the firm-trade union case in which "utility" is transferable via the wage rate. However, in the Edgeworth bargaining problem as well, bargaining can be described as a decision about how much utility to accord each player. Since individual rationality and efficiency of trade have been assumed, a particular level of utility for one player will unambiguously determine a level of utility for the other player. It is often useful to analyze the bargaining process in the space of feasible payoffs, rather than to deal with the allocation in the outcome space directly.

EXAMPLE 9.2 (*resumed*). The set of feasible payoff combinations shown in the diagram is determined by the maximal surplus, $\pi + \nu \leq R(n^*) - C(n^*)$, and by the individual rationality constraints $\pi \geq 0$ and $\nu \geq 0$.

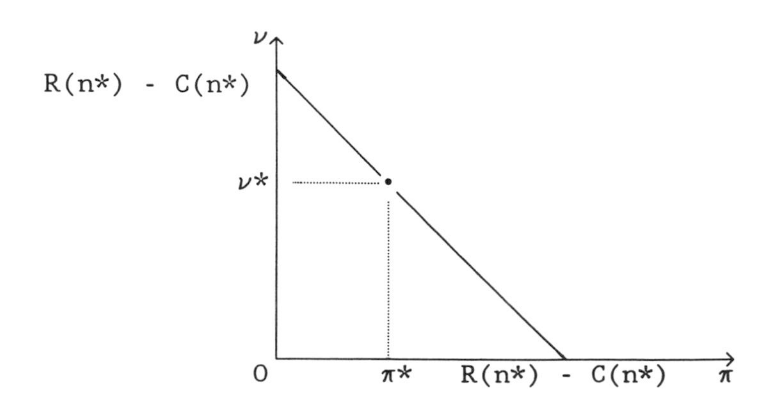

[3]Varian (1984, pp. 278–283) provides an excellent introduction to quasi-linear utility functions.

All individually rational and Pareto-optimal bargaining outcomes lie on the downward sloping line. Payoff combinations below this line are feasible and individually rational but not Pareto-optimal. Any payoff combination (ν^*, π^*) on the Pareto frontier corresponds to a particular wage rate in the interval $[\underline{w}, \overline{w}]$:

$$w^* = R(n^*)/n^* - \pi^*/n^* = c(n^*)/n^* + \nu^*/n^*. \quad ■$$

Similarly, one can transform the Edgeworth bargaining problem into a bargaining problem over utility combinations (u_1, u_2) for consumer 1 and 2, respectively.

EXAMPLE 9.1 (*resumed*). Suppose the consumers have the following utility functions, $\epsilon \leq .5$:

$$u_1(x_1, y_1) = (x_1 \cdot y_1)^{\epsilon} \quad \text{and} \quad u_2(x_2, y_2) = x_2 + 4 \cdot y_2.$$

Solving the optimization problem: Choosing (x_1, y_1) to maximize $(x_1 \cdot y_1)^{\epsilon}$ subject to

$$(\bar{x}_1 + \bar{x}_2 - x_1) + 4 \cdot (\bar{y}_1 + \bar{y}_2 - y_1) \geq u_2,$$

one derives the contract curve, $x_1 = 4 \cdot y_1$, and the maximizers of this problem, $x_1 = .5 \cdot [\Omega - u_2]$ and $y_1 = 2 \cdot [\Omega - u_2]$, respectively, with $\Omega \equiv [(\bar{x}_1 + \bar{x}_2) + 4 \cdot (\bar{y}_1 + \bar{y}_2)]$. Substituting into the objective function and solving for u_2, one obtains the utility possibility frontier (or Pareto frontier) as $\psi(u_1) = \Omega - u_1^{\gamma}$ with $\gamma \equiv 1/(2 \cdot \epsilon)$.

The parameter ϵ measures the concavity of the first consumer's utility function. The smaller ϵ, the more concave $u_1(x_1, y_1)$ will be. Changes in ϵ correspond to monotonic transformations and will not affect the allocations on the contract curve. Furthermore, adding or subtracting any constant number from the utility functions will not affect the behavior of the consumers. Thus, subtracting the utility of the endowments, $u_1(\bar{x}_i, \bar{y}_i)$, one can normalize the utility levels such that the utility from each consumer's endowment equals zero. The following diagram shows the utility possibility frontier of this example.

■

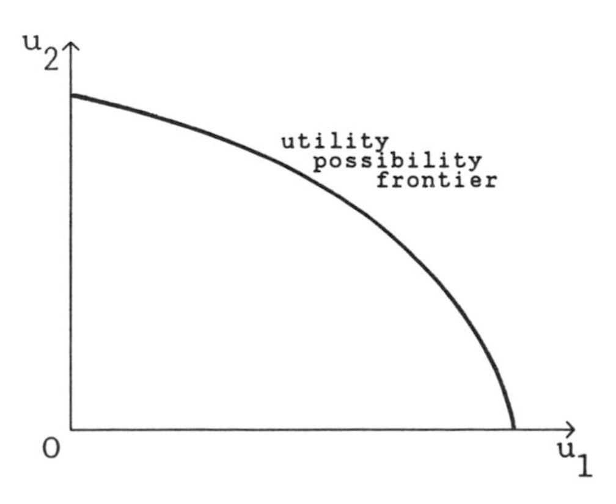

A few remarks are important in this context. Any monotonic transformation of the consumers' utility functions will leave the set of Pareto-optimal and individually rational allocations unchanged, but will lead to a different utility possibility frontier. Furthermore, neither concavity nor convexity of the utility possibility frontier can be guaranteed in general, even if the utility functions of the consumers are concave.

In general, a bargaining problem will be described by a continuous and decreasing function $\psi: \mathbb{R} \to \mathbb{R}$ that associates with any payoff level of player 1, x_1, the maximal possible payoff level of player 2, $x_2 = \psi(x_1)$. The function ψ represents the upper frontier of feasible payoff combinations (x_1, x_2). Payoff combinations below this frontier are feasible but not Pareto-optimal. In addition, one needs to specify the payoff combination $d = (d_1, d_2)$ that arises if bargaining ends without a solution. This payoff combination, d, is often referred to as disagreement payoff combination or threat point. Payoffs that satisfy $x_i \geq d_i$ for $i = 1, 2$ are individually rational. Throughout this chapter, it will be assumed that there exists a pair of payoffs (x_1^o, x_2^o) that satisfy $d_2 = \psi(x_1^o)$ and $x_2^o = \psi(d_1)$. Formally, a bargaining problem is defined as follows.

DEFINITION 9.1. A *bargaining problem* (X, d) specifies a set X of feasible payoff combinations

$$X = \{(x_1, x_2) \in \mathbb{R}^2 | x_2 \leq \psi(x_1)\}$$

and a payoff combination $d = (d_1, d_2) \in X$ that obtains in the case of a break-down of negotiations.

Since ψ is a continuous function, the subset of individually rational payoff combinations in X is a compact (closed and bounded) subset of $\mathbb{R}^2$. If ψ is a concave function, then X is a convex set. The following diagram shows a typical bargaining problem.

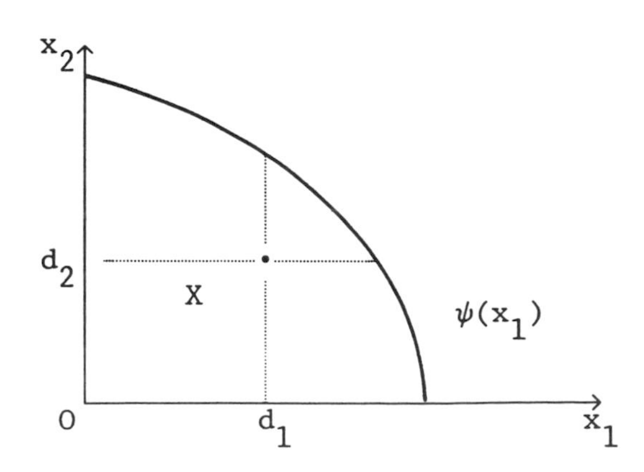

The following sections consider two approaches to finding a solution for a bargaining problem.[4] The first specifies details of the bargaining procedure and applies noncooperative solution concepts to obtain the bargaining outcome as an equilibrium allocation. The second reduces the set of possible bargaining outcomes by suggesting further properties that bargaining solutions should satisfy.

9.1 The Noncooperative Approach: Bargaining Procedures

The obvious way to approach a bargaining problem consists of a detailed description of all possible moves that each player may make. Once possible moves and payoffs are specified, application of an equilibrium concept may determine the outcome of thc bargaining process. The following example illustrates this idea.

EXAMPLE 9.3. Consider two players bargaining about three (indivisible) dollars according to the following procedure: Each player, $i = 1, 2$, suggests a share for herself s_i. If demands are feasible, that is, if $s_1 + s_2 \leq 3$ holds, this outcome is implemented; otherwise each agent gets a payoff equal to zero. Disregarding the strategies 0 and 3 which are weakly dominated, the game can be described by the following payoff matrix.

		Player 2	
		1	2
Player 1	1	1, 1	1, 2
	2	2, 1	0, 0

There are two Nash equilibria in pure strategies, $\{(1,2),(2,1)\}$, in which one player demands one dollar and the other two. Clearly, each player has a preference for a particular equilibrium, but both equilibria are equally rational. This specification of the bargaining procedure is not sufficient to determine the outcome of the bargaining problem. The indeterminateness can be overcome, however, by specifying further details of the bargaining procedure.

Suppose player 1 is allowed to propose a share for herself first. Player 2 can accept this proposal, a, in which case she will obtain the remainder, or reject the offer, r. If she rejects the offer, it is her turn to propose her share and player 1 can either accept or reject the proposal. If the second proposal is unsuccessful as well, each player gets a payoff of zero. The

[4]Edgeworth himself suggested a third solution: as the number of bargainers increases, the set of possible outcomes shrinks.

following extensive form describes this game assuming that both players discount payoffs with the same rate δ.

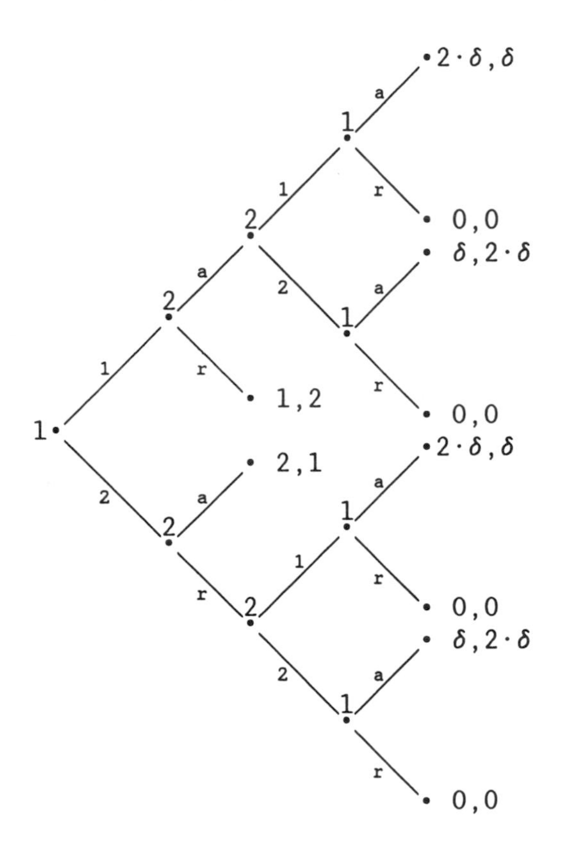

This extensive form game can be analyzed by backward induction. In the second stage, given that her own proposal has been rejected in the first stage, player 1 has to accept any proposal of player 2, since this leaves her better off than the disagreement payoff of zero. Thus it is optimal for player 2 to demand 2 units after rejecting the proposal of player 1. This yields a guaranteed payoff of $2 \cdot \delta$ for player 2 from rejecting the offer of player 1.

Therefore, in stage one, player 2 will accept a proposal s_1 if $3 - s_1 > 2 \cdot \delta$ holds. Hence, a request of 2 from player 1 will be accepted if $2 \cdot \delta < 1$ holds and will be rejected if $2 \cdot \delta > 1$ holds. On the other hand, if player 1 asks for a share of one unit this will always be acceptable.

Since player 1 can foresee this behavior of player 2, her optimal behavior is immediately clear. If $\delta < 1/2$ holds, she will ask for two dollars and her proposal will be accepted immediately. If $\delta > 1/2$ holds, she will demand one dollar which is accepted at once as well. Hence, the game always ends in the first round. ■

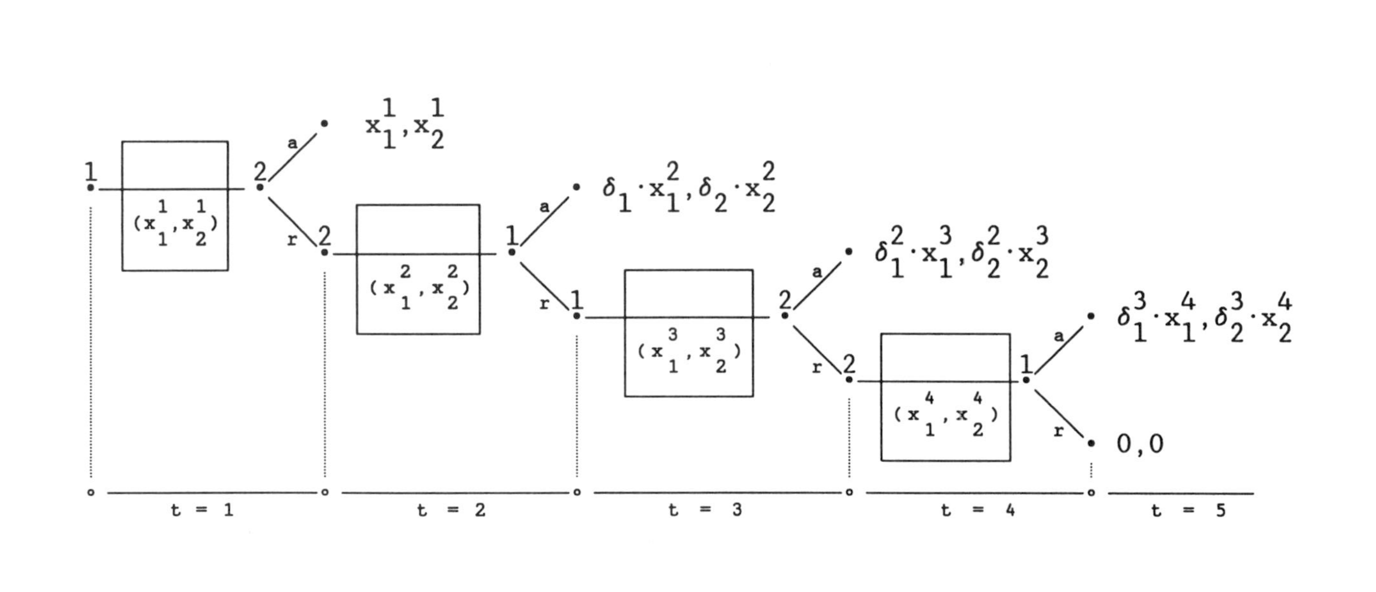

1
(x_1^1, x_2^1)
2
a
r
x_1^1, x_2^1
2
(x_1^2, x_2^2)
1
a
r
$\delta_1 \cdot x_1^2, \delta_2 \cdot x_2^2$
1
(x_1^3, x_2^3)
2
a
r
$\delta_1^2 \cdot x_1^3, \delta_2^2 \cdot x_2^3$
2
(x_1^4, x_2^4)
1
a
r
$\delta_1^3 \cdot x_1^4, \delta_2^3 \cdot x_2^4$
0,0
t = 1
t = 2
t = 3
t = 4
t = 5

Example 9.3 suggests that the solution of a bargaining problem can be determined by a detailed specification of the process of proposals and of delay costs. This idea can be generalized in a number of ways. The most obvious concerns the range of possible outcomes. To apply the bargaining procedure of example 9.3 to a general bargaining problem, one has to describe how bargaining proceeds over a larger set of outcomes:

- bargaining takes place in stages and player 1 begins
- in each stage, one player proposes a payoff allocation (x_1, x_2)
- the opponent responds by accepting, a, or rejecting, r, the offer
- if an offer is accepted the game ends and the accepted allocation is implemented
- if the offer is rejected, a new round begins and the player who rejected the proposal suggests a new allocation
- both players discount each round with a constant rate, δ_1 and δ_2 respectively
- if no agreement has been reached in stage T, then both players receive a payoff of zero.

According to this bargaining rule, player 1 proposes payoff combinations in rounds with odd numbers and player 2 in rounds with even numbers. The diagram on page 241 illustrates this procedure for four rounds of bargaining, $T = 5$. A square is used to indicate a large set of possible proposals; the branch of the tree passing through such a square indicates a representative proposal.

This is a game in extensive form with perfect information. Hence, one can determine subgame perfect equilibria by backward induction. Each stage of delay reduces the set of feasible payoff combinations because the players discount the future. The payoff possibility frontier in any stage t,

$$\phi_t(x_1) \equiv \delta_2^{t-1} \cdot \psi(x_1/\delta_1^{t-1}),$$

relates discounted payoffs of the two players. The next diagram shows how the set of possible payoffs shrinks due to discounting in a bargaining game with five stages ($T = 5$).

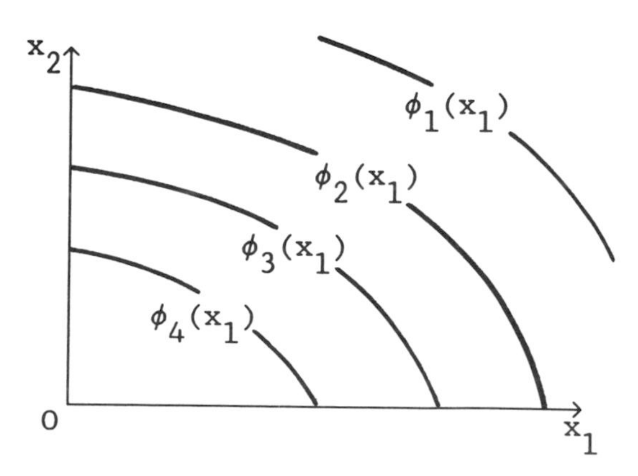

Examples 9.1 and 9.2 illustrate how one can use backward induction to derive the first period proposal that will be accepted immediately.

EXAMPLE 9.1 (*resumed*). Consider the Edgeworth bargaining game with payoff frontier $\psi(u_1) = \Omega - u_1^{\gamma}$. Assume that bargaining takes place over five stages ($T = 5$) according to the procedure just described.

With the help of the backward induction argument, it is easy to determine the unique subgame perfect equilibrium of this game:

Stage 4: Player 2 proposes. Since bargaining ends in the following round and the disagreement payoff equals zero, player 2 will optimally ask for the best possible allocation $u_2^4 = \psi(0) = \Omega$. Player 1 will accept this proposal, since she receives the same payoff as if she rejected the proposal. Note that the present value of this proposal to player 2 is only $\delta_2^3 \cdot \psi(0) = \phi_4(0)$.

Stage 3: Player 1 makes a proposal knowing that player 2 will reject any offer yielding her a payoff with a present value below $\phi_4(0)$. Given this constraint, the best discounted payoff that player 1 can achieve, u_1^3, leaves player 2 exactly with this minimum discounted payoff, $\phi_3(u_1^3) = \phi_4(0)$.

Stage 2: Player 2 will optimally propose an allocation that yields player 1 the minimum discounted payoff that induces her to accept the offer, $\phi_2(u_1^3) = u_2^2$.

Stage 1: Player 1 will ask for a payoff u_1^1 to which player 2 is indifferent between acceptance and rejecting; that is, $\phi_1(u_1^1) = u_2^2 = \phi_2(u_1^3)$.

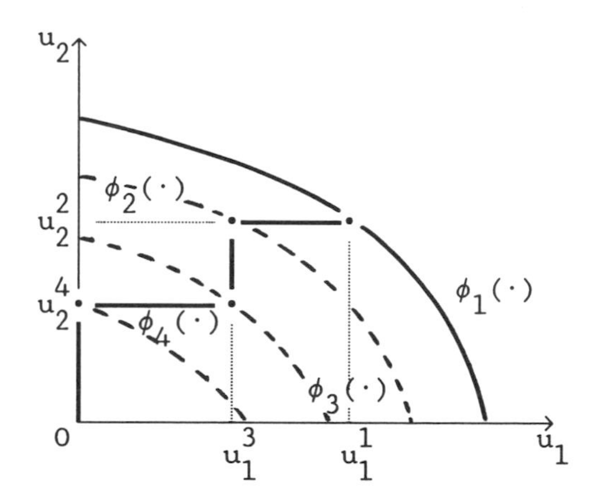

The backward induction argument illustrated in the diagram above, generates a sequence of discounted proposals for player 1 according to the following recursive system of equations:

$$\phi_1(u_1^1) = \phi_2(u_1^3), \qquad \phi_3(u_1^3) = \phi_4(0).$$

Discounted proposals of player 2 in even-numbered periods are simply $u_2^t = \phi_2(u_1^{t+1})$. Given the definition of $\phi_t(u_1)$, one can also express this sequence in terms of the undiscounted payoff possibility frontier $\psi(u_1)$ and the discount factors δ_1 and δ_2:

$$\psi(u_1^1) = \delta_2 \cdot \psi(u_1^3/\delta_1), \qquad \delta_2^2 \cdot \psi(u_1^3/\delta_1^2) = \delta_2^3 \cdot \psi(0).$$

From these equations, it is straightforward to determine the unique subgame perfect equilibrium proposal of player 1,

$$\begin{aligned}(u_1^1, \psi(u_1^1)) = \Big(&[(1+\delta_1^\gamma \cdot \delta_2) \cdot (1-\delta_2) \cdot \Omega]^{1/\gamma},\\ &\Omega - [(1+\delta_1^\gamma \cdot \delta_2) \cdot (1-\delta_2) \cdot \Omega]\Big)\end{aligned}$$

that player 2 will accept without delay. ■

From the argument in example 9.1, it easy to see that the recursive formula that determines the subgame perfect equilibrium proposals has the general form:

$$\delta_2^{t-1} \cdot \psi(x_1^t/\delta_1^{t-1}) = x_2^{t+1} = \delta_2^t \psi(x_1^{t+2}/\delta_1^t) \quad \text{for } t \text{ odd},$$

with the end condition $x_1^T = 0$ if T is odd and $x_2^T = 0$ for T even.

EXAMPLE 9.2 (*resumed*). In this example the payoff possibility frontier is given as $\psi(\pi) \equiv A - \pi$ with $A \equiv [R(n^*) - C(n^*)]$. Assume that bargaining proceeds according to the specified rule for five periods ($T = 5$).

Applying the general formula to derive the sequence of discounted payoffs for player 1,

$$\psi(\pi^1) = \delta_2 \cdot \psi(\pi^3/\delta_1), \qquad \delta_2^2 \cdot \psi(\pi^3/\delta_1^2) = \delta_2^3 \cdot \psi(0),$$

and substituting $\psi(\pi) \equiv A - \pi$ yields the two equations

$$A - \pi^1 = \delta_2 \cdot [A - \pi^3/\delta_1], \qquad \delta_2^2 \cdot [A - \pi^3/\delta_1^2] = \delta_2^3 \cdot [A - 0],$$

which can be easily solved for $\pi^3 = \delta_1^2 \cdot (1 - \delta_2^1) \cdot A$ and

$$\pi^1 = [(1-\delta_2) \cdot (1 + \delta_1 \cdot \delta_2)] \cdot A.$$

The expression in square brackets denotes the share of the surplus A the firm obtains as a result of five rounds of negotiation. Since the trade union gets the last chance to propose a share, it will obtain most of the surplus, provided it is patient enough, that is for a high δ_2. In addition, one checks easily that each player's share is increasing in its discount factor. Hence bargaining strength is related to the patience of players. ■

An undesirable aspect of this bargaining procedure is its dependence on the terminal time T. Whenever T is an odd number, player 2 is allowed to make the last offer giving an advantage over player 1. On the other hand, for T even, player 1 has this last proposer advantage. To avoid such endgame dependence, one is led to investigate the outcome of the bargaining process for the case in which T converges to infinity. An important question concerns the uniqueness of the equilibrium. For each finite bargaining game, there was a unique subgame perfect equilibrium, but a different one depending on whether T was odd or even. Hence the answer to this question is not clear.

THEOREM 9.1. *If the payoff frontier ψ is a concave and differentiable function and if the discount factors δ_1 and δ_2 are both less than one, then there is a unique subgame perfect equilibrium for the bargaining game with $T = \infty$. The subgame perfect equilibrium payoff combination $(x_1^*, \psi(x_1^*))$ is determined by the following equation*:

$$\psi(x_1^*) = \delta_2 \cdot \psi(\delta_1 \cdot x_1^*). \qquad (*)$$

Proof. The proof of this theorem consists of two parts. First, it is shown that there exists a subgame perfect equilibrium by presenting a subgame perfect strategy combination. Second, it is shown that there can be no other subgame perfect equilibrium.

(i) Existence: Since $\psi(\cdot)$ is a concave and differentiable function and $\delta_i < 1$, $i = 1, 2$, holds, equation $(*)$ has a unique solution. This is illustrated in the following diagram.

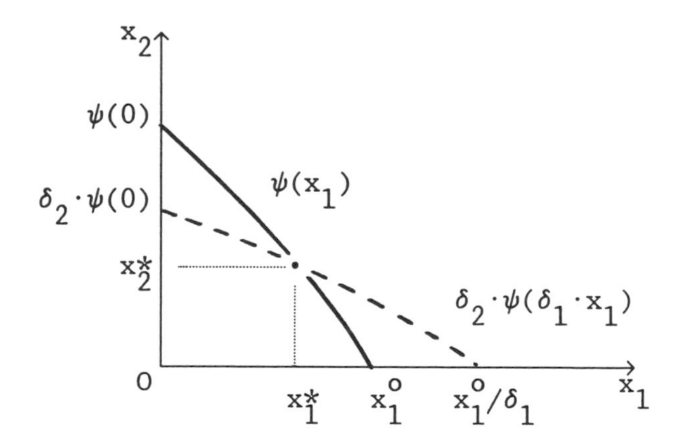

A strategy of a player has to specify which move she would make after any possible history of moves. Note that player 1 proposes a payoff combination in periods with odd numbers and responds to a proposal of player 2 in periods with even numbers. Similarly, player 2 proposes in even-numbered periods and responds in odd-numbered periods. Consider the following strategies:

Strategy of player 1:

$\cdot$propose $(x_1^*, \psi(x_1^*))$ if t is an odd number,

$$\cdot\begin{cases}\text{accept} \\ \text{reject}\end{cases} (x_1, \psi(x_1)) \text{ if } \begin{matrix} x_1 \geq \delta_1 \cdot x_1^* \\ x_1 < \delta_1 \cdot x_1^* \end{matrix}\Bigg\} \text{ if } t \text{ is an even number.}$$

Strategy of player 2:

$\cdot$propose $(\delta_1 \cdot x_1^*, \psi(\delta_1 \cdot x_1^*))$ if t is an even number,

$$\cdot\begin{cases}\text{accept} \\ \text{reject}\end{cases} (x_1, \psi(x_1)) \text{ if } \begin{matrix} \psi(x_1) \geq \delta_2 \cdot \psi(\delta_1 \cdot x_1^*) \\ \psi(x_1) < \delta_2 \cdot \psi(\delta_1 \cdot x_1^*) \end{matrix}\Bigg\}$$

if t is an odd number.

One checks easily that these strategies form a subgame perfect equilibrium. Suppose that t is odd and player 1 follows her equilibrium strategy. Player 2 receives the proposal $(x_1^*, \psi(x_1^*))$. If player 2 follows her equilibrium strategy, she will accept this offer since $\psi(x_1^*) = \delta_2 \cdot \psi(\delta_1 \cdot x_1^*)$ holds. This yields her a payoff of $\delta_2^{t-1} \cdot \psi(x_1^*)$. If player 2 rejects, she can at best hope to have her offer accepted in the next stage. Since player 1 accepts only $x_1 \geq \delta_1 \cdot x_1^*$, her payoff would be

$$\delta_2^t \cdot \psi(x_1) \leq \delta_2^t \cdot \psi(\delta_1 \cdot x_1^*) = \delta_2^{t-1} \cdot [\delta_2 \cdot \psi(\delta_1 \cdot x_1^*)] = \delta_2^{t-1} \cdot \psi(x_1^*).$$

Thus, it is optimal for player 2 to follow her equilibrium strategy in any odd period t.

Suppose now that player 2 follows her equilibrium strategy. If player 1 proposes $(x_1^*, \psi(x_1^*))$ according to her equilibrium strategy, then player 2 will accept, and player 1 obtains a payoff of $\delta_1^{t-1} \cdot x_1^*$. Since player 2 accepts only $\psi(x_1) \geq \delta_2 \cdot \psi(\delta_1 \cdot x_1^*) = \psi(x_1^*)$, the payoff of player 1 from any accepted proposal satisfies $\delta_1^{t-1} \cdot x_1 \leq \delta_1^{t-1} \cdot x_1^*$. Hence, it is optimal for player 1 to follow her equilibrium strategy in odd periods. Similarly, one can show that these strategies are mutually optimal for even periods too.

(ii) Uniqueness: To prove uniqueness, it will be shown that there is a unique subgame perfect equilibrium payoff combination. This is sufficient for uniqueness of the equilibrium strategy combination because each subgame has the same structure as the original game. Therefore, an equilibrium strategy combination of one subgame must be an equilibrium strategy combination of any subgame. Hence, subgame perfect equilibrium strategies must be of the form suggested.

To see that there is a unique equilibrium payoff combination $(x_1^*, \psi(x_1^*))$, assume to the contrary that there is a set of equilibrium payoffs. Denote by $\underline{x}_1$ the smallest equilibrium payoff of player 1, and by $\bar{x}_1$ the highest equilibrium payoff. Clearly, $\underline{x}_1 \leq x_1^* \leq \bar{x}_1$ must hold. Since $\psi(\cdot)$ is a decreasing function, the associated equilibrium payoffs of player 2 must satisfy $\psi(\bar{x}_1) \leq \psi(x_1^*) \leq \psi(\underline{x}_1)$.

If player 2 proposed $(x_1, \psi(x_1))$ with $x_1 > \delta_1 \cdot \bar{x}_1$, then player 1 would accept in any case because $\bar{x}_1$ is the highest equilibrium payoff she can hope to obtain. The minimal equilibrium payoff of player 2, $\psi(\bar{x}_1)$, can therefore be no worse than $\psi(\delta_1 \cdot \bar{x}_1)$. Hence, $\bar{x}_1$ must satisfy

$$\psi(\bar{x}_1) \geq \psi(\delta_1 \cdot \bar{x}_1) \geq \delta_2 \cdot \psi(\delta_1 \cdot \bar{x}_1). \tag{a}$$

Similarly, if player 1 proposes $(x_1, \psi(x_1))$ with $\psi(x_1) \geq \delta_2 \cdot \psi(\delta_1 \cdot \underline{x}_1)$, then player 2 certainly accepts. Therefore, the minimal equilibrium payoff of player 1, $\underline{x}_1$, cannot be less than $\psi^{-1}(\delta_2 \cdot \psi(\delta_1 \cdot \underline{x}_1))$ where $\psi^{-1}(\cdot)$ denotes the inverse of $\psi(\cdot)$. Since $\psi(\cdot)$ is a decreasing function, this implies

$$\psi(\underline{x}_1) \leq \delta_2 \cdot \psi(\delta_1 \cdot \underline{x}_1). \tag{b}$$

It is clear from the preceding diagram that all the values of $\bar{x}_1$ that satisfy the inequality (a) must lie to the left of x_1^*, and all the values of $\underline{x}_1$ satisfying (b) must lie to the right of x_1^*. Hence, (a) and (b) imply $\bar{x}_1 \leq x_1^* \leq \underline{x}_1$. Since $\bar{x}_1$ and $\underline{x}_1$ were upper and lower bounds for the equilibrium payoff of player 1 respectively, $\bar{x}_1 = x_1^* = \underline{x}_1$ follows. ■

Remark 9.1. One can derive equation $(*)$ that defines the subgame perfect equilibrium payoff with the help of the following consideration. It is easy to see that any subgame perfect equilibrium will consist of a proposal x_1^* of player 1 that will be accepted immediately because any delay is costly. Whatever will be accepted in the future must be acceptable in stage 1. In the finite horizon case, backward induction generated the sequence of proposals of player 1 according to the formula

$$\phi_t(x_1^t) \equiv \delta_2^{t-1} \cdot \psi(x_1^t / \delta_1^{t-1}) = \delta_2^t \cdot \psi(x_1^{t+2} / \delta_1^t) \equiv \phi_{t+1}(x_1^{t+2}).$$

Applying this equation for $t = 1$ and $x_1^t = \delta^{t-1} \cdot x_1^*$, one obtains equation $(*)$,

$$\delta_2^0 \cdot \psi(x_1^* / \delta_1^0) = \delta_2^1 \cdot \psi(\delta_1^2 \cdot x_1^* / \delta_1^1). \quad ■$$

Examples 9.1 and 9.2 may be used to illustrate how one can apply theorem 9.1 to particular bargaining games.

EXAMPLE 9.2 (*resumed*). Consider the case in which the firm and the trade union bargain without a definite time limit ($T = \infty$).

From equation $(*)$, one computes easily the subgame perfect equilibrium payoff for the firm as

$$\pi^* = \left[(1 - \delta_2) \cdot (1 - \delta_1 \cdot \delta_2)^{-1}\right] \cdot A$$

and for the trade union as

$$\nu^* = \delta_2 \cdot \left[(1 - \delta_1) \cdot (1 - \delta_1 \cdot \delta_2)^{-1}\right] \cdot A.$$

Note that there remains an advantage for the firm because it is allowed to propose first. ■

EXAMPLE 9.1 (*resumed*). Consider the case with no time limit for the negotiations of the two consumers ($T = \infty$).

In this case, the graph of the function $\psi(u_1) = \Omega - u_1^\gamma$ describes all Pareto-optimal utility pairs that the two consumers can achieve. Hence from

$$\Omega - u_1^\gamma = \delta_2 \cdot \left[\Omega - (\delta_1 \cdot u_1)^\gamma\right]$$

one can determine the bargaining outcome for player 1 as

$$u_1^* = \left[\frac{(1 - \delta_2)}{(1 - \delta_1^\gamma \cdot \delta_2)} \cdot \Omega\right]^{1/\gamma}.$$

Similarly, player 2 obtains

$$u_2^* = \left[\frac{\delta_2 \cdot (1 - \delta_1^\gamma)}{(1 - \delta_1^\gamma \cdot \delta_2)} \cdot \Omega\right].$$

It is easy to check that a monotonic transformation of player 1's utility function will affect the outcome of the bargaining game, even if it leaves the set of Pareto-optimal and individually rational commodity allocations unchanged. To see this, note that a decrease in ϵ will unambiguously increase $\gamma = 1/(2 \cdot \epsilon)$. Differentiating the bargaining result of consumer 2 with respect to γ yields

$$du_2^*/d\gamma = (1 - \delta_1^\gamma \cdot \delta_2)^{-2} \cdot \left[(\delta_2 - 1) \cdot \delta_1^\gamma \cdot \ln \delta_1\right] \cdot \delta_2 \cdot \Omega > 0,$$

because of $\ln \delta_1 < 0$ and $\delta_2 < 1$. An increase in γ, therefore, raises the bargaining outcome of player 2, u_2^*, hence the quantities of both goods that consumer 2 obtains along the unchanged contract curve. Thus, as ϵ decreases, the bargaining result in terms of commodities will increase for consumer 2 and decrease for consumer 1. ■

This section showed that an explicit specification of the bargaining process may determine the outcome of the bargaining problem as a subgame perfect equilibrium. On the other hand, this outcome is very sensitive to all details of the negotiation process as well as to the delay costs of the two players. In particular, one has to keep in mind that the practice of analyzing bargaining in terms of utilities gives the players' payoff functions a cardinal interpretation.

9.2 The Cooperative Approach: Desirable Properties of Outcomes

There is an alternative way to arrive at a prediction for the outcome of a bargaining problem. Instead of describing the bargaining procedure in full detail, one may try to characterize the outcome by requirements (axioms) that an outcome is expected to satisfy. The following example illustrates this point.

EXAMPLE 9.4. Consider two players who have to share \$100. If they reach no agreement they obtain nothing, $d = (0,0)$. Acceptable shares need not sum to \$100 (free disposal). The following diagram shows this bargaining problem.

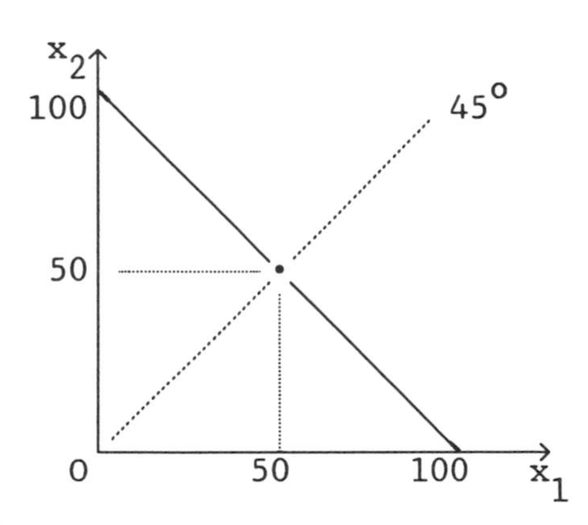

Suppose one requires a solution to satisfy the following two criteria:

(i) An outcome (x_1, x_2) should be Pareto-optimal.

(ii) In a bargaining problem in which feasibility of $(x_1, x_2) = (a, b)$ implies feasibility of $(x_1, x_2) = (b, a)$, both players should get the same shares, $x_1 = x_2$.

The first requirement has been discussed before and seems reasonable for rational and fully informed players. It suggests that a solution lies on the upper frontier of the bargaining problem. An argument for the second property can be based on the principle that all relevant aspects of the bargaining problem should be included in the formulation of the problem. A game with the property that one can swap the payoffs of the players for any feasible allocation is called *symmetric*. Symmetry of a bargaining problem indicates that there are no distinguishing features for any player. Thus one can argue that their shares should not be different.

The bargaining set of this example is obviously symmetric. Hence according to requirement (ii), an outcome must lie on the 45°-line, and by (i), on the Pareto frontier. This identifies $(x_1^*, x_2^*) = (50, 50)$ as the unique outcome that satisfies requirements (i) and (ii). ■

Example 9.4 illustrates the idea of the cooperative approach: a set of "reasonable" requirements for the outcome of the bargaining process may determine the outcome. In fact, the two requirements, Pareto optimality and symmetry, determine a unique outcome in all symmetric bargaining problems. There is however no reason both players should get the same payoff in bargaining problems that are not symmetric. Since the symmetry condition constrains only the solution of symmetric bargaining problems, nonsymmetric bargaining problems would therefore retain all individually rational and Pareto-optimal payoff combinations as possible solutions. Such considerations suggest imposing conditions on the bargaining outcome that apply not only to some particular cases but to a large set of bargaining problems. A bargaining solution is therefore defined as a procedure assigning to each bargaining problem a unique outcome.

DEFINITION 9.2. Denote by $\mathscr{X}$ a set of bargaining problems. A *bargaining solution* is a function $f\colon \mathscr{X} \to \mathbb{R}^2$ that associates with each bargaining problem $(X, d) \in \mathscr{X}$ a particular outcome $(x_1^*, x_2^*) = f(X, d)$.

Note that this definition in itself represents an assumption of how bargaining situations should be viewed. It assumes in particular that only the set of payoff possibilities X and the threat point d matter for a solution and that the same principles should apply for all bargaining problems in $\mathscr{X}$. Given this approach, it is not surprising to find that different systems of axioms will characterize different solutions to the bargaining process. The system of axioms presented here is a generalization of the system proposed by Nash (1953).

The first two axioms require the outcome of any bargaining problem to be (strongly) individually rational and Pareto-optimal. These requirements are very suggestive indeed, and were proposed in the context of the Edgeworth bargaining problem a long time ago.

Axiom 1 (strong individual rationality). The outcome of a bargaining problem, $(x_1^*, x_2^*) = f(X, d)$, shall be strictly better for both players than the disagreement payoff, $d_1 < x_1^*$ and $d_2 < x_2^*$.

Axiom 2 (Pareto optimality). There is no feasible payoff combination (x_1, x_2) that has a higher payoff for both players than the bargaining solution $(x_1^*, x_2^*) = f(X, d)$: $x_1 > x_1^*$ and $x_2 > x_2^*$ implies $(x_1, x_2) \notin X$.

A bargaining outcome that satisfies these two axioms lies on the Pareto frontier of the bargaining problem and is strictly preferred to the respective disagreement payoff by both players.

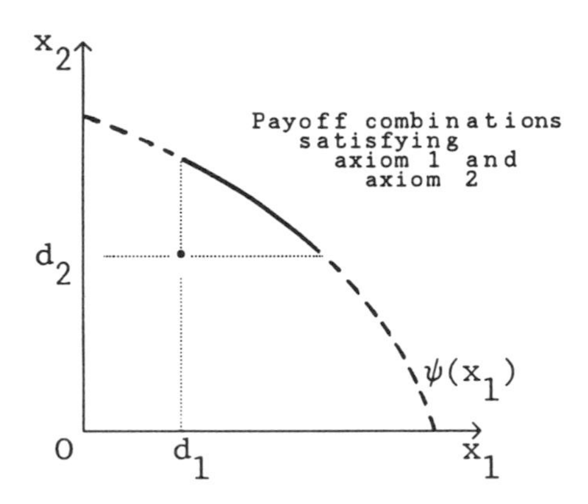

The next axiom identifies bargaining problems as essentially identical if they can be linearly transformed into one another. Except for this linear transformation, such problems shall have the same solution.

Axiom 3 (invariance). If a bargaining problem (Y, d') is related to another bargaining problem (X, d) such that $(y_1, y_2) = (A_1 + B_1 \cdot x_1, A_2 + B_2 \cdot x_2)$ for all $(x_1, x_2) \in X$ and all $(y_1, y_2) \in Y$ holds, then the solutions of the bargaining problems, $(x_1^*, x_2^*) = f(X, d)$ and $(y_1^*, y_2^*) = f(Y, d')$, shall satisfy

$$(y_1^*, y_2^*) = (A_1 + B_1 \cdot x_1^*, A_2 + B_2 \cdot x_2^*).$$

The invariance of the bargaining solution to linear transformations that axiom 3 requires is usually justified by the invariance of expected utility functions to such transformations. If payoffs of both players are considered to be von Neumann–Morgenstern utilities that satisfy the expected utility hypothesis, then they are unique representations of preferences up to a linear-affine transformation only. This means that the same preferences over lotteries will be represented if the payoff x_i is replaced by the payoff $y_i = \beta_i(x_i)$ as long as $\beta_i(x_i)$ is of the form

$$\beta_i(x_i) = A_i + B_i \cdot x_i,$$

where A_i is an arbitrary real number (positive, negative, or zero) and B_i is an arbitrary positive real number. Since utility transformations of this kind do not affect the preferences of an expected-utility-maximizing player, the outcome of the bargaining problem should not depend on them either, no matter what values A_i and B_i may take. Note however that bargaining problems often involve choices under certainty, such as in examples 9.1 and 9.2 in which a justification by the expected-utility hypothesis is inappropriate.

This axiom makes it possible to treat a large number of bargaining problems as essentially equivalent. In particular, all bargaining problems with a linear Pareto frontier are equivalent in this sense. By choosing A_1, A_2, B_1, and B_2 appropriately, one can transform any bargaining problem with a linear Pareto frontier into any other bargaining problem with a linear Pareto frontier. Thus, finding a solution for one bargaining problem with a linear Pareto frontier is by axiom 3 sufficient to assign a solution to any bargaining problem with linear Pareto frontier.

Example 9.5. Consider the following bargaining problem (X, d) with linear Pareto frontier

$$X = \{(x_1, x_2) \in \mathbb{R}^2 | x_2 \leq a - b \cdot x_1\}$$

for some $a > 0$ and $b > 0$.

There is a linear transformation $\beta_i(x_i) = A_i + B_i \cdot x_i$, $i = 1, 2$, that maps the set

$$\Delta_X = \{(x_1, x_2) \in X | x_1 \geq d_1,\ x_2 \geq d_2\}$$

onto the set

$$\Delta = \{(x_1, x_2) \in \mathbb{R}^2 | x_1 \geq 0,\ x_2 \geq 0,\ x_1 + x_2 \leq 1\}.$$

The following diagram illustrates this transformation. Note that (d_1, d_2) is mapped to $(0, 0)$. Formally, this transformation is achieved by setting

(A_1, B_1, A_2, B_2) to

$$A_1 = -\frac{d_1}{a - d_1}, \quad b_1 = \frac{1}{a - d_1}, \quad A_2 = -\frac{d_2}{(a/b) - d_2}, \quad \text{and}$$

$$B_2 = \frac{1}{(a/b) - d_2}.$$

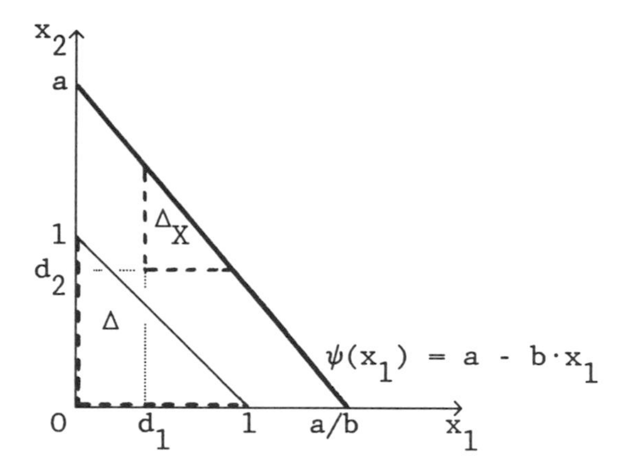

One can check easily that, with these parameters, the function $\beta_i(x_i)$ transforms the bargaining problem (X, d) into the bargaining problem $(\Delta, 0)$. The invariance axiom 3 requires in this case the following: If (x_1^*, x_2^*) is the bargaining outcome for (X, d), then

$$(y_1^*, y_2^*) = \left(\frac{x_1^* - d_1}{a - d_1}, \frac{x_2^* - d_2}{(a/b) - d_2} \right)$$

shall be the solution of $(\Delta, 0)$ and vice versa. ■

The fourth axiom requires the solution of a bargaining problem to be independent of payoff combinations that are feasible and individually rational but do not affect the solution. Removing such irrelevant alternatives shall leave the solution unchanged.

Axiom 4 (independence of irrelevant alternatives). Consider a bargaining problem (X, d) with solution (x_1^*, x_2^*). If another bargaining problem (Y, d) with $Y \subseteq X$ contains the solution (x_1^*, x_2^*), then (x_1^*, x_2^*) shall be the solution for the bargaining problem (Y, d) as well.

The following diagram illustrates the idea of this axiom. Removing payoff combinations from the set X without removing the solution itself leads to a bargaining problem with the same solution.

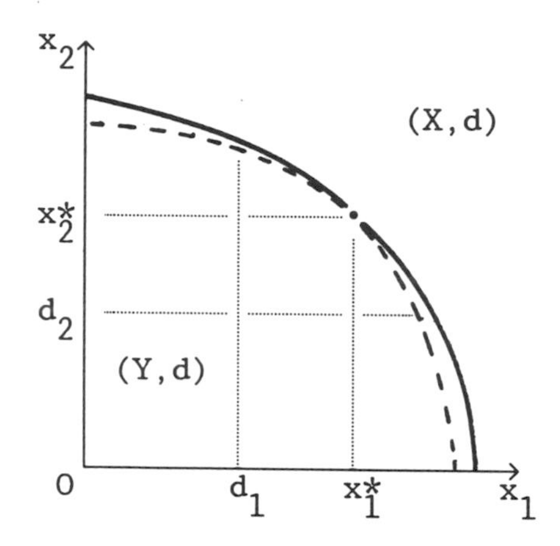

An immediate implication of axiom 4 is that the solution of a bargaining problem with linear Pareto frontier remains the solution of all bargaining problems contained in it and having a Pareto frontier that is tangent at this solution.

EXAMPLE 9.6. Consider a bargaining problem (X, d) where X is defined by a strictly concave and differentiable function $\psi(x_1)$.

Since $\psi(\cdot)$ is differentiable, there is a unique tangent at every point $(x_1, \psi(x_1))$ with slope $\psi'(x_1)$. Since $\psi(\cdot)$ is concave, all points of X must lie below such a tangent. The tangential line $\tilde{\psi}(x_1) = \tilde{x}_2 - \psi'(\tilde{x}_1) \cdot [x_1 - \tilde{x}_1]$ determines another bargaining problem $(\tilde{X}, d)$.

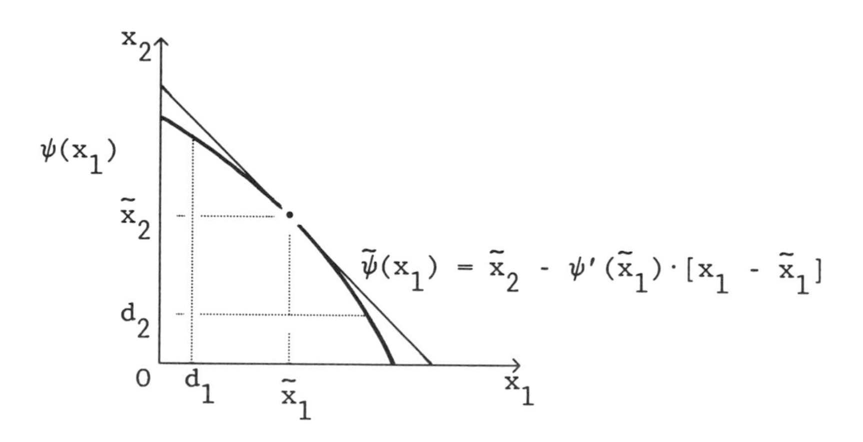

It is obvious from the diagram that $X \subseteq \tilde{X}$ holds. Hence, by axiom 4, if $(\tilde{x}_1, \tilde{x}_2)$ is the solution for the bargaining problem $(\tilde{X}, d)$, then it shall be the solution for (X, d) as well. ■

The final axiom is the symmetry requirement discussed in the context of example 9.4.

Axiom 5 (symmetry). If a bargaining problem (X, d) satisfies

(i) $d_1 = d_2$

(ii) $(x_1, x_2) \in X$ implies $(x_2, x_1) \in X$,

then the solution $(x_1^*, x_2^*) = f(X, d)$ shall satisfy $x_1^* = x_2^*$.

In example 9.4, it was demonstrated that the symmetry axiom 5 together with Pareto optimality (axiom 2) determine a unique solution in symmetric bargaining problems. The following theorem will show that the five axioms yield a unique bargaining outcome for any set $\mathscr{X}$ containing only convex bargaining problems. The idea of the proof is quite simple. The bargaining problem $(\Delta, 0)$ is symmetric with $(.5, .5)$ as a unique bargaining solution. Consider an arbitrary bargaining problem (X, d). Since all bargaining problems are convex, each Pareto optimal point has a tangent that can be mapped to the unit simplex, as in example 9.5. The theorem shows that there is a unique point on the Pareto frontier of each bargaining problem, say $a(X, d)$, that has the property that the linear function that maps the tangent to the unit simplex maps the tangential point to the point $(.5, .5)$ on the simplex. By axiom 3, $a(X, d)$ must be the solution of the bargaining problem with the tangent to (X, d) at $a(X, d)$ as Pareto frontier. Applying axiom 4, as in example 9.6, shows that $a(X, d)$ is the solution of the bargaining problem (X, d).

THEOREM 9.2. *Suppose that, for all $(X, d) \in \mathscr{X}$, $\psi(x_1)$ is a concave function and that there exists $(x_1, x_2) \in X$ with $x_1 > d_1$ and $x_2 > d_2$. If the bargaining solution $f(\cdot)$ satisfies axioms* 1, 2, 3, *and* 4, *then there exists $\alpha \in (0, 1)$ such that*

$$f(X, d) = \text{argmax}\left\{(x_1 - d_1)^{\alpha} \cdot (x_2 - d_2)^{1-\alpha} \middle| (x_1, x_2) \in X\right\}.$$

If the solution also satisfies axiom 5, *then $\alpha = .5$ holds.*

Proof. The proof of this theorem proceeds in three steps. First it is checked that the suggested solution satisfies axioms 1 to 4. Then it is shown that every solution that satisfies axioms 1 to 4 is characterized by some α. Thirdly, it is demonstrated that axiom 5 implies $\alpha = .5$.

(i) Since ψ is concave, the set X must be convex. Hence for any $\alpha \in (0, 1)$, there is a unique maximizer (x_1^*, x_2^*) satisfying

$$(x_1^* - d_1)^{\alpha} \cdot (x_2^* - d_2)^{1-\alpha} \geq (x_1 - d_1)^{\alpha} \cdot (x_2 - d_2)^{1-\alpha}$$

$$\text{for all} \quad (x_1, x_2) \in X.$$

A1: Since there is a payoff combination $(x_1, x_2) \in X$ with $x_1 > d_1$ and $x_2 > d_2$, $(x_1^* - d_1)^\alpha \cdot (x_2^* - d_2)^{1-\alpha} \geq (x_1 - d_1)^\alpha \cdot (x_2 - d_2)^{1-\alpha} > 0$ follows. Hence one has $x_1^* > d_1$ and $x_2^* > d_2$, and axiom 1 is satisfied.

A2: If (x_1^*, x_2^*) were not Pareto optimal, there would be $(x_1, x_2) \in X$ with $x_1 > x_1^*$ and $x_2 > x_2^*$. This would imply

$$(x_1 - d_1)^\alpha \cdot (x_2 - d_2)^{1-\alpha} > (x_1^* - d_1)^\alpha \cdot (x_2^* - d_2)^{1-\alpha},$$

contradicting the fact that (x_1^*, x_2^*) is a maximizer on X. Hence axiom 2 is satisfied.

A3: Suppose that (Y, d') is such that, for all $(y_1, y_2) \in Y$, $y_i = A_i + B_i \cdot x_i$, $i = 1, 2$, holds; in particular, $d_i' = A_i + B_i \cdot d_i$, $i = 1, 2$. Then, for $(y_1^*, y_2^*) = ([A_1 + B_1 \cdot x_1^*], [A_2 + B_2 \cdot x_2^*])$,

$$\begin{aligned}
&(y_1^* - d_1')^\alpha \cdot (y_2^* - d_2')^{1-\alpha} \\
&\quad = ([A_1 + B_1 \cdot x_1^*] - [A_1 + B_1 \cdot d_1])^\alpha \\
&\qquad \cdot ([A_2 + B_2 \cdot x_2^*] - [A_2 + B_2 \cdot d_2])^{1-\alpha} \\
&\quad \geq ([A_1 + B_1 \cdot x_1] - [A_1 + B_1 \cdot d_1])^\alpha \\
&\qquad \cdot ([A_2 + B_2 \cdot x_2] - [A_2 + B_2 \cdot d_2])^{1-\alpha} \\
&\quad = (y_1 - d_1')^\alpha \cdot (y_2 - d_2')^{1-\alpha}
\end{aligned}$$

holds for all $(y_1, y_2) \in Y$. Hence, axiom 3 is satisfied.

A4: Suppose that (Y, d) is a bargaining problem with $Y \subset X$ and $(x_1^*, x_2^*) \in Y$. Hence, from

$$(x_1^* - d_1)^\alpha \cdot (x_2^* - d_2)^{1-\alpha} \geq (x_1 - d_1)^\alpha \cdot (x_2 - d_2)^{1-\alpha} \quad \text{for all} \quad (x_1, x_2) \in X,$$

it follows that

$$(x_1^* - d_1)^\alpha \cdot (x_2^* - d_2)^{1-\alpha} \geq (y_1 - d_1)^\alpha \cdot (y_2 - d_2)^{1-\alpha} \quad \text{for all} \quad (y_1, y_2) \in Y$$

holds. This shows that axiom 4 is satisfied as well.

(ii) Given an arbitrary bargaining solution $f(\cdot)$ satisfying axioms 1, 2, 3, and 4, it will be shown that there is $\alpha \in (0, 1)$ such that, for all bargaining problems (X, d), $f(X, d) = \text{argmax}\{(x_1 - d_1)^\alpha \cdot (x_2 - d_2)^{1-\alpha} | (x_1, x_2) \in X\}$ holds.

Let $(z_1^*, z_2^*) = f(\Delta, 0)$ be the bargaining outcome for the bargaining problem $(\Delta, 0)$ with $\Delta = \{(z_1, z_2) \in \mathbb{R}^2 | z_1 \geq 0,\ z_2 \geq 0,\ z_1 + z_2 \leq 1\}$. Axioms 1 and 2 imply $z_1^* > 0$, $z_2^* > 0$, and $z_1^* + z_2^* = 1$.

Hence, setting $z_1^* = \alpha$, one can write $z_2^* = 1 - \alpha$. It is easy to check that for any parameter γ

$$(\gamma, 1-\gamma) = \operatorname{argmax}\{z_1^{\gamma} \cdot z_2^{1-\gamma} | (z_1, z_2) \in \Delta\}$$

holds. One can therefore view the bargaining outcome $(z_1^*, z_2^*) = (\alpha, 1-\alpha)$ as the solution to the optimization problem:

$$\text{maximise} \quad z_1^{\alpha} \cdot z_2^{1-\alpha} \quad \text{subject to} \quad (z_1, z_2) \in \Delta.$$

For any bargaining problem (X, d), consider (x_1^*, x_2^*),

$$(x_1^*, x_2^*) = \operatorname{argmax}\{(x_1 - d_1)^{\alpha} \cdot (x_2 - d_2)^{1-\alpha} | (x_1, x_2) \in X\},$$

which is well defined since ψ is a concave function. It remains to be shown that (x_1^*, x_2^*) is the bargaining outcome $f(X, d)$.

The tangent through (x_1^*, x_2^*) of the function $\psi(x_1)$ defines a new bargaining problem (Y, d). Since the bargaining problem (Y, d) has a linear Pareto frontier, there is a linear transformation

$$(z_1, z_2) = ([A_1 + B_1 \cdot y_1], [A_2 + B_2 \cdot y_2])$$

that maps $\Delta_Y = \{(y_1, y_2) \in Y | y_1 \geq d_1,\ y_2 \geq d_2\}$ onto Δ, in particular (d_1, d_2) to $(0,0)$ (compare example 9.5). In addition, this transformation maps the set X to

$$\begin{aligned}\Delta_X = \{(z_1, z_2) \in \mathbb{R}^2 | (z_1, z_2) \\ = ([A_1 + B_1 \cdot x_1], [A_2 + B_2 \cdot x_2]), (x_1, x_2) \in X\}.\end{aligned}$$

Furthermore, one has $([A_1 + B_1 \cdot x_1^*], [A_2 + B_2 \cdot x_2^*]) = (\alpha, 1-\alpha)$. To see this, recall that $(x_1^*, x_2^*) = \operatorname{argmax}\{(x_1 - d_1)^{\alpha} \cdot (x_2 - d_2)^{1-\alpha} | (x_1, x_2) \in X\}$ holds. By the construction of $Y, (x_1^*, x_2^*)$ maximizes $(x_1 - d_1)^{\alpha} \cdot (x_2 - d_2)^{1-\alpha}$ on Y as well. Hence,

$$\begin{aligned}
&[A_1 + B_1 \cdot x_1^*]^{\alpha} \cdot [A_2 + B_2 \cdot x_2^*]^{1-\alpha} \\
&= ([A_1 + B_1 \cdot x_1^*] - [A_1 + B_1 \cdot d_1])^{\alpha} \\
&\quad \cdot ([A_2 + B_2 \cdot x_2^*] - [A_2 + B_2 \cdot d_2])^{1-\alpha} \\
&= (B_1^{\alpha} \cdot B_2^{1-\alpha}) \cdot [(x_1^* - d_1)^{\alpha} \cdot (x_2^* - d_2)^{1-\alpha}] \\
&\geq (B_1^{\alpha} \cdot B_2^{1-\alpha}) \cdot [(y_1 - d_1)^{\alpha} \cdot (y_2 - d_2)^{1-\alpha}] \\
&= ([A_1 + B_1 \cdot y_1] - [A_1 + B_1 \cdot d_1])^{\alpha} \\
&\quad \cdot ([A_2 + B_2 \cdot y_2] - [A_2 + B_2 \cdot d_2])^{1-\alpha} \\
&= [A_1 + B_1 \cdot y_1]^{\alpha} \cdot [A_2 + B_2 \cdot y_2]^{1-\alpha} = z_1^{\alpha} \cdot z_2^{1-\alpha}
\end{aligned}$$

for all $(y_1, y_2) \in Y$ and all $(z_1, z_2) \in \Delta$. Thus, $([A_1 + B_1 \cdot x_1^*], [A_2 + B_2 \cdot x_2^*])$ maximizes $z_1^\alpha \cdot z_2^{1-\alpha}$ on Δ. This shows

$$(\alpha, 1-\alpha) \equiv ([A_1 + B_1 \cdot x_1^*], [A_2 + B_2 \cdot x_2^*]).$$

By construction, $(\alpha, 1-\alpha) = ([A_1 + B_1 \cdot x_1^*], [A_2 + B_2 \cdot x_2^*]) \in \Delta_X \subseteq \Delta$ holds. Hence, $(\alpha, 1-\alpha) = f(\Delta, 0) \in \Delta_X$ and axiom 4 imply $f(\Delta, 0) = f(\Delta_X, 0)$. The inverse linear-affine transformation $(x_1, x_2) = ([z_1 - A_1]/B_1, [z_2 - A_2]/B_2)$ maps Δ_X back to X. By axiom 3, $f(X, d) = ([f_1(\Delta_X, 0) - A_1]/B_1, [f_2(\Delta_X, 0) - A_2]/B_2)$ follows and, using $f(\Delta_X, 0) = ([A_1 + B_1 \cdot x_1^*], [A_2 + B_2 \cdot x_2^*])$, one obtains $f(X, d) = (x_1^*, x_2^*)$.

(iii) Suppose the bargaining solution also satisfies axiom 5. $(\Delta, 0)$ is a symmetric bargaining problem. Hence, $(z_1^*, z_2^*) = f(\Delta, 0)$ must satisfy $z_1^* = z_2^*$, and, by Pareto optimality (axiom 2), $z_1^* + z_2^* = 1$. This implies $f(\Delta, 0) = (z_1^*, z_2^*) = (.5, .5)$ (compare example 9.4). Consequently, $\alpha = .5$ must hold in this case. ■

Remark 9.2. Roth (1979) has shown that axioms 1, 3, and 4 imply axiom 2 (Pareto optimality). Hence one can obtain the characterization of theorem 9.2 without axiom 2 at the cost of a somewhat longer proof. ■

Theorem 9.2 makes it possible to derive the bargaining solution of a given bargaining problem (X, d) by solving an optimization problem on the set X. Nash (1953) showed that the five axioms are sufficient to predict a unique payoff combination as the outcome of a bargaining problem. Therefore, it has become customary to call the solution of the optimization problem for $\alpha = .5$ *Nash bargaining solution*. In contrast, for $\alpha \neq .5$, one refers to this solution as the *asymmetric Nash bargaining solution*. In fact, the parameter α that distinguishes the possible bargaining solutions can be viewed as a power index for the bargaining parties: $\alpha > .5$ indicates that player 1 is stronger than player 2, and $\alpha < .5$ that player 1 is weaker than player 2.

EXAMPLE 9.2 (*resumed*). In this example, the bargaining problem $(X, 0)$ is described by the Pareto frontier $\psi(\pi) = A - \pi$ where A denotes the maximal surplus that the trade union and the firm can achieve. The disagreement payoff combination is supposed to be zero for both parties in this case. Let α be the bargaining power of the firm.

To derive the bargaining solution that satisfies axioms 1 to 4, one solves the optimization problem: Choose (π, ν) to maximize $\pi^\alpha \cdot \nu^{1-\alpha}$ subject to $\nu \leq A - \pi$, $\pi \geq 0$, $\nu \geq 0$, to obtain the solution

$$\pi^* = \alpha \cdot A \quad \text{and} \quad \nu^* = (1-\alpha) \cdot A.$$

This payoff combination corresponds to the wage rate

$$w^* = (1-\alpha)\cdot\frac{R(n^*)}{n^*} + \alpha\cdot\frac{C(n^*)}{n^*}.$$

Note that the wage rate that results as a bargaining solution is a weighted average of the average revenue and the average opportunity cost of working. If the bargaining power of the firm approaches one, the wage will become equal to the average opportunity cost. That means that the firm can extract all the surplus and just compensate the trade union for its participation. At the opposite extreme, if the trade union's bargaining power approaches one (α converges to zero), then the firm will earn zero profits. ■

For applications, a disadvantage of the asymmetric Nash bargaining solution lies in the indeterminateness of the bargaining power index α. By choosing α appropriately, every Pareto optimal and individually rational payoff combination can be obtained as a bargaining solution. The Nash bargaining solution provides, however, an opportunity to relate the outcome of an arbitrary bargaining problem to the outcome of bargaining over a single unit of some good. This indicates the possibility of determining the bargaining strength of players in an experimental situation and, based on axioms 1 to 4, of extending the results to arbitrary bargaining situations.

From a normative point of view, the bargaining solution may be used to discuss a bargaining problem on the basis of a "standardized problem." This may facilitate the choice of further criteria for determining an exact solution. The symmetry axiom that Nash used to obtain a unique solution has this flavor because it argues for equal shares in cases in which there are no features of the bargaining problem that make it possible to distinguish the two players.

Example 9.1 (*resumed*). In this example, the bargaining problem (X,d) has the following form:

$$X = \{(u_1,u_2)\in\mathbb{R}^2 \mid u_1^\gamma + u_2 \le \Omega\}, \qquad d=(0,0).$$

Denote by α the bargaining power of consumer 1, then the asymmetric Nash bargaining solution is given by the solution to the following optimization problem: Choose (u_1,u_2) to maximize

$$u_1^\alpha \cdot u_2^{1-\alpha} \quad \text{subject to} \quad u_1^\gamma + u_2 \le \Omega, \quad u_1 \ge 0 \quad \text{and} \quad u_2 \ge 0.$$

The following diagram illustrates this problem.

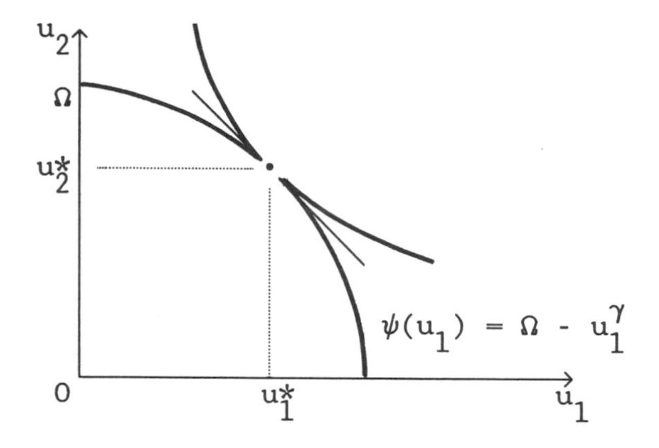

It is straightforward to check that this problem has the following solution:

$$u_1^* = \left[\frac{\alpha \cdot \Omega}{\alpha + \gamma \cdot (1-\alpha)}\right]^{1/\gamma}, \qquad u_2^* = \frac{[\gamma \cdot (1-\alpha)] \cdot \Omega}{\alpha + \gamma \cdot (1-\alpha)}.$$

As in the previous example, the payoff to player 1 is increasing in α, whereas the payoff of player 2 is decreasing in α. Furthermore, this example shows that a monotonic transformation of the utility functions affects the bargaining outcome. ■

9.3 The Nash Program

In the previous sections, two approaches were advanced to arrive at a prediction of the outcome of bargaining situations. One approach relied on a detailed description of the bargaining procedure to obtain the bargaining outcome as a subgame perfect equilibrium of a completely specified game. The main ingredients of this specification were a precise sequence of proposals, counter-proposals, and assumptions about time preferences. The second approach suggested several properties that any bargaining outcome should satisfy. First, Pareto efficiency and individual rationality are demanded together with some distributional rule for the particular case of bargaining over one unit of utility. These properties would suffice to identify a unique outcome in this special case. Second, utilities are interpreted as expected utilities to justify an assumption of invariance against linear transformations. Independence of irrelevant alternatives is invoked to allow a transformation of every bargaining problem into a bargaining problem over the unit interval without affecting the outcome. Combining these assumptions determines a solution to the bargaining problem.

What is predicted as the outcome of the bargaining process depends, in the first case, on the specification of the bargaining procedure and, in the second case, on the degree of confidence that an analyst is willing to put into these axioms. Nash (1953) realized the tension between these fundamentally different approaches and suggested that any system of axioms that predicts a particular outcome should be supplemented by a bargaining procedure that implements this outcome.

It may appear as if the suggestion of the Nash program could be easily carried out. After all, axioms 1 to 4 identify an outcome on the Pareto frontier, and the strategic approach of alternating offers leads to an immediate settlement on some outcome on the frontier as well. All that needs to be done is to match the outcomes of the two approaches. The following diagram illustrates this idea.

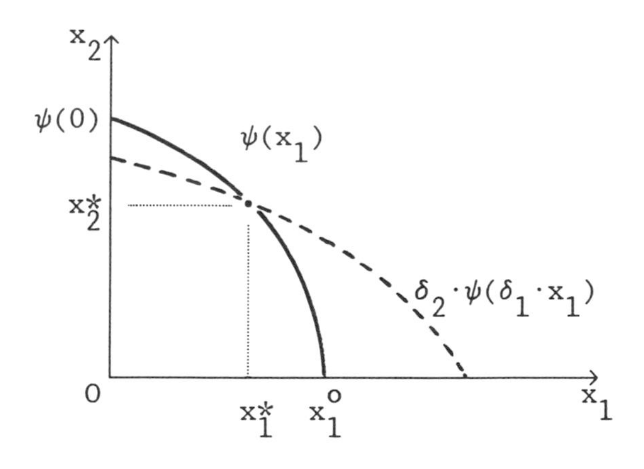

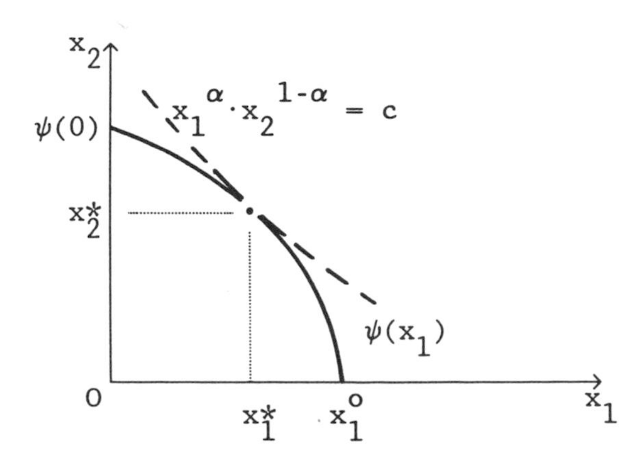

In fact, the solution x_1^* in the upper diagram (page 261) must satisfy

$$\psi(x_1^*) = \delta_2 \cdot \psi(\delta_1 \cdot x_1^*),$$

and the solution in the lower diagram (page 261) must solve

$$\alpha = \frac{-\epsilon_\psi(x_1^*)}{1 - \epsilon_\psi(x_1^*)},$$

where $\epsilon_\psi(x_1) \equiv \psi'(x_1) \cdot x_1/\psi(x_1)$ denotes the elasticity of the function $\psi(\cdot)$. Thus, the bargaining power of player 1, α, can be related to characteristics of the Pareto frontier $\psi(\cdot)$ and to the discount factors of both players.

EXAMPLE 9.7. Consider a bargaining problem $(X, 0)$ with the following Pareto frontier:

$$\psi(x_1) = 1 - x_1.$$

In this case, one easily computes the elasticity as $\epsilon_\psi(x_1) = -x_1/(1-x_1)$, and the solution for $\psi(x_1^*) = \delta_2 \cdot \psi(\delta_1 \cdot x_1^*)$ as $x_1^* = (1-\delta_2)/(1-\delta_1 \cdot \delta_2)$. Hence, one derives

$$\alpha = (1-\delta_2)/(1-\delta_1 \cdot \delta_2)$$

as the bargaining power of player 1. Note that player 1's bargaining power is increasing in δ_2 and decreasing in δ_1. The more patient a player, the higher her bargaining power. ■

Example 9.7 shows how one can relate the outcome of the strategic approach to the outcome of the axiomatic approach. There is, however, a problem in the interpretation of payoffs. Axiom 3 requires the bargaining solution to be invariant in regard to linear-affine transformations. This can be justified if one interprets utilities as von Neumann–Morgenstern utilities and assumes that preferences satisfy expected utility theory. In contrast, the strategic approach to bargaining assumes that discount factors applied to utilities reflect delay costs. Hence both approaches give the utility functions of the players a cardinal interpretation, though for different reasons. In the axiomatic approach, no attempt is made to give utilities an intertemporal interpretation, and in the strategic approach no attempt is made to interpret utilities as von Neumann–Morgenstern utilities. It is, however, extremely restrictive to require players' preferences to satisfy the assumptions of expected utility theory as well as those for a time-separable utility representation.[5]

[5] It is possible to model delay costs by risk of breakdown of negotiations (compare Osborne and Rubinstein, 1990, Chapter 4). This avoids incompatible interpretations of the preference representations.

The two approaches presented in this chapter provided many results on the outcome of bargaining and on the factors influencing these outcomes. Nevertheless, these approaches have been criticized as neglecting important aspects observed in actual negotiations. Unlike many other areas of economics, bargaining theory has been subject to testing. These tests confirm some results of bargaining theory, such as the importance of patience and risk-attitudes, but raise many additional questions.[6] In the social sciences, it is often hard to draw unambiguous conclusions from experimental tests because of the difficulties in controlling the environment during the tests and in assessing unobservable characteristics of the participants. Even with these qualifications, there are strong indications that bargaining behavior is influenced by other aspects of the environment and by the attitudes of the subjects regarding social objectives like fairness.

Furthermore, strategic bargaining theory considers bargaining situations that are completely specified. To apply game-theoretic solution concepts, it is necessary to specify possible moves and their precise sequence completely. Many actual bargaining situations, however, lack such a predetermined structure. If two consumers haggle over the price of some item, there is rarely a clear order of proposals and counter-proposals. But does that mean it is impossible to predict the outcome of a bargaining game except in cases for which one can describe negotiations in full detail? Experimental results[7] indicate that there are rules of bargaining even in incompletely specified situations. Hence for bargaining theory, there remains a challenge to discover and describe such rules.

9.4 Summary

Two approaches for the analysis of bargaining between two players were studied in this chapter. The first relied on the explicit specification of the bargaining procedures and of the bargaining costs of players to derive a subgame perfect equilibrium. If bargaining ends in finite time, then a unique subgame perfect equilibrium exists. This result can be extended to cover the case of an infinite bargaining process. The equilibrium outcomes depend crucially on the time preferences of the players, and the player who proposes first has an advantage over the player who has to respond to that offer.

[6] Roth (1991) provides an excellent survey of the literature on laboratory experiments on bargaining.

[7] Crawford (1990) contains a short review of the literature on experiments on bargaining and further references.

The second approach tries to characterize the outcome of a two-player bargaining problem by a set of axioms. In fact, five axioms are sufficient to determine a unique function that selects a solution for any bargaining problem with a concave payoff possibility frontier. Four of these axioms characterize a bargaining solution in terms of one parameter that can be interpreted as a player's bargaining strength. In this case, the degree of confidence in regard to the predicted outcome depends on the arguments advanced to justify the axioms. Nash (1953) suggested supplementing the axiomatic approach to bargaining with the strategic approach. One can link the two approaches at the cost of strong restrictions on the players' preference representation.

9.5 Remarks on the Literature

The strategic approach to the bargaining problem, presented in section 9.1, was first studied by Ståhl (1972) for the finite horizon case, but became widely known only ten years later with the publication of Rubinstein's (1982) paper treating the infinite horizon case. Since then the literature on bargaining has grown rapidly. Many alternative ways of modeling delay costs have been studied, as well as the influence of outside options and incompleteness of information about the opponent's payoff. Major difficulties were encountered extending the bargaining approach to more than two players. Good surveys on this literature can be found in Binmore and Dasgupta (1987), Osborne and Rubinstein (1990), and Binmore, Osborne, and Rubinstein (1990). The presentation in section 7.1 was inspired by Binmore (1987). The proof of theorem 7.1 is an adaptation of the proof given in Osborne and Rubinstein (1990, p. 45, theorem 3.4).

The axiomatic approach was suggested in Nash (1950, 1953). Many different systems of axioms have been proposed and analyzed. Roth (1979) studies the asymmetric bargaining solution. Binmore's contributions in Chapters 2, 4, 8, and 11 of Binmore and Dasgupta (1987) develop this theory further and discuss its relationship to alternative approaches. A survey on various sets of axioms for bargaining solutions can be found in Roth (1979).

Exercises

9.1 *A consumer leaves home with* \$10 *to buy tomatoes. At the market she meets a seller with* 32 *kg of tomatoes. Assume that both players' preferences over money* m *and tomatoes* x *can be described by the following utility function:* $u(x,m) = \sqrt{x} + m$.

(a) Draw an Edgeworth box and indicate the Pareto-optimal and individually rational allocations. Draw the utility possibility frontier for the two traders and indicate the threat point.

(b) Suppose that both players have the same discount factor $\delta < 1$. Derive the bargaining result if the players alternate with their offers and if the seller begins.
(c) What is the expected price and quantity that the traders can achieve if the first proposer is randomly chosen with probability .5?

9.2 *Consider the wage negotiations of example 9.2. Suppose that the firm proposes first.*
(a) Derive the bargaining result for bargaining over T periods. What is the difference between the cases in which T is even and in which T is odd?
(b) Show that the difference between the bargaining results for T odd and T even disappears as T goes to infinity.

9.3 *Consider the wage bargaining example 9.2 again. Assume that both players have the same discount factor.*
(a) Show that the player who proposes first gets the total surplus if δ converges to zero.
(b) Show that the first-mover advantage vanishes as δ goes to 1.
(c) Analyze the two-stage bargaining game for $\delta = 1$.

9.4 *Consider a trade union negotiating with a firm over a wage-employment contract (w, L). The firm maximizes profits $\pi(w, L) = R(L) - w \cdot L$. The trade union's objective function is the expected utility of its members N given a wage w and the risk of unemployment $(1 - L/N)$ which results from the negotiated contract: $U(w, L) = (L/N) \cdot u(w) + (1 - L/N) \cdot \bar{u}$, where $\bar{u}$ denotes the utility of an unemployed worker. Assume that the revenue function $R(\cdot)$ and the expected utility function $u(\cdot)$ are strictly increasing and strictly concave functions.*
(a) Assuming $\bar{u} = 0$, derive the set of feasible utility-profit combinations (U, π) and draw a diagram showing the Pareto frontier.
(b) Derive the asymmetric Nash bargaining solution with a market power coefficient of α for the firm.
(c) Show how employment and the wage rate change in response to changes in this market power parameter α.

9.5 *Consider two parties bargaining over the sharing of some surplus. Let $x \in [0, 1]$ be the share of player 1 and $y \in [0, 1]$ the share of player 2. Both players' reservation value is zero. Suppose there is a probability q that negotiations break down if an offer is rejected. Both players simultaneously propose a share for themselves. Player 1 rejects the offer of player 2 if $(1 - q) \cdot x \geq (1 - y)$ holds. Similarly, player 2 rejects player 1's offer if $(1 - q) \cdot y \geq (1 - x)$ holds.*
(a) Given a proposal (x, y), determine for each player the probability of breakdown that would make her indifferent between acceptance and rejection; call this probability "willingness to accept a breakdown."

(b) Assume that the player with less willingness to accept a breakdown of negotiations makes a concession by reducing the share demanded for herself. If both have the same willingness to accept a breakdown they both make a concession. Show in a diagram the adjustment process for a few steps of negotiation according to this principle.
(c) Show that this negotiation process leads to the (symmetric) Nash bargaining solution of this game.

10

Cooperative Games

In analyzing social situations, noncooperative game theory views agents as interdependent decision makers who are unable to make binding commitments. This analysis focuses therefore on systems of strategies that are stable in the sense that no agent wants to deviate provided no other agents deviate either. The previous chapters have shown how this program can be successfully applied to various scenarios of complete and incomplete information in strategic form and extensive form games.

The second part of Chapter 9, however, introduced a different perspective on the analysis of social situations. Instead of specifying all moves, or all strategies, and all information states of all agents, the Nash bargaining solution considered the set of all possible outcomes of an interaction and selected a particular outcome based on *a priori* considerations. This alternative approach to the analysis of games will be expanded in this chapter. Neglecting a fully detailed description of strategies may be justified if one assumes that contracts specifying each player's actions can be negotiated and enforced by some outside party, such as a court system. Given the possibility of enforcing contracts, every outcome that agents can achieve jointly becomes a feasible agreement.

The description of a game in coalitional form, introduced at the end of Chapter 1, reflects this change in the method of analysis. A game in coalitional form is completely described by the player set and a function that specifies for each subgroup of players the payoff combinations they can jointly achieve. This description abstracts from all details of the interaction that lead to these payoffs. As the bargaining problem in

Chapter 9 illustrates, many different strategic form games may give rise to the same set of bargaining outcomes. The possibility of binding agreements may therefore overcome the difficulty the analyst encounters in specifying the strategies of all agents in full detail.

It is possible to conduct the analysis of cooperative games in terms of characteristic functions that associate a set of feasible payoff vectors with each group of players. This case is illustrated by the two-player bargaining problem of example 9.1. The analysis of cooperative games is substantially simplified if one considers the special case of transferable utility or games with side payments. Transferable utility means that utilities of players can be summed to a total level of utility for a group and that there is some way to redistribute this sum among the players. Note, however, that even quasi-linear preferences only allow for transfer of utility over a certain range.[1] Nevertheless, transferable utility will be assumed throughout this chapter. This can be justified by the fact that most concepts of cooperative game theory have an appropriately modified analogue for games without side payments (games with nontransferable utility). Restricting the analysis to games with transferable utility, makes a very simple definition of the characteristic function possible.

DEFINITION 10.1. A game with transferable utility in coalitional form $\Gamma = (I, v)$ is described by a player set I and a function v: $\mathscr{P}(I) \to \mathbb{R}$ with $v(\emptyset) = 0$ that indicates the maximal aggregate payoff of a coalition $S \in \mathscr{P}(I)$.

Notice that the set of all subsets of $I, \mathscr{P}(I)$, is the set of coalitions. This implies that the all-player set I, the empty set $\emptyset$, and all sets that contain a single player $\{i\}$, $i \in I$, are treated as coalitions. The following notation will be useful. For any coalition S, denote by $s = \#S$ the number of players in S, in particular, $0 = \#\emptyset$ and $n = \#I$. Any vector $x = (x_1, x_2, \ldots, x_n) \in \mathbb{R}^n$ can be interpreted as a list of payoffs for the n players.

A characteristic function as just defined associates with each coalition the maximum payoff the players of this coalition can achieve. Simple as this definition may appear, even with the possibility of binding agreements, there remains a problem as to what a particular subgroup of players can secure for themselves. Payoffs of the players in a coalition depend in general on the actions chosen by players who do not belong to this group. Thus even if members of a coalition agree to play a certain strategy combination, they will not control their payoff because they cannot commit those players who are not part of the coalition. In a sense, a coalition reduces a multi-player game to a two-player game in which the coalition

[1]Varian (1984, p. 283) discusses quasi-linear utility functions and the range over which utility is transferable.

and its complement play the roles of the two players. But even for two-player games a solution concept is needed to determine an outcome of the game. Von Neumann and Morgenstern (1947) suggest the maximin value as the outcome that a coalition can guarantee itself no matter what other players may do. But, in nonzero-sum games, the maximin value is a pessimistic prediction about the outcome of the game for a coalition. The following example illustrates these considerations.

EXAMPLE 10.1. Consider the following version of the "battle of the sexes", summarized by its payoff matrix.

		Player 2	
		L	R
Player 1	T	2, 1	0, 0
	B	0, 0	1, 2

For the coalition of player 1 and 2, one obtains easily $v(I) = 3$, a joint outcome achieved by playing either (T, L) or (B, R). It is less clear, however, what value to associate with single-player coalitions. A few possibilities are given below:

(i) $v(\{1\}) = \max_{s_1} \min_{s_2} p_1(s_1, s_2) = 0$ and
$v(\{2\}) = \max_{s_2} \min_{s_1} p_2(s_1, s_2) = 0$,

(ii) $v(\{1\}) = p_1(T, L) = 2$ and $v(\{2\}) = p_2(T, L) = 1$,

(iii) $v(\{1\}) = p_1(B, R) = 1$ and $v(\{2\}) = p_2(B, R) = 2$,

(iv) $v(\{1\}) = p_1((\frac{2}{3}, \frac{1}{3}), (\frac{1}{3}, \frac{2}{3})) = 2/3$ and
$v(\{2\}) = p_2((\frac{2}{3}, \frac{1}{3}), (\frac{1}{3}, \frac{2}{3})) = 2/3$.

Version (i) corresponds to the maximin value, an extremely unsatisfactory outcome for any player. Versions (ii) through (iv) represent Nash equilibrium outcomes. Nash equilibria are, however, non-unique and, except for the mixed strategy outcome, nonsymmetric. These ambiguities regarding the appropriate value for a one-player coalition illustrate the difficulties encountered in deriving the characteristic function of a game from its strategic form. The fact that a two-player game has been chosen for this example is not important. Similar problems arise between any two groups of players. ■

In economic applications, however, games in coalitional form are rarely derived directly from games in strategic form. One often finds situations that incompletely describe individual players' strategy sets but for which coalitional outcomes can be assigned in a natural way. The following examples illustrate such cases.

EXAMPLE 10.2. Consider three players who can join forces to earn \$100. Two players together can achieve $\$\xi$, $0 \le \xi \le 100$, and a single player zero.

This game is described by the player set $I = \{1, 2, 3\}$ and the following characteristic function, $v: \mathscr{P}(I) \to \mathbb{R}$: $v(I) = 100$, $v(\{i, j\}) = \xi$ for $i, j \in I$ and $i \ne j$, $v(\{i\}) = 0$ for all $i \in I$. ■

Notice that precise actions of the players remain unspecified in the description of example 10.2. Similarly, one need not know possible bargaining strategies to determine the characteristic function in the following game.

EXAMPLE 10.3 (*Shubik, 1982, page 151*). A farmer's land is worth \$100,000 to him for agricultural use; to a manufacturer it is worth \$200,000 as a plant site; a subdivider would pay up to \$300,000.

Denoting the farmer as player 1, the manufacturer as player 2, and the subdivider as player 3, one obtains the player set $I = \{1, 2, 3\}$. The characteristic function v (in \$100,000 units) follows immediately from the description of the situation:

$$v(I) = 3,$$
$$v(\{1,2\}) = 2, \qquad v(\{1,3\}) = 3, \qquad v(\{2,3\}) = 0,$$
$$v(\{1\}) = 1, \qquad v(\{i\}) = 0 \quad \text{for} \quad i = 2, 3. \qquad ■$$

Even quite complex games about the distribution of costs and benefits among a given set of players can be analyzed as a cooperative game without specifying the strategies of the players.

EXAMPLE 10.4 (*Moulin, 1986, p. 222, modified*). Three communities near a large city consider developing a water treatment system. So far these communities dispose of their sewage by sending it to a central treatment plant operated by city authorities at a monthly fee.

A cost-benefit study estimates the present value of these payments over the usual lifetime of a water treatment plant at \$100 per household. To build and operate a water treatment plant for the same period is estimated at a present value cost of

\$500,000 for up to 5,000 households
\$600,000 for up to 10,000 households
\$700,000 for up to 15,000 households

Community 1 is estimated to serve 5000 households on average during the period under consideration, community 2 has to serve 3000 households, and community 3 has 4000 households.

The decision problem of the three communities can be modeled as a game in characteristic function form with three players, $I = \{1, 2, 3\}$, and the following characteristic function v (measured in \$100,000 units):

$$v(I) = 5,$$
$$v(\{1,2\}) = 2, \qquad v(\{1,3\}) = 3, \qquad v(\{2,3\}) = 1,$$
$$v(\{1\}) = v(\{2\}) = v(\{3\}) = 0.$$

The net benefit of a coalition is calculated as the difference between the joint benefit (number of households times present value of fees saved) minus the cost of a water treatment system of the appropriate size. A single community can of course still send its sewage to the central treatment plant. This guarantees it a net benefit of zero. ■

The games considered in these examples have the property that any coalition of players can obtain an outcome at least as high as the sum of its members' payoffs if they did not join the coalition. In this sense, coalition formation was advantageous for the players. Larger groups enjoy increasing returns to size. The following definition makes this notion precise.

DEFINITION 10.2. A game in coalitional form Γ is called *superadditive*, if, for all $S, T \in \mathscr{P}(I)$ with $S \cap T = \emptyset$, $v(S \cup T) \geq v(S) + v(T)$ follows.

There are, of course, games that are not superadditive. It is sometimes argued, however, that superadditivity is a natural assumption for games in coalitional form because it excludes considering games for which coalition formation does not make sense anyway. Yet superadditivity fails to hold if any two coalitions jointly achieve a lower payoff than the sum of their individual payoffs. This still allows for beneficial cooperation among many coalitions, in particular in the all-player coalition. It is not easy to find meaningful examples of games in coalitional form that are not superadditive. In fact, all examples in this chapter satisfy this condition. On the other hand, there is also no need to impose this condition on any result presented in this chapter.

Before the discussion of solution concepts, three additional examples will indicate the scope of applications for cooperative games in economics. The following example generalizes the Edgeworth pure exchange problem to economies with many traders.

EXAMPLE 10.5 (*market game*). Consider n consumers, $I = \{1, 2, \ldots, n\}$, who consume ℓ commodities $x_i \in \mathbb{R}_+^\ell$. Before trade begins, each consumer $i \in I$ owns a positive endowment of goods $\bar{x}_i \in \mathbb{R}_+^\ell$. Preferences of these consumers are supposed to be given by utility functions $u_i(x_i) = x_i^1 + \alpha_i(x_i^2, \ldots, x_i^\ell)$, $i \in I$, where $\alpha_i(\cdot)$ is a strictly increasing and concave function. This assumption about preferences implies that utility is transferable between agents via payments in terms of the first commodity. The following diagram illustrates such an exchange situation for the case of two consumers, $n = 2$, and two commodities, $\ell = 2$.

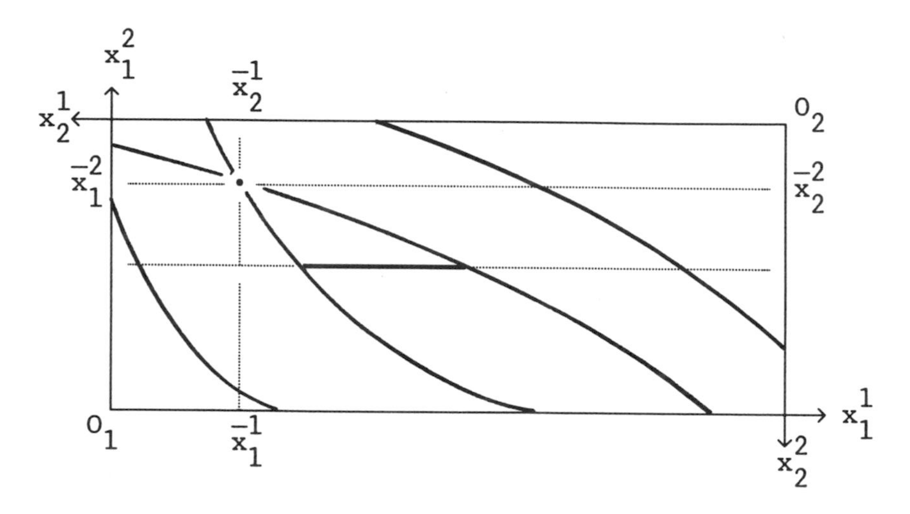

This trade situation can be modeled easily as a game in coalitional form with player set I and characteristic function v:

$$v(S) = \max\left\{ \sum_{i \in S} u_i(x_i) \,\middle|\, \sum_{i \in S} x_i = \sum_{i \in S} \bar{x}_i \right\} \quad \text{for all} \quad S \in \mathscr{P}(I). \qquad \blacksquare$$

Remark 10.1. The pure exchange economy of example 10.5 is a special case of a market game. Market games, studied by Shapley and Shubik (1969), provide a good illustration of the meaning of transferable utility. Shapley and Shubik (1969) assume that side payments are feasible allowing the maximum joint utility to be distributed among the consumers without any restriction. This assumption implies quasi-linear utility functions, as in example 10.5. Given the endowments of the first commodity, it is however impossible to obtain all feasible utility combinations by transfers of good 1 alone. This can be seen in the Edgeworth box by the fact that the contract curve does not reach the lowest indifference curve of a player along the horizontal part where only transfers of good 1 take place. Thus, endowments form a constraint on transfers of utility. Note, however, that all individually rational payoff transfers can be effected through good 1. ■

The next example represents an economic application of a voting game.

Example 10.6 (*weighted majority voting*). Consider a group of n shareholders of a company that has to decide on some investment project.

Denote by w_i the number of shares shareholder i owns. The company charter assigns one vote to each share and requires a minimum of q votes for the adoption of a project. Suppose the investment project yields a return R to be distributed equally per share.

This situation can be modeled as a game in characteristic function form with player set $I = \{1, 2, \ldots, n\}$ and characteristic function v:

$$v(S) = \begin{cases} \left(\sum_{i \in S} w_i\right) \cdot \dfrac{R}{W} & \text{for} \quad \sum_{i \in S} w_i \geq q \\ 0 & \text{for} \quad \sum_{i \in S} w_i < q \end{cases},$$

with $W \equiv \sum_{i \in I} w_i$. ■

The last example considers the problem of profit distribution among investment fund managers.

Example 10.7 (*Lemaire, 1981, p. 18, modified*). Three investment fund managers consider investment possibilities for a year. Fund manager 1 has \$3,000,000 to invest, manager 2 has \$1,000,000, and manager 3 has \$2,000,000. There is an investment opportunity available with the following interest scheme.

Deposit	Interest rate
less than \$2,000,000	8 percent
\$2,000,000 up to \$5,000,000	9 percent
\$5,000,000 and more	10 percent

This situation can be readily translated into a characteristic function, v, for a three-player game, $I = \{1, 2, 3\}$, (in \$10,000 units):

$$v(I) = 60,$$

$$v(\{1,2\}) = 36, \qquad v(\{1,3\}) = 50, \qquad v(\{2,3\}) = 27,$$

$$v(\{1\}) = 27, \qquad v(\{2\}) = 8, \qquad v(\{3\}) = 18. \quad ■$$

All examples given so far have in common that they describe distributional problems. In each case, a set of players has to decide how to share a joint payoff. Once an agreement has been reached, a contract can be written that obliges the players to carry out the actions necessary to achieve the outcome. A solution to a game in coalitional form is, therefore, a proposed distribution of the joint payoff to the players that is accepted. The following two sections present different approaches for characterizing payoff vectors as "acceptable."

10.1 The Core

This section studies payoff distributions that are stable against secession of subgroups. A natural requirement for any payoff vector to be mutually agreeable seems to be that no subgroup could do better by breaking the agreement and operating on its own. The core concept tries to capture this idea formally.

Any solution vector must, of course, be feasible. Furthermore, in order to describe what a coalition S can achieve on its own, it is necessary to specify what is feasible for a coalitions S.

DEFINITION 10.3. A payoff vector $x \in \mathbb{R}^n$ is called *feasible for coalition* S if $\Sigma_{i \in S} x_i \leq v(S)$ holds. A payoff vector $x \in \mathbb{R}^n$ is called *feasible* if it is feasible for I, the coalition of all players.

Since the set of coalitions includes single-player coalitions $\{i\}$, $i \in I$, and the grand coalition I, it follows from the principle of stability against secession of a coalition, that an agreeable payoff vector x must

- give each player at least as much as he could obtain on his own, $x_i \geq v(\{i\})$ for all $i \in I$, (individual rationality)
- exhaust the potential of the grand coalition's payoff, $\Sigma_{i \in I} x_i = v(I)$ (Pareto optimality, group rationality)

An *imputation* is a payoff vector that is individually rational and group rational (Pareto-optimal).

DEFINITION 10.4. A payoff vector x is called an *imputation* if it belongs to the set

$$I(v) = \left\{ x \in \mathbb{R}^n | \sum_{i \in I} x_i = v(I),\ x_i \geq v(\{i\}) \text{ for all } i \in I \right\}.$$

All payoff vectors in $I(v)$ are stable against breakaway of a single player. A larger subgroup may, however, be willing to block an agreement $x \in I(v)$ because it can produce a better payoff for its members.

EXAMPLE 10.2 (*resumed*). Consider the case in which the two-player coalitions can achieve an outcome of $\xi = 20$.

The set of imputations of this game,

$$I(v) = \{(x_1, x_2, x_3) \in \mathbb{R}^3 | x_1 + x_2 + x_3 = 100,\ x_1 \geq 0,\ x_2 \geq 0,\ x_3 \geq 0\},$$

is a three-dimensional simplex. An imputation like $(x_1, x_2, x_3) = (10, 5, 85)$ is however not acceptable to a coalition of player 1 and 2. By breaking away, player 1 and 2 can share an outcome of 20 making possible a payoff for these players that strictly dominates their payoff according to the suggested imputation, (10, 5). ■

Example 10.2 makes it quite clear that an acceptable payoff vector should assign payoffs so that the sum of the payoffs for any coalition is larger or at least equal to what this coalition can achieve on its own. Any suggestion of a vector x that has $\Sigma_{i \in S} x_i < v(S)$ for some coalition S can be blocked by this coalition. By breaking away from the suggested agreement and for example, splitting the difference $v(S) - \Sigma_{i \in S} x_i$ equally among its members, all members of the coalition S would be better off. A stable agreement x should therefore give each coalition at least as much as it could obtain alone, $\Sigma_{i \in S} x_i \geq v(S)$ for all $S \in \mathscr{P}(I)$.

Definition 10.5. The core of a game $\Gamma = (I, v)$, is the set of all feasible payoff vectors that cannot be blocked by any coalition,

$$C(v) = \left\{ x \in \mathbb{R}^n | x \text{ feasible}, \sum_{i \in S} x_i \geq v(S) \quad \text{for all} \quad S \in \mathscr{P}(I) \right\}.$$

The core is by definition a subset of the set of imputations, $C(v) \subseteq I(v)$, since, for the grand coalition I, a payoff vector with $\Sigma_{i \in I} x_i > v(I)$ is not feasible. The following examples show the geometry of the core.

Example 10.2 (*resumed*). Consider the following three cases in which the two-player coalitions can achieve an outcome of (i) $\xi = 20$, (ii) $\xi = 50$, or (iii) $\xi = 80$.

In each of these cases the core is the set of payoff vectors (x_1, x_2, x_3) satisfying the set of inequalities

$$x_1 + x_2 \geq \xi, \qquad x_1 + x_3 \geq \xi, \qquad x_2 + x_3 \geq \xi,$$
$$x_1 \geq 0, \qquad x_2 \geq 0, \qquad x_3 \geq 0$$

plus the equality

$$x_1 + x_2 + x_3 = 100.$$

The following three diagrams show the core for the three cases. Since $x_3 = 100 - x_1 - x_2$ must hold for an imputation, one can represent the core of a three-player game in the (x_1, x_2)-plane.

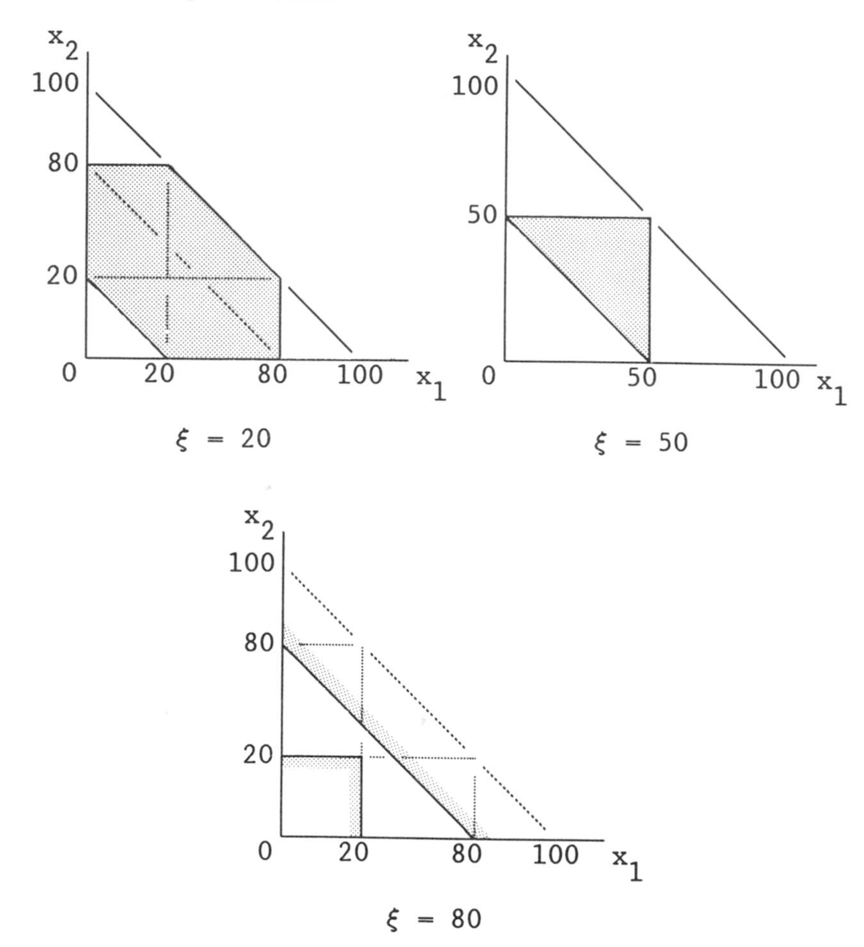

As one can see from these diagrams, the core shrinks as ξ gets bigger. For $\xi = 80$ the core is empty. Thus there is no payoff vector that is robust against blocking by a coalition. On the other hand, for $\xi = 20$ the core is a rather large set. Note that, for $\xi = 200/3$ the core has just one element, $C(v) = \{(100/3, 100/3, 100/3)\}$, corresponding to an equal sharing of the aggregate payoff 100. ■

Example 10.2 treats two-player coalitions symmetrically. This provides an opportunity to study the impact of intermediate coalitions on the size of the core. If the payoff to the set of all players, I, is big compared with the outcome that smaller coalitions can achieve, then the core will be rather large. On the other hand, if the smaller coalitions obtain an outcome close enough to the outcome of the all-player set I, then the core will be empty. In the latter case, each coalition has an incentive to break away from a general agreement and to secure for its members a larger payoff than the grand coalition could give them. The following examples use the same type of diagram to illustrate the core for three-player games that treat two-player coalitions asymmetrically.

EXAMPLE 10.3 (*resumed*). Recall the characteristic function of this example:

$$v(I)=3,\qquad v(\{1,2\})=2,\qquad v(\{1,3\})=3,\qquad v(\{2,3\})=0,$$
$$v(\{1\})=1,\qquad v(\{i\})=0$$

for $i=2,3$.

After substituting for x_3, the core of this game is given by the following set of inequalities,

$$x_1+x_2\geq 2,\quad x_2\leq 0,\quad x_1\leq 3,\quad x_1\geq 1,\quad x_2\geq 0,\quad x_1+x_2\leq 3.$$

Obviously, player 2 has no power in this example. Each allocation in the core will assign zero to player 2. Player 1, however, plays a major role, since the potential coalition of player 3 with player 1 guarantees player 1 at least 2. Hence, $C(v)=\{(\alpha,0,3-\alpha)|\alpha\in[2,3]\}$ is the set of core allocations.

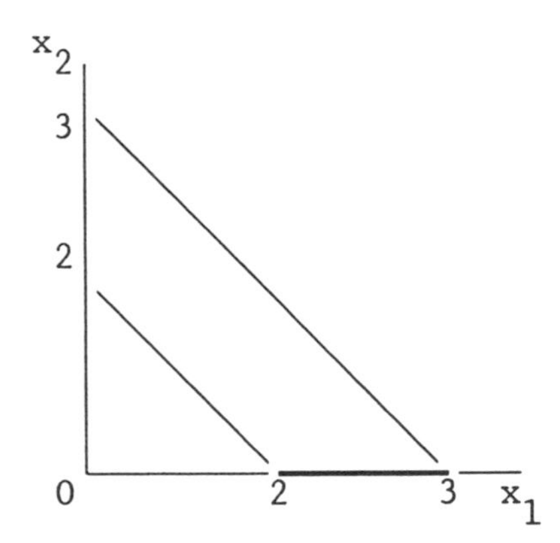

■

EXAMPLE 10.4 (*resumed*). In this problem of cost sharing the following characteristic function v had been established: $v(I)=5$, $v(\{1,2\})=2$, $v(\{1,3\})=3$, $v(\{2,3\})=1$, $v(\{1\})=v(\{2\})=v(\{3\})=0$.

Substituting $x_3=5-x_1-x_2$ into the inequalities defining the core, one derives the following set of inequalities in x_1 and x_2:

$$x_1+x_2\geq 2,\quad x_2\leq 2,\quad x_1\leq 4,\quad x_1\geq 0,\quad x_2\geq 0,\quad x_1+x_2\leq 5.$$

The following diagram shows this set.

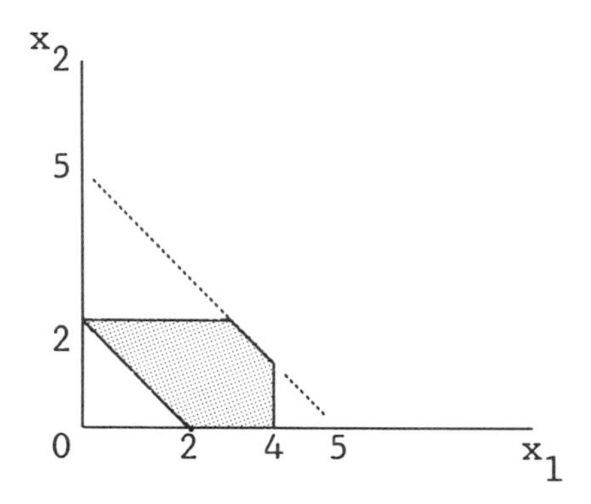

■

Example 10.4 makes clear that the core is a negative selection criterion. It rules out certain allocations as unstable given the blocking power of the coalitions, but it does not suggest directly a particular allocation.

EXAMPLE 10.7 (*resumed*). The cooperation problem of the investment fund managers was modeled by the following characteristic function: $v(I) = 60$, $v(\{1,2\}) = 36$, $v(\{1,3\}) = 50$, $v(\{2,3\}) = 27$, $v(\{1\}) = 27$, $v(\{2\}) = 8$, $v(\{3\}) = 18$.

Noting that $x_3 = 60 - x_1 - x_2$ must hold, the core is given by the following system of inequalities:

$$x_1 + x_2 \geq 36, \quad x_2 \leq 10, \quad x_1 \leq 33, \quad x_1 \geq 27, \quad x_2 \geq 8, \quad x_1 + x_2 \leq 60.$$

The set of allocations satisfying these constraints can be seen in the following diagram.

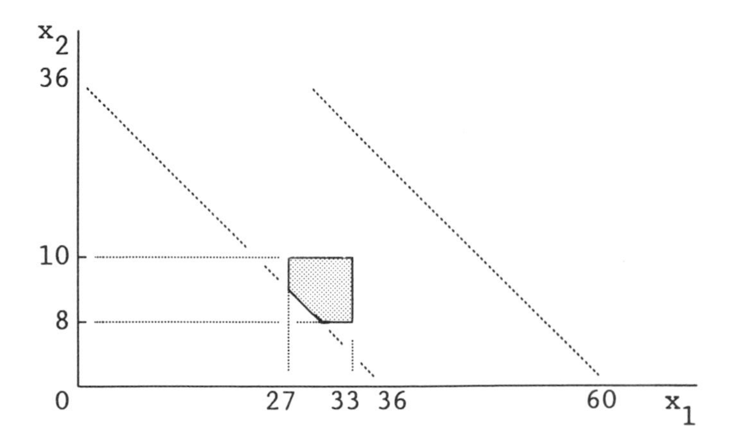

■

These examples show the structure of the core for three-player games. The set of allocations that cannot be blocked by any coalition may contain a single allocation, many allocations, or no allocation at all. The latter case in particular poses a problem for any application of the core concept. This problem is made harder by the fact that it is usually difficult to check for an arbitrary game whether it has an empty core or not, as examples 10.5 and 10.6 illustrate.

From the definition and from the examples discussed so far, it is clear that the core is described by a large number of inequalities together with one equality for the sum of all payoffs. Since these inequalities bound the core set of allocations from below, the core is empty if and only if the smallest payoff allocation satisfying all inequalities sums to more than the outcome the set of all players can achieve. These considerations link the question of the nonemptiness of the core to a linear programing problem. Therefore, one can make use of the structure of such problems to arrive at a test method for the nonemptiness of the core.

The following definitions make it possible to test whether the core of a game is empty by considering a specific set of allocations only. The concept of a balanced family of coalitions is particularly useful in this context.

Denote by $\mathscr{S}(i) = \{S \in \mathscr{P}(I) | i \in S\}$ the set of all coalitions that contain player i.

DEFINITION 10.6. A *balanced family of coalitions* is defined by a function γ: $\mathscr{P}(I) \to [0,1]$ with $\Sigma_{S \in \mathscr{S}(i)} \gamma(S) = 1$ for all $i \in I$.

Remark 10.2. The "balanced family of coalitions" is of course the set of coalitions for which $\gamma(S) > 0$ holds. In definition 10.6, the balanced family of coalitions is implicitly defined by the function γ. For the following result, only the functions γ are important. The term balanced family of coalitions will therefore be used for the function γ defined in 10.6 as well. ■

Notice that the notion of a balanced family of coalitions makes no reference to the characteristic function v of a particular game in coalitional form. Balanced families provide a useful way of organizing the set $\mathscr{P}(I)$ of all coalitions. A few examples of balanced families of coalitions may help to clarify this claim.

EXAMPLE 10.8 (*balanced families of coalitions*). Consider a player set with just three players $I = \{1, 2, 3\}$.

In this case, it is possible to list all subsets:

$$\mathscr{P}(I) = \{\varnothing, \{1\}, \{2\}, \{3\}, \{1,2\}, \{1,3\}, \{2,3\}, I\}.$$

Furthermore, the sets of all subsets containing a particular player are given as:

$\mathscr{S}(1) = \{\{1\},\{1,2\},\{1,3\},I\}$, $\mathscr{S}(2) = \{\{2\},\{1,2\},\{2,3\},I\}$, and
$\mathscr{S}(3) = \{\{3\},\{1,3\},\{2,3\},I\}$.

It is easy to check that the following functions represent balanced families of coalitions:

(a) $\gamma(I) = 1$, $\gamma(S) = 0$ for all $S \neq I$
(b) $\gamma(I) = 0$, $\gamma(\{i,j\}) = .5$ for all $i \neq j$, $\gamma(S) = 0$ for all other S
(c) $\gamma(\{i\}) = 1$ for all $i \in I$, $\gamma(S) = 0$ for all other S
(d) $\gamma(I) = .5$, $\gamma(\{i,j\}) = .25$ for all $i \neq j$, $\gamma(S) = 0$ for all other S
(e) $\gamma(\{i\}) = \gamma(\{j,k\}) = 1$ for any $i, j, k \in I$ with $i \neq j \neq k$, $\gamma(S) = 0$ for all other S. ■

The balanced family (e) in example 10.8 is particularly useful because it is true for any player set I and any coalition $S \subset I$, that

$$\gamma(S) = \gamma(I \backslash S) = 1 \quad \text{and} \quad \gamma(T) = 0$$

for any other coalition T forms a balanced family of coalitions. With the help of the concept of a balanced family of coalitions, one can characterize games in coalitional form in the following way.

DEFINITION 10.7. A game in coalitional form $\Gamma = (I, v)$ is called *balanced* if

$$\sum_{S \in \mathscr{P}(I)} \gamma(S) \cdot v(S) \leq v(I)$$

for all balanced families of coalitions holds.

Notice that it is sufficient to find a single balanced family of coalitions for which $\sum_{S \in \mathscr{P}(I)} \gamma(S) \cdot v(S) > v(I)$ holds, to conclude that a game is not balanced.

THEOREM 10.1 *A game has a nonempty core if and only if it is balanced.*

Proof.

(i) *Necessity:* $C(v) \neq \varnothing \Rightarrow \Gamma = (I, v)$ is balanced.
By the definition of the core, any $x \in C(v)$ satisfies

$$\sum_{i \in S} x_i \geq v(S) \quad \text{for all} \quad S \in \mathscr{P}(I).$$

Consider an arbitrary balanced family of coalitions γ. Since $\gamma(S) \geq 0$ for all $S \in \mathscr{P}(I)$ holds,

$$\sum_{S \in \mathscr{P}(I)} \gamma(S) \cdot v(S) \leq \sum_{S \in \mathscr{P}(I)} \gamma(S) \cdot \left(\sum_{i \in S} x_i \right) \qquad \text{(a)}$$

follows. Note that

$$\sum_{S \in \mathscr{P}(I)} \sum_{i \in S} \gamma(S) \cdot x_i = \sum_{i \in I} \sum_{S \in \mathscr{S}(i)} \gamma(S) \cdot x_i$$

holds. Hence,

$$\sum_{S \in \mathscr{P}(I)} \gamma(S) \cdot \left(\sum_{i \in S} x_i \right) = \sum_{i \in I} x_i \cdot \left[\sum_{S \in \mathscr{S}(i)} \gamma(S) \right] = v(I), \qquad \text{(b)}$$

because $[\Sigma_{S \in \mathscr{S}(i)} \gamma(S)] = 1$ for any balanced family of coalitions. Combining (a) and (b) yields $\Sigma_{S \in \mathscr{P}(I)} \gamma(S) \cdot v(S) \leq v(I)$. Since this holds for an arbitrary balanced family γ, Γ is a balanced game.

(ii) *Sufficiency:* $\Gamma = (I, v)$ balanced $\Rightarrow C(v) \neq \varnothing$.

It will be shown that a game Γ is not balanced if the core is empty. This means that one can find a balanced family of coalitions for which

$$\sum_{S \in \mathscr{P}(I)} \gamma(S) \cdot v(S) > v(I)$$

holds, whenever $C(v) = \varnothing$.

Consider the following two sets: $A \equiv \{x \in \mathbb{R}^n | \Sigma_{i \in I} x_i = v(I)\}$ and

$$B \equiv \left\{ x \in \mathbb{R}^n | \sum_{i \in S} x_i \geq v(S) \quad \text{for all} \quad S \in \mathscr{P}(I),\, S \neq I \right\}$$

Note that $A \cap B = C(v)$ holds. Hence, $A \cap B = \varnothing$ if and only if $C(v) = \varnothing$. Note that A is a hyperplane in $\mathbb{R}^n$ and B is a closed and convex subset of $\mathbb{R}^n$. If the core is empty, then $C(v) = A \cap B = \varnothing$ implies that the sets A and B can be separated by a hyperplane. Hence, for any $x \in A$

$$v(I) = \sum_{i \in I} x_i < \sum_{i \in I} y_i \quad \text{for all} \quad y \in B \qquad (*)$$

holds. The following diagram illustrates this argument for the case $n = 2$.

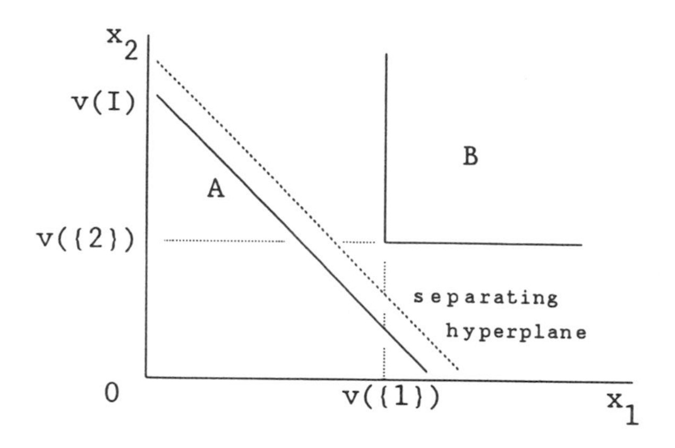

In particular, the inequality ($*$) must hold for the vector y^* that solves the following optimization problem:

$$\text{Choose } y^* \in B \text{ that minimizes } \sum_{i \in I} y_i.$$

Since B is closed, convex, and bounded below, a minimizer must exist and be characterized by the first-order conditions of the Lagrangian function

$$\mathscr{L}(y, \lambda) = -\sum_{i \in I} y_i + \sum_{S \in \mathscr{P}(I)} \lambda_S \cdot \left[\sum_{i \in S} y_i - v(S)\right].$$

At the optimum (y^*, λ^*),

$$\text{for all } \quad i \in I, \quad \frac{\partial \mathscr{L}(\cdot)}{\partial y_i} = 0 \quad \text{and} \quad \mathscr{L}(y^*, \lambda^*) = -\sum_{i \in I} y_i^*$$

must hold. This yields the following system of equations:

$$-1 + \sum_{S \in \mathscr{S}(i)} \lambda_S^* = 0 \tag{i}$$

$$\sum_{\substack{S \in \mathscr{P}(I) \\ S \neq I}} \lambda_S^* \cdot \left[\sum_{i \in S} y_i^* - v(S)\right] = 0. \tag{ii}$$

Setting $\lambda_I^* = 0$, by (i), λ^* is a balanced family of coalitions. From (ii),

$$\sum_{S \in \mathscr{P}(I)} \lambda_S^* \cdot v(S) = \sum_{S \in \mathscr{P}(I)} \lambda_S^* \cdot \left[\sum_{i \in S} y_i^*\right] = \sum_{i \in I} y_i^* \cdot \left[\sum_{S \in \mathscr{S}(i)} \lambda_S^*\right] = \sum_{i \in I} y_i^*$$

follows, where the second equality requires a reordering of terms, and (i) has been used to obtain the last equality. Hence, there is a balanced family λ^* such that $\Sigma_{S \in \mathscr{P}(I)} \lambda_S^* \cdot v(S) = \Sigma_{i \in I} y_i^* < v(I)$. The strong inequality follows from equation $(*)$. Thus a game cannot be balanced if the core is empty. ■

Theorem 10.1 is very useful for checking whether the core of a game in coalitional form is empty, as the following examples will show.

EXAMPLE 10.2 (*resumed*). Consider the game $\Gamma = (I, v)$,

$$v(I) = 100, \quad v(\{i,j\}) = \xi \quad \text{for} \quad i, j \in I \quad \text{and} \quad i \neq j,$$
$$v(\{i\}) = 0 \quad \text{for all} \quad i \in I.$$

It has been shown before that the core is empty for $\xi = 80$.

To prove the core of this game is empty, it suffices to find a single balanced family of coalitions for which $\Sigma_{S \in \mathscr{P}(I)} \gamma(S) \cdot v(S) > v(I)$ holds. Since the two-player coalitions' payoff plays a major role in this case, it is worth trying the balanced family (b) from example 10.8,

$$\gamma(I) = 0, \quad \gamma(\{i,j\}) = .5 \quad \text{for all} \quad i \neq j, \qquad \gamma(S) = 0 \quad \text{for all other } S.$$

One checks easily that

$$\sum_{S \in \mathscr{P}(I)} \gamma(S) \cdot v(S) = .5 \cdot \xi + .5 \cdot \xi + .5 \cdot \xi = 1.5 \cdot \xi > 100 = v(I)$$

for $\xi > 200/3$ holds for this family. Hence one can conclude that, for $\xi > 200/3$ the game is not balanced and, by theorem 10.1, the core is empty. ■

With the help of the concept of a balanced game, it is not difficult to investigate whether the core is empty in examples 10.5 and 10.6. Note that these games have an arbitrary number of players. This question cannot be answered therefore by inspecting a diagram.

EXAMPLE 10.6 (*resumed*). In this game shareholders with differing numbers of shares w_i have to make a decision about investment according to a one-share one-vote principle and a required number of share-votes equal to q. Given a net return R in case of the adoption of the project, the following characteristic function arises:

$$v(S) = \left(\sum_{i \in S} w_i \right) \cdot \frac{R}{W} \quad \text{for} \quad \sum_{i \in S} w_i \geq q, \qquad v(S) = 0 \quad \text{for} \quad \sum_{i \in S} w_i < q,$$

with $W \equiv \Sigma_{i \in I} w_i$.

The concept of a balanced game makes it possible to prove that the core of this game is not empty by showing that the game is balanced. Denote by $\alpha(S)$ the function indicating whether S is a winning coalition or not; $\alpha(S) = 1$ for $\Sigma_{i \in S} w_i \geq q$, $\alpha(S) = 0$ for $\Sigma_{i \in S} w_i < q$. Then one can write the characteristic function as $v(S) = \alpha(S) \cdot (\sum_{i \in S} w_i) \cdot R/W$. To see that this game is balanced, take an arbitrary balanced family $\gamma(\cdot)$ and check the following equalities and inequalities:

$$\begin{aligned}
&\sum_{S \in \mathscr{P}(I)} \gamma(S) \cdot v(S) \\
&\quad = \sum_{S \in \mathscr{P}(I)} \gamma(S) \cdot \alpha(S) \cdot \left(\sum_{i \in S} w_i \right) \cdot \frac{R}{W} = \sum_{i \in I} \sum_{S \in \mathscr{S}(i)} \gamma(S) \cdot \alpha(S) \cdot w_i \cdot \frac{R}{W} \\
&\quad = \sum_{i \in I} w_i \cdot \frac{R}{W} \cdot \left[\sum_{S \in \mathscr{S}(i)} \gamma(S) \cdot \alpha(S) \right] \leq \sum_{i \in I} w_i \cdot \frac{R}{W} \cdot \left[\sum_{S \in \mathscr{S}(i)} \gamma(S) \right] \\
&\quad = \sum_{i \in I} w_i \cdot \frac{R}{W} = R = v(I).
\end{aligned}$$

The equalities result from simple rearrangements of the summands, and the definitions of $v(S)$ and of a balanced family of coalitions. The inequality follows since $0 \leq \gamma(S) \cdot \alpha(S) \leq \gamma(S)$ holds. Since this is true for an arbitrary balanced family, it must be true for any balanced family. Hence the game is balanced and, by theorem 10.1, the core is nonempty. ■

This section concludes with a discussion of the pure exchange economy given in example 10.5.

EXAMPLE 10.5 (*resumed*). The characteristic function of this game v was given as

$$v(S) = \max \left\{ \sum_{i \in S} u_i(x_i) \middle| \sum_{i \in S} x_i = \sum_{i \in S} \bar{x}_i \right\} \quad \text{for all} \quad S \in \mathscr{P}(I).$$

It is a well-known fact from microeconomic theory that the core of a pure exchange economy is nonempty.[2] The usual way to show this result consists of demonstrating that a Walrasian equilibrium exists and that the equilibrium allocation lies in the core of the game. A second way to prove that the core of such an economy is nonempty shows that the market game of the pure exchange economy is balanced. Because of the importance of this example for economic theory, both claims will be proved here.

Claim A. A Walrasian equilibrium exists and the equilibrium allocation lies in the core.

To prove the existence of a Walrasian equilibrium, it is shown that the following optimization problem has a solution, and that this solution forms

[2] Varian (1984, pp. 235–239) provides an introduction to the Walrasian equilibrium and the core.

a Walrasian equilibrium

$$\text{Choose} \quad x = (x_1, \ldots, x_n) \quad \text{to maximize} \quad \sum_{i \in I} u_i(x_i)$$

$$\text{subject to} \quad \sum_{i \in I} x_i = \sum_{i \in I} \bar{x}_i, \quad \text{and} \quad x_i \geq 0 \quad \text{for all} \quad i \in I.$$

Since the utility functions u_i are continuous and the set of feasible allocations is compact, a maximizer x^* exists. By the Kuhn-Tucker theorem,[3] there exists a vector of multipliers $p^* \in \mathbb{R}_+^\ell$ such that

$$\mathscr{L}(x, p^*) = \sum_{i \in I} u_i(x_i) + p^* \cdot \left[\sum_{i \in I} \bar{x}_i - \sum_{i \in I} x_i \right] = \sum_{i \in I} \left(u_i(x_i) + p^* \cdot [\bar{x}_i - x_i] \right)$$

is maximized by x^*. Furthermore, $p^* \cdot [\Sigma_{i \in I} \bar{x}_i - \Sigma_{i \in I} x_i^*] = 0$ is a necessary condition for a maximum. Since all utility functions u_i are strictly increasing $p_k^* > 0$ must hold for all $k = 1, \ldots, \ell$. It is easy to see that these conditions imply that (p^*, x^*) satisfies

(i) $\Sigma_{i \in I} x_i^* = \Sigma_{i \in I} \bar{x}_i$ (market clearing)
(ii) x_i^* maximizes $u_i(x_i)$ subject to $p^* \cdot x_i = p^* \cdot \bar{x}_i$ for all $i \in I$ (optimal behavior of consumers subject to a budget constraint).

Hence (p^*, x^*) is a Walrasian equilibrium. This establishes existence of an equilibrium.

To see that x^* is an allocation in the core, assume to the contrary that x^* is not in the core. Then there must be a coalition S and another feasible allocation $(x_i)_{i \in S}$ yielding a strictly greater utility for every consumer in this coalition S than $(x_i^*)_{i \in S}$ does; formally,

$$u_i(x_i) > u_i(x_i^*) \quad \text{for all} \quad i \in S \quad \text{and} \quad \sum_{i \in S} x_i = \sum_{i \in S} \bar{x}_I \quad \text{must hold.}$$

Since consumers chose x_i^* to maximize their utility subject to their budget constraint they must have been unable to afford the preferred x_i. Hence $p^* \cdot x_i > p^* \cdot \bar{x}_i$ must hold for all $i \in S$. Summing over the agents of coalition S, $\Sigma_{i \in S} p^* \cdot x_i > \Sigma_{i \in S} p^* \cdot \bar{x}_i$ follows. This contradicts feasibility $\Sigma_{i \in S} x_i = \Sigma_{i \in S} \bar{x}_i$. Thus, x^* must be an element of the core.

Claim B. The pure exchange game $\Gamma = (I, v)$, just given is balanced.

To prove this game is balanced, consider an arbitrary balanced family of coalitions γ. For any coalition $S \in \mathscr{P}(I)$, let $(y_i^S)_{i \in S}$ be an allocation

[3]See Varian (1984, p. 319) for an exposition of this theorem.

that achieves

$$v(S) = \sum_{i \in S} u_i(y_i^S) = \max\left\{ \sum_{i \in S} u_i(x_i) \mid \sum_{i \in S} x_i = \sum_{i \in S} \bar{x}_i \right\}.$$

A simple rearrangement of summands yields

$$\sum_{S \in \mathscr{P}(I)} \gamma(S) \cdot v(S) = \sum_{S \in \mathscr{P}(I)} \gamma(S) \cdot \sum_{i \in S} u_i(y_i^S) = \sum_{i \in I} \sum_{S \in \mathscr{S}(i)} \gamma(S) \cdot u_i(y_i^S).$$

Note that any balanced family of coalitions specifies a convex combination for any set of coalitions containing a particular player i, $\mathscr{S}(i)$. Therefore, by concavity of u_i, one has in addition,

$$u_i\left(\sum_{S \in \mathscr{S}(i)} \gamma(S) \cdot y_i^S \right) \geq \sum_{S \in \mathscr{S}(i)} \gamma(S) \cdot u_i(y_i^S).$$

Summing this inequality over all $i \in I$ yields the inequality

$$\sum_{i \in I} u_i\left(\sum_{S \in \mathscr{S}(i)} \gamma(S) \cdot y_i^S \right) \geq \sum_{i \in I} \sum_{S \in \mathscr{S}(i)} \gamma(S) \cdot u_i(y_i^S) = \sum_{S \in \mathscr{P}(I)} \gamma(S) \cdot v(S). \tag{A}$$

Consider the average consumption of a player i over all coalitions in which this player participates $z_i \equiv \sum_{S \in \mathscr{S}(i)} \gamma(S) \cdot y_i^S$. The allocation $(z_1, \ldots, z_n)$ is feasible for the grand coalition I, as the following equalities show:

$$\begin{aligned} \sum_{i \in I} z_i &\equiv \sum_{i \in I} \sum_{S \in \mathscr{S}(i)} \gamma(S) \cdot y_i^S = \sum_{S \in \mathscr{P}(I)} \gamma(S) \cdot \sum_{i \in S} y_i^S \\ &= \sum_{S \in \mathscr{P}(I)} \gamma(S) \cdot \sum_{i \in S} \bar{x}_i = \sum_{i \in I} \bar{x}_i \cdot \left[\sum_{S \in \mathscr{S}(i)} \gamma(S) \right] = \sum_{i \in I} \bar{x}_i. \end{aligned}$$

Since $(z_1, \ldots, z_n)$ is feasible.,

$$\begin{aligned} v(I) &= \max\left\{ \sum_{i \in I} u_i(x_i) \mid \sum_{i \in I} x_i = \sum_{i \in I} \bar{x}_i \right\} \\ &\geq \sum_{i \in I} u_i(z_i) = \sum_{i \in I} u_i\left(\sum_{S \in \mathscr{S}(i)} \gamma(S) \cdot y_i^S \right) \geq \sum_{S \in \mathscr{P}(I)} \gamma(S) \cdot v(S) \end{aligned}$$

follows. The first inequality is a consequence of $v(S)$ being the highest aggregate payoff that coalition S can achieve; the second inequality has been derived before in (A). The last two inequalities prove the game is balanced. ■

The concept of the core is based on a negative selection criterion because it rules out allocations that can be reasonably opposed by some subgroup of players. The condition that no coalition could improve on the proposed payoff allocation by reallocating its total payoff among its members separately, may appear to be a necessary requirement for any agreement. However, as some of the examples in this section have shown, the core may be empty. Thus there may be no allocation that is stable against deviations of subgroups. The following section presents an alternative approach for finding a solution for games in coalitional form.

10.2 The Shapley Value

The solution for a game in coalitional form presented in this section uses the axiomatic method encountered in Chapter 9 in the context of the Nash bargaining solution. This approach advances criteria, called axioms, that an acceptable allocation should satisfy. The surprising result of this method is the small number of requirements that are sufficient to determine a unique allocation, called *value* by Shapley (1953) or *Shapley value* thereafter, to avoid confusion with other "value" concepts, for example, the maximin-value in Chapter 2.

Most economists would probably agree that a payoff allocation that rational players accept should satisfy efficiency (Pareto optimality) and symmetry. The first requirement rejects allocations that provide opportunities for making someone better off without harming anyone else. The second condition imposes equal treatment of players who play identical roles in the game. The role of players in a cooperative game is determined by the contributions they make to the different coalitions of the game. In this perspective, players are *symmetric*, play identical roles, if they contribute in the same way to any coalition they join. Given the focus on a player's contribution to various coalitions, a third condition appears quite natural: a player who contributes to the outcome of any coalition exactly his individual payoff $v(\{i\})$ shall be assigned this individual payoff in the value allocation. Denoting by $\phi(v) \equiv (\phi_1(v), \phi_2(v), \ldots, \phi_n(v))$ a payoff allocation for the n players of the game in coalitional form $\Gamma = (I, v)$, one can state these requirements as three assumptions on the allocation $\phi(v)$.

Assumption 1 (efficiency). The payoff allocation $\phi(v)$ distributes the total payoff of the game,

$$\sum_{i \in I} \phi_i(v) = v(I).$$

Assumption 2 (symmetry). For any two players $i, j \in I$ who contribute the same to every coalition, formally for whom $v(S \cup \{i\}) = v(S \cup \{j\})$ *for all* $S \in \mathscr{P}(I)$ *with* $i, j \notin S$ *holds,*

$$\phi_i(v) = \phi_j(v).$$

Assumption 3 (dummy player). For any player $i \in I$ who contributes his own value to any coalition, formally for whom $v(S) = v(S \setminus \{i\}) + v(\{i\})$ for all $S \in \mathscr{S}(i)$ holds,

$$\phi_i(v) = v(\{i\}).$$

For some games these three assumptions are sufficient to determine the allocation $\phi(v)$ uniquely.

EXAMPLE 10.2 (*resumed*). This three-player game has the following characteristic function:

$$v(I) = 100, \quad v(\{i,j\}) = \xi \text{ for } i, j \in I \quad \text{and} \quad i \neq j, v(\{i\}) = 0 \text{ for all } i \in I.$$

By assumption 1 (efficiency), the suggested allocation $\phi(v)$ must satisfy

$$\phi_1(v) + \phi_2(v) + \phi_3(v) = 100.$$

Next, consider the only coalition that does not contain players 1 and 2, namely $\{3\}$. One has $v(\{3\} \cup \{1\}) = \xi = v(\{3\} \cup \{2\})$. Hence, player 1 and 2 contribute in the same way to any coalition to which they do not belong. By assumption 2 (symmetry), $\phi_1(v) = \phi_2(v)$ must hold. A similar argument establishes $\phi_3(v) = \phi_2(v)$. Hence

$$\phi_1(v) = \phi_2(v) = \phi_3(v)$$

follows from the symmetry assumption. Combining these two conditions yields $\phi(v) = (100/3, 100/3, 100/3)$ as the unique allocation satisfying assumptions 1 and 2 in this game. ■

It is important to note that the value allocation exists no matter what value ξ takes. For $\xi \in [0, 200/3]$, the core is nonempty and the allocation $\phi(v)$ belongs to the core. For all other ξ, the core is empty but the allocation $\phi(v)$ remains unchanged. This highlights an essential difference between the two approaches: the joint outcome of the two-player coalition, ξ, determines the influence of a two-player coalition in the game. For the allocation $\phi(v)$ the influence of a coalition is completely irrelevant. What matters is an individual player's contribution to various coalitions. As long as these contributions remain unchanged the allocation $\phi(v)$ will remain unaffected. Thus for the core, absolute outcomes of coalitions matter; for the value allocation $\phi(v)$, the relative contribution of different players to a coalition's outcome is essential.

In general, however, these three assumptions are not sufficient to determine a value allocation. To obtain a unique value allocation a consistency requirement is necessary. Suppose the same set of players I plays two games $\Gamma = (I, v)$ and $\Gamma' = (I, w)$ successively. The overall payoff that each coalition $S \in \mathscr{P}(I)$ receives from playing both games is $v(S) +$

$w(S)$. On the other hand, $(v+w)(S) \equiv v(S)+w(S)$ for all coalitions $S \in \mathscr{P}(I)$ defines another characteristic function. Thus, one can consider the two games as a single game in which each coalition S achieves $(v+w)(S)=v(S)+w(S)$. The two situations, sequential play of v and w and compound $v+w$, yield the same joint payoff for every coalition. The fourth assumption maintains that the distribution $\phi(v+w)$ for the compound game $v+w$ should equal the sum of the distributions $\phi(v)+\phi(w)$ of the constituent games. This means the distribution ϕ shall be invariant to an arbitrary decomposition of a game. Note that the value allocation rules for ϕ are assumed to be applicable to all games with the same player set. The value ϕ is therefore a function associating particular payoff allocations with coalitional games that have the same player set.

Denote by $\Gamma(I)$ the set of all games in coalitional form with the same player set I. Note that for any two games $v, w \in \Gamma(I)$ the characteristic function $v+w$ is also an element of $\Gamma(I)$.

Assumption 4 (additivity). For any two games $v, w \in \Gamma(I)$, $\phi(v+w) = \phi(v)+\phi(w)$.

Given assumptions 1 to 4, the following characterization of the value function ϕ can be given.

THEOREM 10.2. *The unique value* $\phi = (\phi_1, \ldots, \phi_n)$ *satisfying assumptions* 1, 2, 3, *and* 4 *is*

$$\phi_i(v) = \sum_{S \in \mathscr{P}(I)} q(s) \cdot [v(S) - v(S \setminus \{i\})]$$

with $q(s) \equiv (s-1)!(n-s)!/n!$.

Before proving this theorem, two remarks will provide interpretations of its meaning, and some examples will illustrate how to apply it. An interesting aspect of the value is its interpretation as the "expected value of each players contribution to coalitions of the game." The following remark shows that the weights $q(s)$ can be viewed as probabilities.

Remark 10.3. The weights $q(s)$ in the formula of the value ϕ satisfy

$$q(s) \geq 0, \quad \text{for all} \quad s = 1, 2, \ldots, n, \quad \text{and} \quad \sum_{S \in \mathscr{S}(i)} q(s) = 1.$$

The non-negativity of $q(s)$ is obvious. To establish that the sum of the weights equals one, consider the following equations:

$$\begin{aligned}
\sum_{S \in \mathscr{S}(i)} q(s) &= \sum_{s=1}^{n} \frac{(n-1)!}{(s-1)!(n-s)!} \cdot q(s) \\
&= \sum_{s=1}^{n} \frac{(n-1)!}{(s-1)!(n-s)!} \cdot \frac{(s-1)!(n-s)!}{n!} = \sum_{s=1}^{n} \frac{1}{n} = 1.
\end{aligned}$$

To obtain the first equality, note that the same weight $q(s)$ applies to each coalition that contains i and has s members. The number of coalitions that contain player i and have s members is equal to the number of coalitions with $(s-1)$ members that one can form from $(n-1)$ players. This follows, since each coalition of $(s-1)$ members formed from the set of players $I \backslash \{i\}$ yields a different coalition of s members containing player i. It is a standard result of combinatorics that one can form $\binom{m}{k} = m!/k!(m-k)!$ combinations, that is k-player coalitions, from a set of m players. Hence, with $m = (n-1)$ and $k = (s-1)$, one concludes that there are $(n-1)!/(s-1)!(n-s)!$ coalitions with s players that contain player i. ■

In light of this remark, the probabilities $q(s)$ can be written as

$$q(s) = \frac{(s-1)!(n-s)!}{n!} = \frac{1}{n} \cdot \frac{(s-1)!(n-s)!}{(n-1)!} = \frac{1}{n} \cdot \frac{1}{\binom{n-1}{s-1}}.$$

The expression $\binom{n-1}{s-1}$ gives the number of coalitions with s members that include a particular player. Hence, $q(s)$ can be interpreted as the probability that a particular player joins a coalition with $s-1$ members times the *ex-ante* probability of becoming this particular player. This implies that each coalition is considered equally likely. The allocation that the value assigns to player $i \in I$ is, in this interpretation, the expected value of the marginal contributions that this player can make to the coalitions of this game.

Remark 10.4. Note that, for all coalitions $S \in \mathscr{P}(I)$ that do not contain player i, $S = S \backslash \{i\}$ holds and, therefore, $v(S) - v(S \backslash \{i\}) = 0$. Consequently, one can sum over $\mathscr{S}(i)$ instead of $\mathscr{P}(I)$ in the formula of the value,

$$\sum_{S \in \mathscr{P}(I)} q(s) \cdot [v(S) - v(S \backslash \{i\})] = \sum_{S \in \mathscr{S}(i)} q(s) \cdot [v(S) - v(sS \backslash \{i\})].$$

■

Remark 10.4 makes the computation of the value much easier. For a game with player set $I = \{i, j, k\}$, one can compute the value ϕ as follows:

$$\begin{aligned} \phi_i(v) &= q(3) \cdot [v(I) - v(\{j,k\})] + q(2) \cdot [v(\{i,j\}) - v(\{j\})] \\ &\quad + q(2) \cdot [v(\{i,k\}) - v(\{k\})] + q(1) \cdot [v(\{i\}) - v(\varnothing)] \\ &= \tfrac{1}{3} \cdot [v(I) - v(\{j,k\})] + \tfrac{1}{6} \cdot [v(\{i,j\}) - c(\{j\})] \\ &\quad + \tfrac{1}{6} \cdot [v(\{i,k\}) - v(\{k\})] + \tfrac{1}{3} \cdot [v(\{i\}) - v(\varnothing)] \end{aligned}$$

since $q(3) = 0!2!/3! = \frac{1}{3}$, $q(2) = 1!1!/3! = \frac{1}{6}$, and $q(1) = 2!0!/3! = \frac{1}{6}$ holds. The following three examples apply this formula to compute the value allocation.

Example 10.3 (*resumed*). The characteristic function of the land use problem was given as $v(I) = 3$, $v(\{1,2\}) = 2$, $v(\{1,3\}) = 3$, $v(\{1\}) = 1$ and $v(S) = 0$ for all other $S \in \mathscr{P}(I)$.

No players are symmetric or null players and one has to apply the value formula to all players to obtain

$$\phi_1(v) = \tfrac{1}{3} \cdot [3-0] + \tfrac{1}{6} \cdot [2-0] + \tfrac{1}{6} \cdot [3-0] + \tfrac{1}{3} \cdot [1-0] = 13/6.$$

Computing similarly the value for the other players yields

$$\phi(v) \equiv (\phi_1(v), \phi_2(v), \phi_3(v)) = (13/6, 1/6, 4/6).$$

Player 1, the farmer who owns the land, gets the largest share because no joint payoff is possible without his participation. More surprisingly, player 2, the manufacturer whose offer will not be taken up by the grand coalition, receives a positive share. Note that the value allocation could be blocked by a coalition of player 1 and player 3. Such a coalition could distribute the share of player 2 between them and make them better off than they are in the value allocation. Hence, ϕ is not a core allocation. ■

Example 10.4 (*resumed*). The following characteristic function represents the cost sharing problem for water treatment facilities of the three communities:

$$v(I) = 5, \quad v(\{1,2\}) = 2, \quad v(\{1,3\}) = 3,$$
$$v(\{2,3\}) = 1, \quad v(\{1\}) = v(\{2\}) = v(\{3\}) = 0.$$

Using the value formula, one computes easily

$$\phi(v) = (\phi_1(v), \phi_2(v), \phi_3(v)) = (13/6, 7/6, 10/6).$$

It is worth reflecting on this result. If the three communities decide to jointly set up their own water treatment system, they will make a net surplus of \$500,000. The most immediate proposal for how to share this surplus would probably be to proportion it to the population of these communities. Since community 1 has a population of 5000 out of a total of 12,000 inhabitants in the three communities, it would be entitled to 5/12 of the surplus. Similarly, community 2's share would be 3/12 and community 3's 4/12. Thus, the surplus allocation (in \$100,000 units) would be $5 \cdot (5/12, 3/12, 4/12) = (25/12, 15/12, 20/12)$ for a proportional distribution scheme.

The value allocation, however, distributes the surplus as (13/6, 7/6, 10/6) = (26/12, 14/12, 20/12). Obviously, community 1 gets a higher share in the value allocation than corresponds to its population share, and community 2 obtains a smaller share, whereas community 3 receives exactly the same share in both schemes. Since the benefit (costs of $100 per head saved) was proportional, the difference between the value allocation and the per-head shares must be sought in the cost structure. Indeed, the technology behind the cost function shows strong increasing returns to scale. Community 1 with its larger population contributes considerably more to the cost savings per head than community 2 with the smallest population. This is reflected in a higher than proportional share of community 1 and a lower than proportional share of community 2. The value adjusts for asymmetric contributions to the joint surplus. It is easy to check that in this case the value allocation lies in the core of the game. ■

Example 10.7 is a further game with a value allocation in the core.

EXAMPLE 10.7 (*resumed*). In this game the investment opportunities of three fund managers gave rise to the following characteristic function: $v(I) = 60$, $v(\{1,2\}) = 36$, $v(\{1,3\}) = 50$, $v(\{2,3\}) = 27$, $v(\{1\}) = 27$, $v(\{2\}) = 8$, $v(\{3\}) = 18$.

Applying the value formula yields

$$\phi(v) \equiv (\phi_1(v), \phi_2(v), \phi_3(v)) = (30, 9, 21)$$

for this game. This allocation cannot be blocked by any coalition. Note the difference to the division (30, 10, 20) that would result if interest income were distributed in proportion to each player's investment. ■

Consider the proof of theorem 10.2. Firstly, it is checked that the value ϕ satisfies assumptions 1 to 4. Secondly, it will be shown that this value must be unique.

Proof.

Part 1: ϕ satisfies axioms 1 to 4.

A1 (efficiency): To establish efficiency, consider the following equalities:

$$\begin{aligned}
\sum_{i \in I} \phi_i(v) &= \sum_{i \in I} \sum_{S \in \mathscr{S}(i)} q(s) \cdot [v(S) - v(S \setminus \{i\})] \\
&= \sum_{S \in \mathscr{P}(I)} \sum_{i \in S} q(s) \cdot [v(S) - v(S \setminus \{i\})] \\
&= \sum_{S \in \mathscr{P}(I)} s \cdot q(s) \cdot v(S) - \sum_{S \in \mathscr{P}(I)} q(s) \cdot \left[\sum_{i \in S} v(S \setminus \{i\}) \right] \\
&= \sum_{S \in \mathscr{P}(I)} s \cdot q(s) \cdot v(S) - \sum_{S \in \mathscr{P}(I) \setminus I} q(s) \cdot s \cdot v(S) \\
&= q(n) \cdot n \cdot v(I) = v(I).
\end{aligned}$$

The first equality follows by the definition of the value ϕ in the theorem. The second equality holds because it just rearranges the order of summation, the third equation uses the fact that the first summand does not depend on i. The fourth equation follows from the following considerations: one sums first over all subsets of a given set S that contain $s-1$ agents, $[\sum_{i \in S} v(S \setminus \{i\})]$, and, for every S, there are exactly s of these subsets; summing over all subsets of I, except I itself, and correcting for double-counting by multiplying $v(S)$ by s, one establishes the fourth equation.

A2 (symmetry): Consider the following subsets of $\mathscr{P}(I)$:

$$\mathscr{S}_0 = \{S \in \mathscr{P}(I) | i \notin S \text{ and } j \notin S\}, \quad \mathscr{S}_{ij} = \{S \in \mathscr{P}(I) | i \in S \text{ and } j \in S\},$$

$$\mathscr{S}_i = \{S \in \mathscr{P}(I) | i \in S \text{ and } j \notin S\}, \text{ and } \mathscr{S}_j = \{S \in \mathscr{P}(I) | i \notin S \text{ and } j \in S\}.$$

Note that $S \in \mathscr{S}_i$ if and only if $S \setminus \{i\} \in \mathscr{S}_0$ holds. Furthermore, for symmetric players, one has $v(S \setminus \{i\}) = v(S \setminus \{i, j\} \cup \{j\}) = v(S \setminus \{i, j\} \cup \{i\}) = v(S \setminus \{j\})$ for all $S \in \mathscr{S}_{ij}$. These considerations justify the following equations:

$$\begin{aligned}
\phi_i(v) &= \sum_{S \in \mathscr{S}(i)} q(s) \cdot [v(S) - v(S \setminus \{i\})] \\
&= \sum_{S \in \mathscr{S}_{ij}} q(s) \cdot [v(S) - v(S \setminus \{i\})] + \sum_{S \in \mathscr{S}_i} q(s) \cdot [v(S) - v(S \setminus \{i\})] \\
&= \sum_{S \in \mathscr{S}_{ij}} q(s) \cdot [v(S) - v(S \setminus \{i\})] + \sum_{S \in \mathscr{S}_0} q(s) \cdot [v(S \cup \{i\}) - v(S)] \\
&= \sum_{S \in \mathscr{S}_{ij}} q(s) \cdot [v(S) - v(S \setminus \{j\})] + \sum_{S \in \mathscr{S}_0} q(s) \cdot [v(S \cup \{j\}) - v(S)] \\
&= \sum_{S \in \mathscr{S}_{ij}} q(s) \cdot [v(S) - v(S \setminus \{j\})] + \sum_{S \in \mathscr{S}_j} q(s) \cdot [v(S) - v(S \setminus \{j\})] \\
&= \sum_{S \in \mathscr{S}(j)} q(s) \cdot [v(S) - v(S \setminus \{j\})] = \phi_j(v).
\end{aligned}$$

A3 (dummy player): Suppose for some $i \in I$, $v(S) - v(S \setminus \{i\}) = v(\{i\})$ for all $S \in \mathscr{S}(i)$ holds, then

$$\phi_i(v) = \sum_{S \in \mathscr{S}(i)} q(s) \cdot [v(S) - v(S \setminus \{i\})] = v(\{i\}) \cdot \left[\sum_{S \in \mathscr{S}(i)} q(s) \right] = v(\{i\})$$

follows by remark 10.3.

A4 (additivity): For any two games $v, w \in \Gamma(I)$, $v + w \in \Gamma(I)$ follows, since $v + w$ is just another characteristic function. For any coalition $S \in \mathscr{P}(I)$, the payoff in the game $(v+w), (v+w)(S)$, equals $v(S) + w(S)$. Hence one has

$$\begin{aligned}
&[(v+w)(S) - (v+w)(S \setminus \{i\})] \\
&\quad = [v(S) - v(S \setminus \{i\})] + [w(S) - w(S \setminus \{i\})].
\end{aligned}$$

Therefore, for all $i \in I$,

$$\phi_i(v+w) = \sum_{S \in \mathscr{P}(I)} q(s) \cdot [(v+w)(S) - (v+w)(S \setminus \{i\})]$$
$$= \sum_{S \in \mathscr{P}(I)} q(s) \cdot [v(S) - v(S \setminus \{i\})]$$
$$+ \sum_{S \in \mathscr{P}(I)} q(s) \cdot [w(S) - w(S \setminus \{i\})] = \phi_i(v) + \phi_i(w).$$

This establishes that the value ϕ satisfies assumptions 1 to 4.

Part 2. ϕ is unique.

To show that ϕ is unique, it is demonstrated (a) that assumptions 1 to 3 determine ϕ uniquely for a particular set of games, and (b) that, by assumption 4, the value of any other game in $\Gamma(I)$ can be obtained from the value on the particular set in a unique way.

(a) For any set $T \in \mathscr{P}(I)$ one can define the following simple game:

$$v_T(S) = \begin{cases} 1 & \text{if } T \subseteq S \\ 0 & \text{otherwise} \end{cases}.$$

One can interpret coalition T in this game as the "winning coalition." Any group of players containing T wins, all other coalitions lose. Similarly, for any $\lambda \in \mathbb{R}$, the game $\lambda \cdot v_T$ assigns a joint payoff of λ to the winning coalition and nothing to the losing coalitions.

Firstly, any player $i \notin T$ has $v_T(\{i\}) = 0$. Furthermore, a player $i \notin T$ is a null player because, for any $S \in \mathscr{S}(i)$, either $T \subseteq S$ and, therefore, $v_T(S) - v_T(S \setminus \{i\}) = 0$ holds, or T is not a subset of S and therefore $v_T(S) = v_T(S \setminus \{i\}) = 0$ follows. By assumption 3, $\phi_i(v_T) = 0$ has to be assigned to a null player.

Next consider two players $i, j \in T$. For any coalition S such that $i, j \notin S$ holds, T is neither a subset of $S \cup \{i\}$ nor of $S \cup \{j\}$. Hence, $v_T(S \cup \{i\}) = v_T(S \cup \{j\})$ holds and the symmetry assumption 2 implies $\phi_i(v_T) = \phi_j(v_T)$.

Finally, by assumption 1,

$$1 = v_T(I) = \sum_{i \in I} \phi_i(v_T) = \sum_{i \in T} \phi_i(v_T) + \sum_{i \notin T} \phi_i(v_T) = t \cdot \phi_i(v_T)$$

for any $i \in T$.

The last equality follows by symmetry of the players in T. Thus, $\phi_i(v_T) = 1/t$ for all $i \in T$ is established. Hence, assumptions 1, 2, and 3 imply that there is a unique value ϕ for any simple game, namely

$$\phi_i(v_T) = \begin{cases} 1/t & \text{for } i \in T \\ 0 & \text{for } i \notin T \end{cases}.$$

By repeating the steps made to establish $\phi_i(v_T)$, one checks easily that, for any $\lambda \in \mathbb{R}$, $\phi_i(\lambda \cdot v_T) = \lambda/t$ for $i \in T$ and $\phi_i(\lambda \cdot v_T) = 0$ for $i \notin T$ holds.

(b) To see that there is a unique extension for the value from the set of simple games to all games $\Gamma(I)$, note that every game $v \in \Gamma(I)$ is a vector of real numbers with $m = 2^n - 1$ components. There are m subsets of players in a set of n players, I, (excluding the empty set with a payoff of zero in every game). Hence, every component of the m-vector v corresponds to the joint payoff of some coalition. Therefore, $\Gamma(I) = \mathbb{R}^m$ holds.

Consider the set of simple games $B = \{v_T \in \Gamma(I) = \mathbb{R}^m | T \in \mathscr{P}(I) \setminus \varnothing\}$. Since there are $m = 2^n - 1$ subsets in $\mathscr{P}(I) \setminus \varnothing$, the set B contains m vectors. It is a standard result of linear algebra that any vector of $\mathbb{R}^m$ can be obtained uniquely by a linear combination of any m linearly independent m-vectors.[4] Hence, any game $v \in \Gamma(I) = \mathbb{R}^m$ can be uniquely obtained as a linear combination of the simple games in B, if all vectors in B are linearly independent.

The following argument shows by contradiction that the vectors in B are linearly independent. Suppose the vectors in B are not linearly independent. Then there must exist a vector $\alpha \in \mathbb{R}^m$ with $\alpha \neq 0$ such that $\sum_{T \in \mathscr{P}(I) \setminus \varnothing} \alpha_T \cdot v_T = 0$ holds. Let $K \in \mathscr{P}(I) \setminus \varnothing$ be the coalition with the smallest number of members k among the coalitions with $\alpha_T \neq 0$. By linear dependence,

$$v_K(K) = -(1/\alpha_K) \cdot \left[\sum_{T \in \mathscr{P}(I) \setminus K} \alpha_T \cdot v_T(K) \right]$$

must hold. But $v_K(K) = 1$, by the definition of a simple game, and $[\sum_{T \in \mathscr{P}(I) \setminus K} \alpha_T \cdot v_T(K)] = 0$, since no coalition $T \neq K$ with $t \geq k$ can be contained in K. Hence, assuming linear dependence of the elements in B leads to a contradiction. Thus, for any $v \in \Gamma(I) = \mathbb{R}^m$, there is a unique $\alpha \in \mathbb{R}^m$ such that

$$v = \sum_{T \in \mathscr{P}(I) \setminus \varnothing} \alpha_T \cdot v_T$$

[4]This result can be found in any linear algebra textbook; for example, Grossman (1984, pp. 192, 193).

holds. By repeated application of assumption 4 (additivity), one obtains

$$\phi_i(v) = \sum_{T \in \mathscr{P}(I) \setminus \varnothing} \phi_i(\alpha_T \cdot v_T)$$

which determines the unique value for the game v. ■

This section concludes with two examples and a final remark. The first example derives the Shapley value for a special case of the shareholder problem of example 10.6. The second applies the Shapley value to a cost sharing problem. The final remark shows that the value allocation is individually rational if the game is superadditive.

EXAMPLE 10.6 (*modified*). Reconsider the weighted majority voting game for the case of a company with five shareholders, $I = \{1, 2, 3, 4, 5\}$. Shareholder 1 owns 80 shares and the others hold 30 each. Let R, the return on the investment project under consideration, be 100. A decision is reached with 50% of the voting shares. This yields the following characteristic function:

$$v(S) = .5 \cdot \left(\sum_{i \in S} w_i \right) \quad \text{for} \quad \sum_{i \in S} w_i \geq .5 \quad \text{and} \quad v(S) = 0 \quad \text{for} \quad \sum_{i \in S} w_i < .5.$$

The following considerations will simplify computations. Firstly, note that all players but shareholder 1 contribute in the same way to each coalition that does not contain them. Hence, by assumption 2, they are symmetric and must obtain equal allocations. Denote this allocation by α, then $\phi_2(v) = \phi_3(v) = \phi_4(v) = \phi_5(v) = \alpha$ must hold and by assumption 1 $\alpha = [v(I) - \phi_1(v)]/4$. Secondly, to compute $\phi_1(v)$, note that for $S \in \mathscr{S}(1)$

$$[v(S) - v(S \setminus \{1\})] = \begin{cases} 0 & s = 1 \\ 40 + 15 \cdot (s-1) \quad \text{for} & 2 \leq s < 5 \\ 40 & s = 5 \end{cases}$$

holds, since player 1 needs at least one other sharehoalder to obtain a majority and leaving a coalition will make the remaining group unable to reach a decision, except for the case of the all-player coalition ($s = 5$). Recall that there are $(n-1)!/((s-1)!(n-s)!)$ coalitions with s players that contain player i. This implies $q(s) \cdot (n-1)!/((s-1)!(n-s)!) = 1/n$. Therefore, one obtains $\phi_1(v) = (1/5) \cdot [(40 + 15 \cdot 1) + (40 + 15 \cdot 2) + (40 + 15 \cdot 3) + 40] = 250/5 = 50$. And, for all other shareholders, $\phi_i(v) = \alpha = (100 - 50)/4 = 12.5$.

Note that shareholder 1 receives a much greater allocation than corresponds to his relative shareholdings. This reflects his greater influence in this situation. This influence depends as much on the distribution of shares among the other shareholders as on his own holdings. Since 60% of the shares are held by small shareholders, player 1 is the decisive player in nearly all coalitions. This dominant position is reflected in the value allocation.

The importance of the distribution of the rest of the shares becomes obvious by modifying the example such that there are only two shareholders beside shareholder 1 holding 60 shares each. The player set is now $I = \{1, 2, 3\}$ and repeating the argument made before (or computing the value by the formula) yields

$$\begin{aligned}\phi(v) \equiv (\phi_1(v), \phi_2(v), \phi_3(v)) &= (110/3, 95/3, 95/3)\\ &= (36.6, 31.6, 31.6).\end{aligned}$$

Thus player 1 gets a considerably smaller allocation in this case, though his share in the company is unaltered. ■

The last example in this chapter shows an application of cooperative game theory to an economic problem that does not involve "players" in the usual sense. The Shapley value has been used to attribute costs of multi-product firms to the different products. Such an application needs justification. It can be shown, however, that all the assumptions that determine the value allocation have a natural interpretation in the context of cost sharing problems.[5] The following example indicates the necessary reinterpretation of the coalitional form for cost sharing applications.

EXAMPLE 10.9. Consider a firm jointly producing two outputs in quantities x_1 and x_2 according to the following cost function:

$$C(x_1, x_2) = \ln(1 + a_1 \cdot x_1 + a_2 \cdot x_2), \qquad a_1 > 0, a_2 > 0.$$

Note that this cost function has decreasing marginal costs for any output combination. This implies that the average costs are always higher than the marginal costs of production. Therefore, the firm would make losses in a competitive market.

Denote by $c_i(x_1, x_2)$, $i = 1, 2$, the costs attributed to output i and consider the following cost distribution scheme:

$$\begin{aligned}c_1(x_1, x_2) &= .5 \cdot \ln[(1 + a_1 \cdot x_1 + a_2 \cdot x_2) \cdot (1 + a_1 \cdot x_1)/(1 + a_2 \cdot x_2)],\\ c_2(x_1, x_2) &= .5 \cdot \ln[(1 + a_1 \cdot x_1 + a_2 \cdot x_2) \cdot (1 + a_2 \cdot x_2)/(1 + a_1 \cdot x_1)].\end{aligned}$$

It is easy to check that $C(x_1, x_2) = c_1(x_1, x_2) + c_2(x_1, x_2)$ holds. Given inverse demand functions for the two goods $p_1(x_1)$ and $p_2(x_2)$, one can determine a market clearing price system (p_1^*, p_2^*) that covers the cost of

[5] Mirman, Tauman, and Zang (1985) provide an introduction to the cost sharing literature and further references.

production by solving the following two equations for (x_1^*, x_2^*):

$$p_1(x_1^*) = c_1(x_1^*, x_2^*)/x_1^* \quad \text{and} \quad p_2(x_2^*) = c_2(x_1^*, x_2^*)/x_2^*.$$

The cost distribution scheme $c_i(x_1, x_2)$, $i = 1,2$, is the Shapley value applied to the following game in coalitional form: the set of players are the two commodities, $I = \{1,2\}$, and, for a given output combination (x_1, x_2), the characteristic function is defined as

$$v(\{1,2\}) = C(x_1, x_2), v(\{1\}) = C(x_1, 0), v(\{2\}) = C(0, x_2).$$

The worth of an output combination is simply the cost of producing this output combination. If some product is not part of the coalition, then its output is set to zero. It is straightforward to apply the formula of the Shapley value to this game. The cost shares $c_1(x_1, x_2)$ and $c_2(x_1, x_2)$ are the value for the given output combination (x_1, x_2). ■

The previous examples have shown that the value allocation may or may not be part of the core. This raises the question of whether it is at least an imputation. By assumption 1 it is group rational, but individual rationality is not required directly. The following remark answers this question.

Remark 10.5. For superadditive games, the value allocation is individually rational. By superadditivity, $v(S \cup \{i\}) \geq v(S) + v(\{i\})$ must hold for all coalitions that do not contain player i. Hence, $v(S) - v(S \setminus \{i\}) \geq v(\{i\})$ follows for all coalitions S that do contain player i, $S \in \mathscr{S}(i)$. Hence, one obtains from the formula of the Shapley value

$$\phi_i(v) = \sum_{S \in \mathscr{S}(i)} q(s) \cdot [v(S) - v(S \setminus \{i\})]$$

$$\geq v(\{i\}) \cdot \left[\sum_{S \in \mathscr{S}(i)} q(s) \right] = v(\{i\}). \quad ■$$

10.3 Summary

This chapter presented a brief introduction to the analysis of games in coalitional form. In contrast to the description of games in strategic or extensive form, the characteristic function records only jointly feasible outcomes for all possible coalitions of players. This can be justified if two conditions are satisfied:

- there exists a possibility for members of a coalition to make binding agreements regarding their actions
- a "reasonable" concept is available to determine a coalition's value in the resulting game between this coalition and those players not participating in the coalition

Two solution concepts were introduced and discussed in this framework: the core and the Shapley value. They reflect different approaches to the problem of determining a distribution of the joint outcome of the all-player coalition that is "acceptable" for all players. The core requires such an allocation to be stable against secession by subgroups. If a subgroup can achieve a better outcome for its members than the proposed allocation provides for them, then the proposed allocation is not considered viable. This constraint on possible allocations may be too strong, however, and rule out any allocation, or it may prove too weak and leave a large number of acceptable allocations. In fact, there are very few examples in which the core contains only a single point. In contrast, the Shapley value approach assumes several properties that an acceptable allocation should have and demonstrates that there is a unique allocation satisfying these conditions. The value is easy to compute and exists for all games. The value allocation is, however, not necessarily stable against secession of coalitions. Thus, the Shapley value and the core are not always compatible solution concepts for games in coalitional form.

10.4 Remarks on the Literature

The concept of the characteristic function goes back to von Neumann and Morgenstern (1947). Many solution concepts for games in coalitional form have been proposed that are not treated here. Aumann (1987a, pp. 31, 32) provides a short description of these concepts and further references. None of these concepts has found as wide a field of applications in economics as the core and the Shapley value. This is the main reason for not considering them here.

The importance of the core for economic theory was recognized by Shubik (1959). Shubik showed that the Walrasian equilibrium allocation forms an allocation of the core of an appropriately specified game. Shapley (1967) proved that the core of a game is nonempty if and only if the game is balanced (theorem 10.1). Example 10.5 is based on the paper "On Market Games" by Shapley and Shubik (1969). An economic application of the core concept that could not be covered in this book can be found in the context of job matching problems (Crawford and Knoer (1981) and Roth (1984)).

The Shapley value was suggested by Shapley (1953). Since then numerous axiomatizations and modifications have been considered in the literature, and the relationship between the value and other solution concepts has been extensively studied (for further references see Hart (1987)). Because of its economic applications, the extension of the Shapley value to continuum sets of players (Aumann and Shapley (1974)) is particularly worth mentioning. This extension makes it possible to apply the value to general cost sharing problems. Mirman, Tauman and Zang (1985) provide a good introduction to this literature and further references.

Finally, it is important to note that there are extensions of the core and the Shapley value to games without transferable utility (Shapley (1969)). These extensions make it possible to show that for pure exchange economies with a continuum of consumers the core and value allocations coincide with the set of Walrasian equilibrium allocations (Aumann (1964), (1975)). This shows a remarkable coincidence between Walrasian equilibrium outcomes and two important solution concepts for games in coalitional form in economies in which individual agents are insignificant.

Exercises

10.1 *Consider a car dealer, player 1, negotiating with two potential buyers about the sale of a used car. The seller has a reservation price of* \$2000, *and both buyers value the car at* \$3000. *If one allows buyers to jointly purchase the car, then any distribution of the surplus of* \$1000 *is feasible.*

Derive the characteristic function of this game, show in a diagram the core, and determine the Shapley value of this game.

10.2 *Consider three countries negotiating a trade agreement. The joint benefit of an agreement for the different coalitions is as follows:*

$$v(\{1,2,3\}) = x,\ v(\{1,2\}) = v(\{1,3\}) = .6 \cdot x,\ v(\{2,3\}) = .3 \cdot x, \text{ and } v(\{i\}) = 0 \quad \text{for} \quad i = 1,2,3.$$

(a) Draw a diagram of the core for this game. Does the size of the core depend on x?

(b) Derive the Shapley value for this game. Is the value allocation in the core? Does the answer to this question depend on x?

10.3 *Consider an airport authority which has to decide how to charge 5 airlines for the daily use of the check-in facilities. The following table shows the daily number of flights per airline:*

airline	1	2	3	4	5
flights	1	5	1	2	1

The airport authority uses the following cost function for providing check-in services for x flights: $C(x) = 10 \cdot x + x^2$.

Derive the cost allocation according to the Shapley value and compare it with a proportional cost distribution scheme. Which airlines gain and which lose compared with the proportional scheme? Explain your result.

10.4 Prove that the Shapley value of a market game (example 10.5) is individually rational by showing that the market game is superadditive.

10.5 *Consider a parliament in which three parties share the seats in the following proportions*: $(s_1, s_2, s_3) = (.4, .4, .2)$. *A proposed bill will pass if it obtains at least* 50% *of the votes in the parliament. Suppose that the benefit from winning is one for a winning coalition and zero for a losing coalition.*

(a) Derive the characteristic function of this game.

(b) Derive the Shapley value of the game. Compare the weights of the Shapley value with the proportions of seats that each party holds and explain the differences.

(c) Is the Shapley value an element of the core in this game?

Bibliography

Abreu, D. (1983). "Repeated Games with Discounting." Ph.D. dissertation, Princeton University.

Akerloff, G. (1970). The market for lemons: Qualitative uncertainty and the market mechanism. *Quarterly Journal of Economics 84*, 488–500.

Aumann, R. J. (1987a). Game theory. *In* Eatwell, J., Milgate, M, Newman, P. (Eds.) (1989), "The New Palgrave. Game Theory." London: Macmillan, pp. 1–53.

Aumann, R. J. (1987b). Correlated equilibrium as an expression of Bayesian rationality. *Econometrica 55*, 1–18.

Aumann, R. J. (1975). Values of markets with a continuum of traders. *Econometrica 43*, 611–646.

Aumann, R. (1975a). Lectures on game theory. Unpublished manuscript by Imai, H., Cordoba, J., and Osborne, M.

Aumann, R. J. (1964). Markets with a continuum of traders. *Econometrica 32*, 39–50.

Aumann, R. J., Shapley, L. (1976). Long-term competition: A game-theoretic analysis. Unpublished paper.

Aumann, R. J., Shapley, L. S. (1974). "Values of Non-Atomic Games." Princeton, N.J.: Princeton University Press.

Banks, J. S., Sobel, J. (1987). Equilibrium selection in signalling games. *Econometrica 55*, 647–661.

Benoit, J. P., Krishna V. (1985). Finitely repeated games. *Econometrica 53*, 905–922.

Berge, C. (1963). "Topological Spaces Including a Treatment of Multi-valued Functions, Vector Spaces, and Convexity." New York: Macmillan Company.

Bernheim, D. (1984). Rationalizable strategic behavior. *Econometrica 52*, 1007–1028.

Binmore, K. (1992). "Fun and Games: A Text on Game Theory." Lexington, Mass.: Heath.

Binmore, K. (1987). Perfect equilibria in bargaining models. *In* Binmore, K., Dasgupta, P. (Eds.) (1987), "The Economics of Bargaining." Oxford, U.K.: Blackwell, pp. 77–105.

Binmore, K., Dasgupta, P. (eds.) (1987). "The Economics of Bargaining." Oxford, U.K.: Blackwell.

Binmore, K., Osborne, M. J., Rubinstein, A. (1990). Noncooperative models of bargaining. Unpublished paper, Foerder Institute for Economic Research, Tel-Aviv University, Tel-Aviv.

Cho, I. K., Kreps, D. M. (1987). Signalling games and stable equilibria. *Quarterly Journal of Economics 102*, 179–221.

Cournot, A. (1838). "Recherches sur les Principes Mathématiques de la Théorie des Richesses." Reprinted Calmann-Lévy, Paris.

Crawford, V. P. (1990). Explicit communication and bargaining outcomes. *American Economic Review (Papers and Proceedings) 80*, 213–219.

Crawford, V. P., Knoer, E. M. (1981). Job matching with heterogeneous firms and workers. *Econometrica 49*, 437–450.

Crawford, V. P., Sobel, J. (1982). Strategic information transmission. *Econometrica 50*, 579–594.

Dasgupta, P., Maskin, E. (1986). The existence of equilibrium in discontinuous economic games. *Review of Economic Studies 53*, 1–41.

Dixit, A., Nalebuff, B. (1991). "Thinking Strategically: The Competitive Edge in Business, Politics, and Everyday Life." New York: Norton.

Friedman, J. W. (1990). "Game Theory with Applications to Economics." Second Edition. New York, Oxford: Oxford University Press.

Friedman, J. W. (1986). "Game Theory with Applications to Economics." New York, Oxford: Oxford University Press.

Friedman, J. (1971). A noncooperative equilibrium for supergames." *Review of Economic Studies 38*, 1–12.

Fudenberg, D., Maskin, E. (1986). The folk theorem in repeated games with discounting and with incomplete information. *Econometrica 54*, 533–554.

Fudenberg, D., Tirole, J. (1991). "Game Theory." Cambridge, Mass.: MIT Press.

Fudenberg, D., Tirole, J. (1991a). Perfect Bayesian equilibrium and sequential equilibrium. *Journal of Economic Theory 53*, 236–260.

Fudenberg, D., Tirole, J. (1989). Noncooperative game theory for industrial organization: An introduction and overview. *In* Schmalensee, R., Willig, R. D. (Eds.), "Handbook of Industrial Organization." Volume I, pp. 259–327.

Glicksberg, I., (1952). A further generalization of Kakutani's fixed point theorem. *Proceedings of the American Mathematical Society 3*, 170–174.

Green, J., Heller, W. P. (1981). Mathematical analysis and convexity with applications to economics. *In* Arrow, K. J., Intriligator, M. D. (Eds.), "Handbook of Mathematical Economics." Volume I. Amsterdam, New York: North-Holland Publishing Company, pp. 15–52.

Grossman, St. J. (1984). "Elementary Linear Algebra." Second Edition. Belmont: Wardsworth Publishing Company.

Harsanyi, J. C. (1973). Games with randomly disturbed payoffs. *International Journal of Game Theory 2*, 1–23.

Harsanyi, J. C. (1967). Games with incomplete information played by Bayesian players. Parts I, II, III, *Management Science 14*, 159–182, 320–334, 486–502.

Hart, S. (1987). Shapley value. *In* Eatwell, J., Milgate, M, Newman, P. (Eds.) (1989), "The New Palgrave: Game Theory," London: Macmillan, pp. 210–216.

Kakutani, S. (1941). A generalization of Brower's fixed point theorem. *Duke Mathematical Journal 8*, 457–458.

Kalai, E., Samet, D. (1984). Persistent equilibrium. *International Journal of Game Theory 13*, 129–141.

Kohlberg, E., Mertens, J.-F. (1986). On the strategic stability of equilibria, *Econometrica 54*, 1003–1037.

Kreps, D. M. (1990a). "A Course in Microeconomic Theory." New York: Harvester and Wheatsheaf.

Kreps, D. M. (1990b). "Game Theory and Economic Modelling." Oxford: Clarendon Press.

Kreps, D., Wilson, R. (1982). Sequential equilibria. *Econometrica 50*, 863–894.

Kuhn, H. W. (1953). Extensive games and the problem of information. *In* H. W. Kuhn and A. W. Tucker (Eds.), "Contributions to the Theory of Games, II." Princeton, N.J.: Princeton University Press.

Lang, S. (1968). "Analysis I." Reading (Mass.), London: Addison-Wesley Publishing Company.

Lemaire, J. (1981). Cooperative game theory and its insurance applications. *ASTIN Bulletin 21*, 17–40.

Machina, M. (1987). Choice under uncertainty: problems solved and unsolved. *Economic Perspectives 1*, 121–154.

McDonald, I. M., Solow, R. M. (1981). Wage bargaining and employment. *American Economic Review 71*, 896–908.

McMillan, J. (1992). "Games, Strategies, and Managers." New York: Oxford University Press.

Mertens, J.-F. (1989). Stable equilibria: A reformulation. Part I. Definitions and basic properties. *Mathematics of Operations Research 14*, 575–625.

Milgrom, P. R., Weber, R. J. (1985). Distributional strategies for games with incomplete information. *Mathematics of Operations Research 10*, 619–632.

Mirman, L. J., Tauman, Y., Zang, I. (1985). On the use of game-theoretic concepts in cost accounting. *In* Peyton Young, H. (Ed.), "Cost Allocation: Methods, Principles, Applications." Amsterdam: Elsevier (North-Holland).

Moulin, H. (1986). "Game Theory for the Social Sciences" (2nd and revised edition). New York: New York University Press.

Myerson, R. B. (1991). "Game Theory. Analysis of Conflict." Cambridge, Mass.: Harvard University Press.

Myerson, R. B. (1978). Refinements of the Nash equilibrium concept. *International Journal of Game Theory 7*, 73–80.

Nash, J. F. (1953). Two-person cooperative games. *Econometrica 21*, 128–140.

Nash, J. F. (1951). Noncooperative games. *Annals of Mathematics 54*, 289–295.

Nash, J. F. (1950). The bargaining problem. *Econometrica 18*, 155–162.

Newman, P. (1987). Francis Ysidro Edgeworth. *In* Eatwell, J., Milgate, M, Newman, P. (Eds.) (1990), "The New Palgrave: Utility and Probability." London: Macmillan. pp. 38–69.

Osborne, M. J., Rubinstein, A. (1990). "Bargaining and Markets. San Diego: Academic Press.

Pearce, D. G. (1984). Rationalizable strategic behavior and the problem of perfection. *Econometrica 52*, 1029–1050.

Rapoport, A. (1989). Prisoner's dilemma. *In* Eatwell, J., Milgate, M, Newman, P.

(Eds.) (1989), "The New Palgrave: Game Theory." London: Macmillan, pp. 199–204.

Rasmusen, E. (1989). "Games and Information: An Introduction to Game Theory." Oxford, UK: Blackwell.

Riley, J. G. (1987). Signalling. *In* Eatwell, J., Milgate, M., Newman, P., (Eds.), "The New Palgrave Allocation, Information and Markets." London: Macmillan, 287–294.

Rosen, J. B. (1965). Existence and uniqueness of equilibrium points for concave *n*-person games. *Econometrica 33*, 520–534.

Roth, A. E. (1991). Laboratory experimentation in economics: A methodological overview. *In* Oswald, A. J. (Ed.) (1991), "Surveys in Economics." Oxford, Blackwell.

Roth, A. E. (1984). Stability and polarization of interests in job matching. *Econometrica 52*, 47–57.

Roth, A. E. (1979). "Axiomatic Models of Bargaining." Berlin: Springer-Verlag.

Rubinstein, A. (1989). The electronic mail game: Strategic behavior under almost common knowledge. *American Economic Review 79*, 385–391.

Rubinstein, A. (1982). Perfect equilibrium in a bargaining model. *Econometrica 50*, 97–109.

Rubinstein, A. (1979). Equilibrium in supergames with the overtaking criterion. *Journal of Economic Theory 21*, 1–9.

Sabourian (1989). Repeated games: A survey. *In* Hahn, F. (Ed.) (1990), "The Economics of Missing Markets, Information, and Games." Oxford: Clarendon Press. Chapter 4, pp. 62–105.

Samuelson, L. (1990). Dominated strategies and common knowledge. Unpublished paper. Department of Economics, University of Wisconsin.

Selten, R. (1975). Reexamination of the perfectness concept for equilibrium points in extensive games. *International Journal of Game Theory 4*, 25–55.

Selten, R. (1965). Spieltheoretische Behandlung eines Oligopolmodels mit Nachfrageträgheit. *Zeitschrift für die gesamte Staatswissenschaft 12*, 301–324, 667–689.

Shapiro, C. (1989). Theories of oligopoly behavior. *In* Schmalensee R., Willig R. D. (Eds.) (1989), "Handbook of Industrial Organization." Volume 1. Amsterdam, New York: North Holland, Chapter 6, pp. 329–414.

Shapley, L. S. (1969). Utility comparison and the theory of games. *In* La Décision: Agrégation et dynamique des ordres de préférance. Paris: Éditions du CNRS, pp. 251–263.

Shapley, L. S. (1967). On balanced sets and cores. *Naval Research Logistics Quarterly 14*, 453–460.

Shapley, L. S. (1953). A value for *n*-person games. *In* Kuhn, H. W., Tucker, A. W. (Eds.), "Contributions to the Theory of Games II." *Annals of Mathematical Studies Series 28*. Princeton, N.J.: Princeton University Press pp. 307–317.

Shapley, L. S., Shubik, M. (1969). On market games. *Journal of Economic Theory 1*, 9–25.

Shubik, M. (1982). "Game Theory in the Social Sciences: Concepts and Solutions." Cambridge, Mass.: MIT Press.

Shubik, M. (1959). Edgeworth market games. *In* Luce, R. D., Tucker, A. W.

(Eds.), "Contributions to the Theory of Games IV." *Annals of Mathematical Studies Series 40*. Princeton, N.J.: Princeton University Press, pp. 267–278.

Spence, M. (1973). "Market Signalling: Information Transfer in Hiring and Related Processes." Cambridge, Mass.: Harvard University Press.

Ståhl, I. (1972). "Bargaining Theory." Economics Research Institute, Stockholm School of Economics, Stockholm.

Takayama, A. (1988). "Mathematical Economics." 2nd Edition. Cambridge, Mass.: Cambridge University Press.

Tirole, J. (1988). The Theory of Industrial Organization. Cambridge, Mass.: MIT Press.

Van Damme, E. (1991). "Stability and Perfection of Nash Equilibria." Second, Revised and Enlarged Edition. Heidelberg, New York: Springer Verlag.

Van Damme, E. (1990). Refinements of Nash equilibrium. Discussion paper, Center for Economic Research. Tilburg University, Tilburg.

Van Damme, E. (1983). Refinements of the Nash Equilibrium Concept. Heidelberg, New York: Springer-Verlag.

Varian, H. R. (1984). "Microeconomic Analysis." Second Edition. New York: Norton.

Von Neumann, J. (1928). Zur Theorie der Gesellschaftsspiele. *Mathematische Annalen 100*, 295–320.

Von Neumann, J., Morgenstern, O. (1947). "Theory of Games and Economic Behavior." 2nd edition. Princeton, N.J.: Princeton University Press.

Zermelo, E. (1913). Über eine Anwendung der Mengenlehre auf die Theorie des Schachspiels. *Proceedings of the Fifth International Congress of Mathematicians 2*, 501–504.

Index